Researches in Global Ecosystem and Renewable Energy Resources

Researches in Global Ecosystem and Renewable Energy Resources

Edited by **Shania Gomes**

SYRAWOOD
PUBLISHING HOUSE

New York

Published by Syrawood Publishing House,
750 Third Avenue, 9th Floor,
New York, NY 10017, USA
www.syrawoodpublishinghouse.com

Researches in Global Ecosystem and Renewable Energy Resources
Edited by Shania Gomes

International Standard Book Number: 978-1-68286-034-2 (Hardback)

Printed in the United States of America.

Contents

Preface

There has been an emphasis on using renewable energy resources worldwide for addressing the emerging energy crisis and preservation of global ecosystems. This book brings forth some of the most innovative concepts and elucidates the unexplored aspects of this field. The topics included herein are monitoring carbon emissions from different sources, climate variability, evaluation of various fuel and energy sources and their impact on environment, etc. The book is aimed at providing a comprehensive understanding of the field to the students and researchers engaged in the field.

Various studies have approached the subject by analyzing it with a single perspective, but the present book provides diverse methodologies and techniques to address this field. This book contains theories and applications needed for understanding the subject from different perspectives. The aim is to keep the readers informed about the progress in the field; therefore, the contributions were carefully examined to compile novel researches by specialists from across the globe.

Indeed, the job of the editor is the most crucial and challenging in compiling all chapters into a single book. In the end, I would extend my sincere thanks to the chapter authors for their profound work. I am also thankful for the support provided by my family and colleagues during the compilation of this book.

Editor

Influence of Crude Oil Spillage on the γ-Radiation Status of Water and Soil in Ogba/Egbema/Ndoni Area, Nigeria

Yehuwdah E. Chad-Umoren[1] & Efe Ohwekevwo[2]

[1] Department of Physics, University of Port Harcourt, Rivers State, Nigeria

[2] Department of Physics, Rivers State University of Science and Technology, Port Harcourt, Rivers State, Nigeria

Correspondence: Yehuwdah E. Chad-Umoren, Department of Physics, University of Port Harcourt, Rivers State, Nigeria. E-mail: echadumoren@yahoo.com

Abstract

Crude oil spillage is one of the variables in the hydrocarbon industry responsible for the impact the industry makes on the ionizing radiation status of its host environment. A study was carried out to investigate the level of impact of oil spillage on the ionizing radiation profile of three communities in the Niger Delta region of Nigeria reputed for its abundant oil wealth. For the purpose, 15 water samples (five from each community) and 15 soil samples (five from each community) were collected and analyzed using the Gamma Scout γ-spectrometer and Na(TI) detector. The lowest dose rate for both water and soil was the same 1.14 ± 0.07 mSv/yr (0.014 ± 0.001 mR/hr), while the highest dose rate was also the same for both water and soil 1.58 ± 0.11 mSv/yr (0.019 ± 0.001 mR/hr). However, the minimum dose rate, maximum dose rate and mean dose rate computed for water differed from those computed for soil. The study showed that all computed radiation parameters for both water and soil, including the minimum dose rates measured for the three communities, exceeded international regulatory standards for the general populace indicating that oil spill had resulted in the elevation of the radiation levels of the affected communities and that appropriate steps need to be taken to protect those living in the study area from radiation hazards.

Keywords: Radiation dose, hydrocarbon industry, radiation exposure, water samples, soil samples, oil spillage, Niger Delta region

1. Introduction

The Niger Delta region of Nigeria is recognized as critical to global oil security and indispensable to the nation's economic wellbeing. But the Niger delta has also become a victim of its hydrocarbon wealth. The exploitation of its abundant hydrocarbon resources has led to the despoliation and pollution of its environment. Inability to properly harness the abundant gas wealth of the region has led to the environmentally destructive practice of gas flaring which releases harmful gaseous pollutants into the atmosphere. A study of the ambient air pollutants in the region indicates that existing Federal environmental standards were grossly exceeded for particulate matter, oxides of carbon, nitrogen and sulphur (Oluwole et al., 1996).

Activities of the hydrocarbon industry also result in the elevation of the ionizing radiation levels of its host locations in the region. In a study of the external environmental radiation status of some industrial locations in Port Harcourt, which included oil and gas facilities, an average value of 0.014 mR h^{-1} was reported for the background, showing an elevation from the standard background radiation level of 0.013mR h^{-1} (Avwiri & Ebeniro, 1998). Arogunjo et al. (2004) studied the impact of the oil and gas industry on the natural radioactivity distribution in the region and showed that the mean activity concentration of ^{40}K, ^{238}U and ^{234}Th were 34.8 ± 2.4, 16.2 ± 3.7 and 24.4 ± 4.7 Bqkg^{-1}, respectively. It was further observed that areas with oil extraction activities had higher activity concentrations than those without any known oil extraction activity in the region. Stanislav and Elena (1998) studied the environmental impact of offshore oil and gas facilities and showed that produced water from oil and gas production contained naturally occurring radioactive elements (Uranium and Thorium) and their daughter products (^{226}Ra and ^{228}Ra).

Ononugbo et al. (2011) carried out a study of the terrestrial radioactivity in the industrial areas of Ogba/Egbema/Ndoni Local Government Area (ONELGA) of Rivers state. The industrial areas and their host communities were divided into six (6) zones, each having an oil and gas facility. An *in situ* measurement was done using two well calibrated nuclear radiation meters, a digilert-100 and a digilert-50 and a geographical positioning system (GPS). Ten readings were taken in each of the six zones and the host community at randomly selected sites, making a total of 60 sampling points. The mean site radiation levels that were obtained ranged from 0.014 ±0.001 mRh^{-1} (1.183± 0.060 mSvyr^{-1}) to 0.018 ± 0.002 mRh^{-1} (1.183± 0.085 mSvyr^{-1}), while the mean community radiation levels ranged from 0.014 ± 0.001 mRh^{-1} (1.183 ± 0.06 mSvyr$^{-1)}$ to 0.017 ± 0.001 mRh^{-1} (1.435± 0.072 mSvyr^{-1}). The equivalent dose had an average range of 1.056 mSvyr^{-1} to 2.871 mSvyr^{-1}, which is much lower than the International Commission on Radiological Protection recommended 20 mSvyr^{-1} dose limit for radiological workers, but above the permissible level of 1 mSv/yr recommended for the general public (ICRP, 1990). Further analysis showed that 43 sites or 72% of the sampling sites exceeded the normal background level of 0.013mRh^{-1} indicating a certain level of radiation risk for the communities hosting the facilities.

Ebong and Alagoa(1992) have shown that the nature of input raw materials, effluents from the production process and the output production determine the effect of an industrial operation on the radiation levels of its host environment. In the case of the hydrocarbon industry there are a number of variables associated with the industry that contribute to the impact the industry makes on the ionizing radiation profile of its areas of operation. These include the use of radiation generators, sealed and unsealed sources of radioactive materials, naturally occurring radioactive materials (NORMs) originating from reservoir rocks, scales, sludge, oil spillage and gas flaring (Chad-Umoren, 2012; Meindinyo & Agbalagba, 2012; OGP, 2008; Sigalo & Briggs-Kamara, 2004).

However, in some locations, particular variables may be more critical than others. This present study investigates the impact of crude oil spillage as an isolated variable and as the only industry variable affecting the radiation patterns of three communities - Ebegoro, Ebocha and Obrikom in the Ogba/Egbema/Ndoni Local Government Area (ONELGA) in Nigeria's Niger Delta state of Rivers where the operations of a multinational oil firm had led to the release of enormous quantities of crude oil into the environment.

Studies on the impact of the oil and gas industry on the ionizing radiation profile of the Niger Delta region had tended to focus on the contribution of oil and gas facilities (Chad-Umoren & Briggs-Kamara, 2010; Chukwuocha & Enyinna, 2009; Avwiri et al., 2007). Furthermore, there has been only one previous study in ONELGA (Ononugbo et al., 2011) and that had followed the traditional mode of focusing mainly on the impact of facilities such as gas processing plants, flow stations, natural gas compressor stations. There is therefore the need to properly document the effect incessant oil spillage has had on the level of background ionizing radiation of the area. This work provides base-line data for future studies.

Oil spillage is often an unintended release of crude oil into the environment as a result of human activity. Such accidents may involve a refinery, an oil storage facility, barges, oil tankers or oil pipelines. Oil spillage has been described as a major source of water pollution in the Niger Delta region and the increase in its frequency has been attributed to the growth of the industry and the prevalence of ageing oil pipelines. A total of 10,260 oil spills is estimated to have occurred in the region between 1976 and 2007 resulting in the loss of about seven million barrels of oil. Of this quantity, 6% was spilled on land, 25% in swamps and 69% in offshore environment (NREP, 2008).

Crude oil is both toxic due to its toxic chemical content and radioactive due to its radionuclide content, specifically the presence of Uranium and Thorium. These radiouclides and members of their decay chains are found in the earth's crust and therefore become incorporated into the crude oil during the process of oil drilling and production (Chad-Umoren, 2012; OGP, 2008). The prevalence of these radionuclides is depth-dependent so that the level of radioactivity of given crude is determined by the depth of the well that is its source. The deeper the oil well, the more radioactive the crude oil. Consequently, depending on the radioactive nature of the crude, spillage can lead to harmful elevation of the ionizing radiation profile of the environment.

2. Study Area

Ebegoro, Ebocha and Obrikom, the three communities chosen for this study are communities in the Ogba/Egbema/Ndoni Local Government Area (ONELGA) in Nigeria's Niger Delta state of Rivers. ONELGA is located within latitudes 5^{0}23'N and 5^{0}26'N and longitudes 6^{0}33'E and 6^{0}42' North West and has a topography of flat plains in a network of rivers - the Niger, Sombreiro (Nkissa), Orashi and their tributaries along with a series of creeks. It is one of the onshore oil producing areas of Rivers state and has one of the highest oil and gas production onshore of the Niger-Delta with over 900 oil wells, more than thirteen active oil fields and playing host to a number of multinational companies (Abali, 2009). The area has a labyrinth of pipelines carrying oil or gas to flow stations from the different oil wells (UNDP, 2006). Oil and gas activities started in ONELGA in 1964 with actual production commencing in 1966. Since then, production activities have been continuous with increase in the number of drilled oil wells. Gas flaring and oil spillage due to rupture of pipes have been the major sources of environmental pollution in the area.

3. Sample Collection, Preparation and Analyses

The study involved two sets of evaluations – on-site evaluation at the three communities and laboratory analyses. The RadEye Gamma survey meter with a non linear pulse to dose rate conversion capability and a geographical positioning system (GPS) were used for the on-site radiation monitoring. Measurement technique adopted followed the usual procedure for *in situ* measurements (Chad-Umoren et al., 2006; Avwiri et al., 2009).

The laboratory study was carried out using water and soil samples collected from spill sites in the three selected communities. In each community, five sampling locations were randomly chosen, but to as much as possible cover the community. Fifteen soil samples (five from each of the three communities) and fifteen water samples (five also from each of the three communities) were collected from the delineated spill sites. About 2 kg of soil samples were collected from the spill sites and stored in black plastic bags directly after collection to prevent contamination from atmospheric humidity. The volume of water samples collected from the spill sites was about one litre and also stored in a manner that preserved sample integrity.

The samples were later analysed at the laboratory of the Centre for Energy Research and Development, Obafemi Awolowo University, Ile Ife, Nigeria. The soil samples were first air dried, sieved through 2 mm and stored in plastic bags for a period of 28 days to allow them attain secular equilibrium before analysis.

The following conversion factors were used:

1 μSv/hr = 365 x 24 x 10^{-3} mSv/yr

1 mSv/yr = 0.0119 mR/hr

4. Results and Discussions

Table 1. Dose rate and geographical location for water samples

Community	Sampling Site	Geographical Field Location		Dose Rate Laboratory result (μSv/hr)	Equivalent Dose(mSv/yr)	Radiation Level(mR/hr)
Ebegoro	1	N05°23'04.4"	E006°39'27"	0.18±0.018	1.58±0.16	0.019±0.002
	2	N05°23'9"	E006°39'49"	0.18±0.018	1.58±0.16	0.019±0.002
	3	N05°23'49"	E006°39'35''	0.16±0.016	1.40±0.14	0.017±0.002
	4	N05°23'30.6"	E006°39'36''	0.16±0.011	1.40±0.10	0.017±0.001
	5	N05°23'0"	E006°40'37.5"	0.17±0.017	1.49±0.15	0.018±0.002
Ebocha	1	N05°27'21.8"	E006°41'33.3"	0.16±0.014	1.40±0.12	0.017±0.001
	2	N05°27'22.2"	E006°41'34"	0.16±0.013	1.40±0.11	0.017±0.001
	3	N05°26'1"	E006°40'23.8"	0.15±0.015	1.31±0.13	0.016±0.002
	4	N05°27'2"	E006°40'58"	0.14±0.014	1.23±0.12	0.015±0.001
	5	N05°23'21.5"	E006°45'35.5"	0.16±0.013	1.40±0.11	0.017±0.001
Obrikom	1	N05°23'30"	E006°39'23"	0.15±0.011	1.31±0.10	0.016±0.001
	2	N05°23'30"	E006°40'33"	0.14±0.014	1.23±0.12	0.015±0.001
	3	N05°28'23	E006°42'53"	0.13±0.013	1.14±0.11	0.014±0.001
	4	N05°28'40"	E006°35'6"	0.14±0.014	1.23±0.12	0.015±0.001
	5	N05°23'35"	E006°27'40"	0.17±0.014	1.49±0.12	0.018±0.001

Table1 shows the radiation profile of the water samples and the GPS co-ordinates. These results show that the highest radiation dose was obtained at Ebegoro, while the lowest was recorded at Obrikom. Table 2 shows the maximum dose rate, minimum dose rate and mean dose rate for the water from the three communities and the ICRP maximum exposure level for the general public. A consideration of the mean radiation dose values show that the highest value was obtained at Ebegoro with a value of 1.49±0.14 mSv/yr (0.018±0.002 mR/hr) while the lowest was recorded at Obrikom with a value of 1.28±0.11 mSv/yr (0.016±0.001 mR/hr). The radiation contamination of the water from the spill sites is nearly homogeneous across the three communities as the difference between the largest value and the least is only 0.35 mSv/yr or 22%. The implication will be that the spillage was uniformly distributed in the community water sources.

Also, a comparison of the maximum dose rate, minimum dose rate and mean dose rate for the water from the three communities with the ICRP reference level shows that these quantities have values that are higher than the ICRP standard. As shown in Figures 1 and 2, the radiation profile is such that even the lowest radiation levels in each of the spill site exceed the ICRP reference level. The minimum exposure level in Obrikom is 14% above the ICRP level, while the maximum exposure level is 49% higher. In Ebocha the minimum dose rate is 23% above the ICRP standard and the maximum 40%. The minimum exposure level in Ebegoro is 40% above the ICRP value and the maximum exposure level 58%.

Table 2. Maximum, minimum, mean and icrp radiation dose and exposure rates (water samples)

Spill Site	Maximum		Minimum		Mean		ICRP	
	mSv/yr	mR/hr	mSv/yr	mR/hr	mSv/yr	mR/hr	mSv/yr	mR/hr
Ebegoro	1.58±0.16	0.019±0.002	1.40±0.10	0.017±0.001	1.49±0.14	0.018±0.002	1.000	0.013
Ebocha	1.40±0.12	0.017±0.001	1.23±0.12	0.015±0.001	1.35±0.12	0.016±0.001	1.000	0.013
Obrikom	1.49±0.12	0.018±0.001	1.14±0.11	0.014±0.001	1.28±0.11	0.016±0.001	1.000	0.013

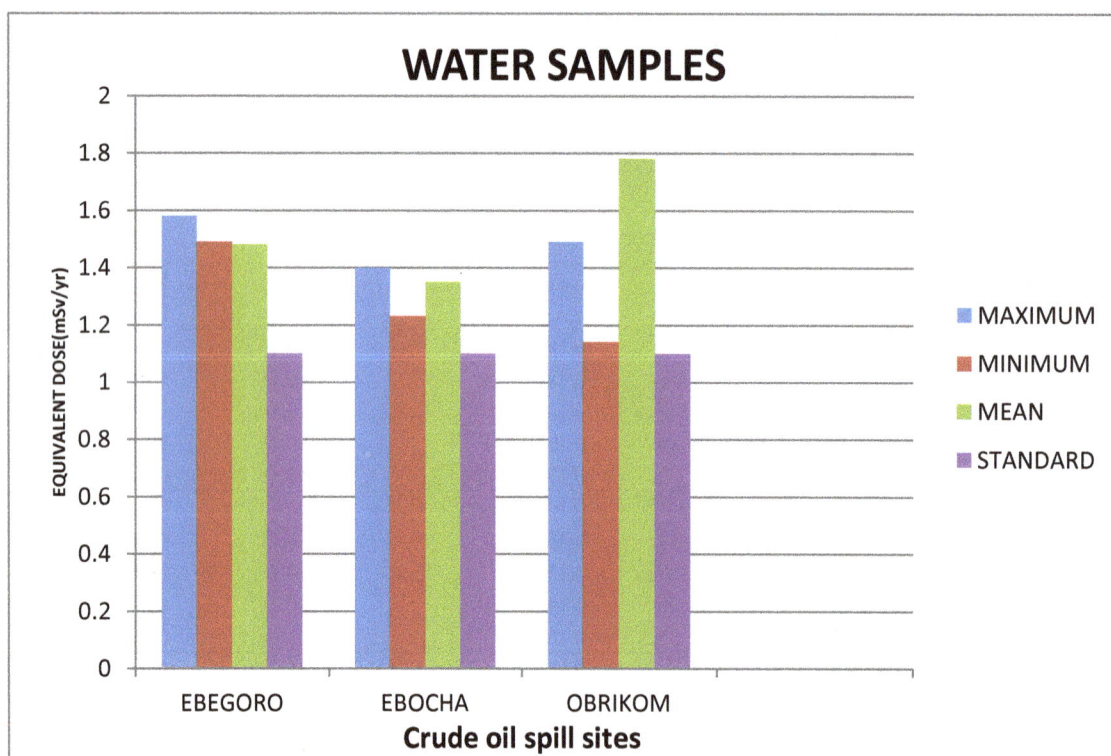

Figure 1. Comparison of maximum, minimum, mean and permissible (ICRP) radiation dose rate (water)

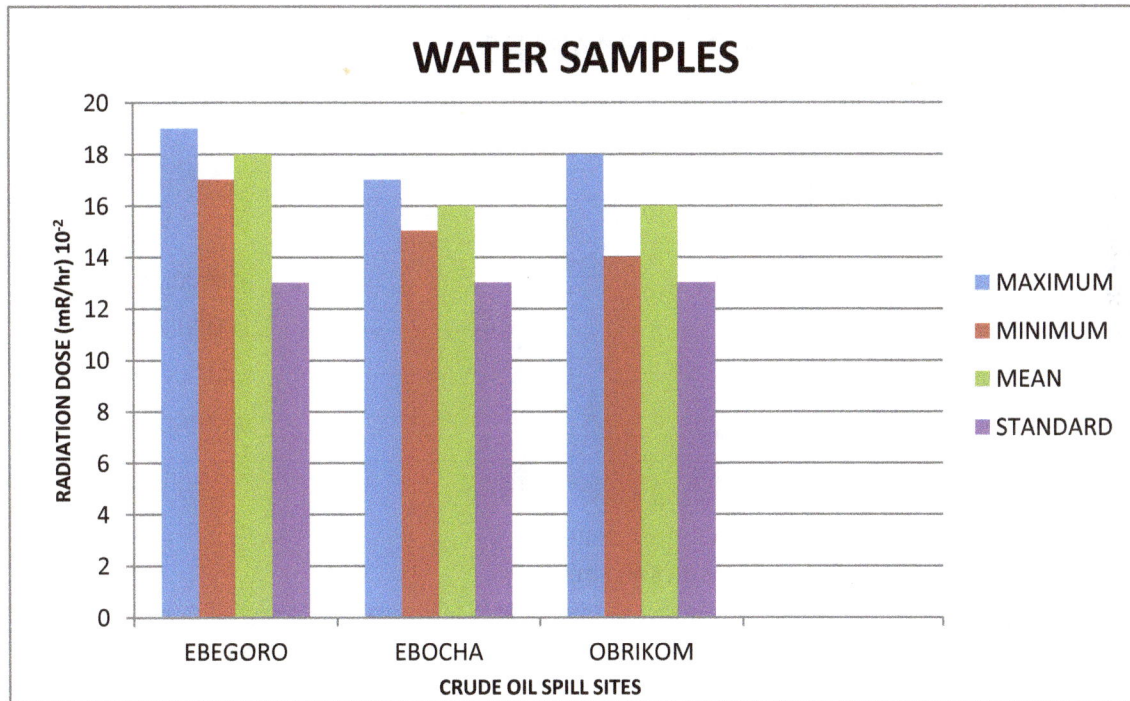

Figure 2. Comparison of maximum, minimum, mean and permissible (ICRP) radiation exposure rates for water

Table 3. Dose rate and geographical location for soil samples

Community	Site	Geographical Field Location	Dose Rate Laboratory result(μSv/hr)	Equivalent Dose(mSv/yr)	Radiation Level(mR/h)
Ebegoro	1	N05^{0}23$^{'}$04.4$^{''}$ E006^{0}39$^{'}$27$^{''}$	0.17±0.017	1.49±0.15	0.018±0.002
	2	N05^{0}23$^{'}$9$^{''}$ E006^{0}39$^{'}$49$^{''}$	0.18±0.013	1.58±0.11	0.019±0.001
	3	N05^{0}23$^{'}$49$^{''}$ E006^{0}39$^{'}$35$^{''}$	0.17±0.017	1.49±0.15	0.018±0.002
	4	N05^{0}23$^{'}$30.6$^{''}$E006^{0}39$^{'}$36$^{''}$	0.16±0.014	1.40±0.12	0.017±0.001
	5	N05^{0}23$^{'}$0$^{''}$ E006^{0}40$^{'}$37.5$^{''}$	0.18±0.018	1.58±0.16	0.019±0.002
Ebocha	1	N05^{0}27$^{'}$21.8$^{''}$ E006^{0}41$^{'}$33.3$^{''}$	0.14±0.014	1.23±0.12	0.015±0.001
	2	N05^{0}27$^{'}$22.2$^{''}$ E006^{0}41$^{'}$34$^{''}$	0.14±0.014	1.23±0.12	0.015±0.001
	3	N05^{0}26$^{'}$1$^{''}$ E006^{0}40$^{'}$23.8$^{''}$	0.16±0.016	1.40±0.14	0.017±0.002
	4	N05^{0}27$^{'}$2$^{''}$ E006^{0}40$^{'}$58$^{''}$	0.17±0.017	1.49±0.15	0.018±0.002
	5	N05^{0}23$^{'}$21.5$^{''}$ E006^{0}45$^{'}$35.5$^{''}$	0.16±0.016	1.40±0.14	0.017±0.002
Obrikom	1	N05^{0}23$^{'}$30$^{''}$ E006^{0}39$^{'}$23$^{''}$	0.16±0.016	1.40±0.14	0.017±0.002
	2	N05^{0}23$^{'}$30$^{''}$ E006^{0}40$^{'}$33$^{''}$	0.17±0.014	1.49±0.12	0.018±0.001
	3	N05^{0}28$^{'}$23$^{''}$ E006^{0}42$^{'}$53$^{''}$	0.17±0.015	1.49±0.13	0.018±0.002
	4	N05^{0}28$^{'}$40$^{''}$E006^{0}35$^{'}$6$^{''}$	0.15±0.012	1.31±0.11	0.016±0.001
	5	N05^{0}23$^{'}$35$^{''}$ E006^{0}27$^{'}$40$^{''}$	0.13±0.008	1.14±0.07	0.014±0.001

Table 4. Minimum, maximum, mean and icrp radiation dose and exposure rates (soil samples)

Community	Maximum		Minimum		Mean		ICRP	
	mSv/yr	mR/hr	mSv/yr	mR/hr	mSv/yr	mR/hr	mSv/yr	mR/hr
EBEGORO	1.58±0.11	0.019±0.001	1.40±0.12	0.017±0.001	1.51±0.14	0.018±0.002	1.00	0.013
EBOCHA	1.49±0.15	0.018±0.002	1.23±0.12	0.015±0.001	1.35±0.13	0.016±0.002	1.00	0.013
OBRIKOM	1.49±0.12	0.018±0.001	1.14±0.07	0.014±0.001	1.37±0.11	0.017±0.001	1.00	0.013

SOIL SAMPLES

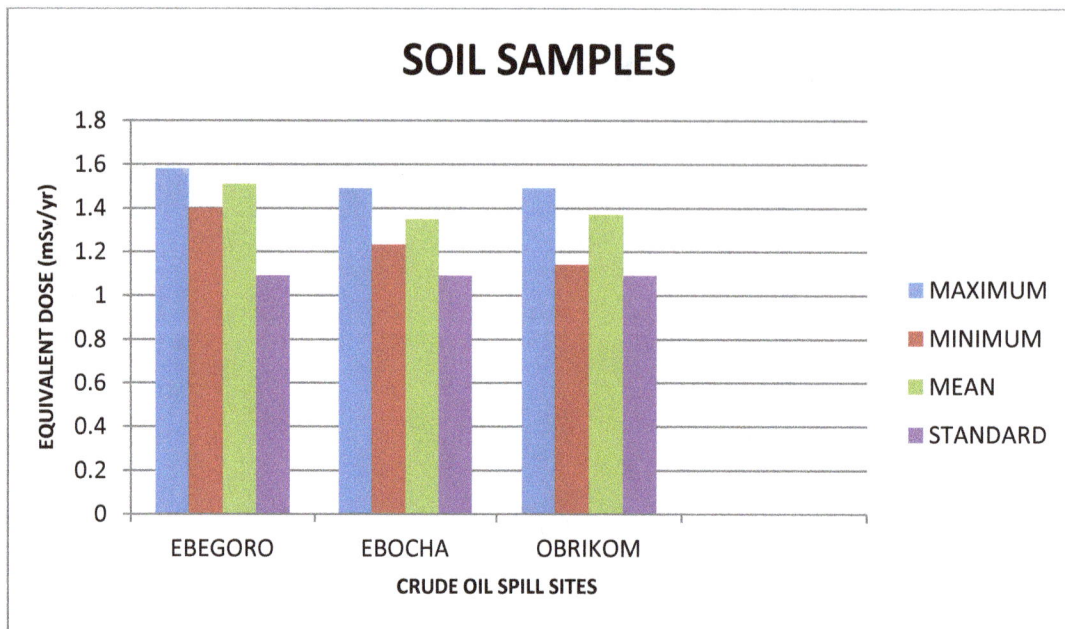

Figure 3. Comparison of minimum, maximum, mean and permissible radiation dose rate

SOIL SAMPLES

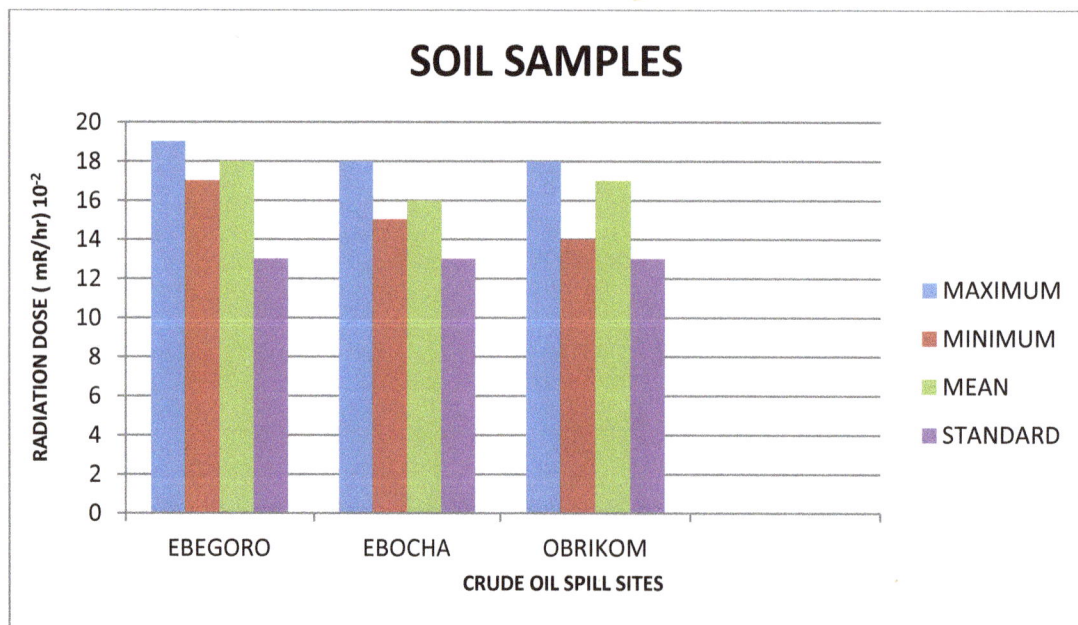

Figure 4. Comparison of radiation dose rate of maximum, minimum, mean and permissible radiation exposure rate

The radiation profile of the soil samples (Table 3) shows the following ranges for the various communities: for Ebegoro community the radiation level ranges from 1.40±0.12 mSv/yr (0.017±0.001 mR/hr) to 1.58±0.11 mSv/yr (0.019±0.001 mR/hr); while it goes from 1.23±0.12 mSv/yr (0.015±0.001 mR/hr) to 1.49±0.15 mSv/yr (0.018±0.002 mR/hr) at Ebocha. For Obrikom, the range is from 1.14±0.07 mSv/yr (0.014±0.001 mR/hr) to 1.49±0.12 mSv/yr (0.018±0.001 mR/hr). The highest value was recorded at Ebegoro, while the least was obtained from Obrikom.

Table 4 shows the maximum dose rate, minimum dose rate, mean dose rate for the soil from the three communities and the ICRP permissible maximum exposure level for the general public. The highest mean radiation value is 1.51±0.14 mSv/yr (0.018±0.002 mR/hr), obtained at Ebegoro; while the lowest is 1.35±0.13 mSv/yr (0.016±0.002 mR/hr), recorded at Ebocha. The radiation contamination of the soil from the spill sites is near evenly spread over the three communities as the difference between the largest value and the least is only 0.16 mSv/yr or 10.6%.

Figures 3 and 4 show that as in the case of the water samples, the minimum dose rate, maximum dose rate and mean dose rate computed for soil samples from the various communities exceed the ICRP reference limit for the general public. The minimum exposure level in Obrikom is 14% higher than the ICRP permissible level, while the maximum exposure level is 49%. In Ebocha, the minimum exposure level is 23% above the ICRP level and the maximum exposure 49%; while in Ebegoro the minimum exposure level is 40% above the ICRP standard and the maximum exposure level is 58% higher.

Comparing the radiation profile of both water and soil it is seen that the lowest radiation dose rate for both is the same (1.14±0.014 mSv/yr (0.014± 0.001)). This result was obtained in the same community, Obrikom. Similarly, it is also seen that the highest radiation dose rate for water is the same with that for soil (1.58±0.16 mSv/yr (0.019±0.002 mR/hr)). This result was also recorded in one community, Ebegoro.

Furthermore, the results show that the impact of the oil spillage on the radiation status of the three communities is more pronounced in Ebegoro than in Obrikom.

A study of the impact of oil spillage in parts of Ughelli another Niger Delta community, showed that radiation levels within the areas affected by the spillage and the host communities were 55% and 33.3%, respectively above the normal background level of 0.013 mR/hr indicating elevations attributable to the spillage (Agalagba & Meindinyo, 2010). Other studies in the region have shown that the hydrocarbon industry is contributing to an increasingly worrisome ionizing radiation climate in the region (Arogunjo et al., 2004; Chad-Umoren, 2012; Ononugbo et al., 2011; Sigalo & Briggs-Kamara, 2004).

5. Conclusion and Recommendation

(1)   A comparison of the radiation profiles of the water with that of the soil indicates that the impact of the spillage on the radiation levels is generally the same for both as the measured values are basically the same for both.

(2)   Although the community least affected by the impact of the spillage on the radiation profile is the Obrikom community as it has the lowest radiation levels for both water and soil, however the observed radiation values for this community exceeds the ICRP permissible level for the general public.

(3)   The results of the present study are significantly higher than those of earlier studies carried out in parts of the Niger Delta at locations that had not experienced oil spillage (Chad-Umoren & Obinoma, 2007; Briggs-Kamara et al., 2009; Chad-Umoren et al., 2006), indicating that oil spillage has resulted in the elevation of the radiation profile of the study area.

(4)   Regular maintenance of oil pipelines and prompt replacement of old and obsolete ones should be carried out. Lack of regular maintenance of storage tanks and oil pipelines has been identified as a major cause of oil spills in Nigeria.

(5)   It may not be possible to completely eliminate oil spillage in the Niger Delta region since it is connected to the beneficial exploitation of the hydrocarbon resources of the region and it is not economically prudent to stop exploiting the region's abundant natural endowments; however, the indigenous peoples of the region and other residents should be provided with facilities that will enable them receive prompt medical help, especially with respect to those ailments that are related to radiation exposure.

(6)   Prompt clean up and remediation exercises should be carried out where spillage has occurred.

References

Abali, B. K. (2009). *Oil and Gas Exploration: What ONELGA suffers*. Port Harcourt, B'Alive publications co.

Agbalagba, O. E., & Meindinyo, R. K. (2010). Radiological impact of oil spilled environment: A Case study of the Eriemu well 13 and 19 oil spillage in Ughelli region of delta state, Nigeria. *Indian Journal of Science and Technology, 3*(9), 1001-1005.

Arogunjo, A. M., Farai, I. P., & Fuwape, I. A. (2004). Impact of oil and gas industry to the natural radioactivity distribution in the delta region of Nigeria. *Nigerian Journal Physics, 16*, 131-136.

Avwiri, G. O., Chad-Umoren, Y. E., Enyinna, P. I., & Agbalagba, E. O. (2009). Occupational Radiation Profile of Oil and Gas Facilities During Production and Off Production Periods in Ughelli, Nigeria. *Journal Facta Universitatis:Working and Living Environmental Protection, 6*(1), 11-19.

Avwiri, G. O., Enyinna, P. I., & Agbalagba, E. O. (2007). Terrestrial Radiation around oil and gas facilities in Ughelli, Nigeria. *Journal of Applied Sciences, 7*, 1543-1546. http://dx.doi.org/10.3923/jas.2007.1543.1546

Avwiri, G. O., & Ebeniro, J. O. (1998). External environmental radiation in an industrial area of rivers state. *Nigerian Journal of Physics, 10*, 105-107.

Briggs-Kamara, M. A., Sigalo, F. B., Chad-Umoren, Y. E., & Kamgba, F. A. (2009). Terrestrial Radiation Profile of a Nigerian University Campus: Impact of Computer and Photocopier Operations. *Journal Facta Universitatis: Working and Living Environmental Protection, 6*(1), 1-9.

Chad-Umoren, Y. E. (2012). Ionizing Radiation Profile of the Hydrocarbon Belt of Nigeria. In Mitsuru Nenoi (Ed.), *Current Topics in Ionizing Radiation Research*, InTech Publications, Janeza Trdine 9, 51000 Rijeka, Croatia.

Chad-Umoren, Y. E., Adekanmbi, M., & Harry, S. O. (2006). Evaluation of Indoor Background Ionizing Radiation Profile of a Physics Laboratory. *Journal Facta Universitatis: Working and Living Environmental Protection., 3*(1), 1-8.

Chad-Umoren, Y. E., & Obinoma, O. (2007). Determination of Ionizing Radiation Level of the Main Campus of the College of Education, Rumuolumeni, Rivers State, Nigeria. *Int'l Journal of Environmental Issues, 5*(1&2), 5-10.

Chad-Umoren, Y. E., & Briggs-Kamara, M. A. (2010). Environmental Ionizing Radiation Distribution in Rivers State, Nigeria. *Journal of Environmental Engineering and Landscape Management, 18*(2), 154-161. http://dx.doi.org/10.3846/jeelm.2010.18

Chukwuocha, E., & Enyinna, P. I. (2009). Radiation monitoring of facilities in some oil wells in Bayelsa and Rivers states. *Scientia Africana, 9*(1), 98-102.

Ebong, I. D. U., & Alagoa, K. D. (1992). Estimates of gamma-ray background air exposure at a fertilizer plant. *Discovery and Innovation, 4*, 25-28.

ICRP. (1990). *Recommendations of International Commission of Radiological Protection.* Oxford: Pergamon press.

Meindinyo, R. K., & Agbalagba, O. E. (2012). Radioactivity concentration and heavy metal assessment of soil and water, in and around Imirigin oil field, Bayelsa state, Nigeria. *Journal of environmental chemistry and Ecotoxicology, 42*, 29-34.

OGP. (2008). Guidelines for the management of Naturally Occurring Radioactive Material (NORM) in the oil & gas industry, International Association of Oil & Gas Producers.

Oluwole, A. F., Olaniyi, H. B., Akeredolu, F. A., Ogunsola, O. J., & Obioh, I. B. (1996). Impact of the Petroleum Industry on Air Quality in Nigeria, Presented at the 8th Biennial International Seminar on the Petroleum Industry and the Nigerian Environment, Port Harcourt, 17-21 November, 1996.

Ononugbo, C. P., Avwiri, G. O., & Chad-Umoren, Y. E. (2011). Impact of Gas Exploitation on the Environmental Radioactivity of Ogba/Egbema/Ndoni Area, Nigeria. *Energy and Environment, 22*(8), 1017-1028. http://dx.doi.org/10.1260/0958-305X.22.8.1017

Sigalo, F. B., & Briggs-Kamara, M. A. (2004). Estimate of Ionizing Radiation Levels within Selected Riverine Communities of the Niger Delta. *Journal of Nigerian Envirironmental Society, 2*, 159-162.

Stanislav, P., & Elena, C. (1998). *Environmental Impact of the Off-shore Oil and Gas Industries.* East Northport, USA.

United Nations Development Programme (UNDP). (2006). Niger Delta Human Development report: Environmental and Social Challenges in the Niger Delta. UN House, Abuja, Nigeria.

Evaluation of Asphaltene Stability During CO$_2$ Flooding at Different Miscible Conditions and Presence of Light Components

Vahid Alipour Tabrizy[1] & Aly A. Hamouda[2]

[1] Department of Reserve Replacement, ASG PTC, Statoil ASA, Norway

[2] Department of Petroleum Engineering, University of Stavanger, 4036 Stavanger, Norway

Correspondence: Vahid Alipour Tabrizy, Department of Reserve Replacement, ASG PTC, Statoil ASA, Norway. E-mail: vtab@statoil.com

Abstract

The negative side effect of the flooding with CO$_2$ is asphaltene deposition; while little work was reported in the literature on asphaltene precipitation due to CO$_2$ flooding in presence of light components. The main objective in this paper is to address asphaltene precipitation for oil containing methane and propane due to CO$_2$ flooding at different miscibility conditions. Experimental measured asphaltene deposition due to miscible CO$_2$ injection is compared with corresponding values estimated by proposed model. It is shown that there is a critical concentration of CO$_2$, where below it; solubility parameter of the liquid is enhanced, hence preventing asphaltene from depositing. The first objective of the paper is to address an approach which is based on solubility parameters/CO$_2$ fraction in the liquid to qualitatively assess stability/instability region for the asphaltene. The second objective is to quantitatively compare the predicted and experimental results. It is shown that the higher CO$_2$ flooding pressure and temperature, the more deposited asphaltene. It was also shown that a higher risk for asphaltene deposition in case of chalk cores than for sandstone cores.

Keywords: miscible CO$_2$ flooding, asphaltene, solubility parameter, light components

1. Introduction

Carbone dioxide flooding in Enhanced Oil Recovery (EOR) processes has been encouraging; however it may result to asphaltene deposition which it turns affect reservoir rock and fluid properties (Moritis, 2006; Chukwudeme & Hamouda, 2009; Hamouda et al., 2008; Idem & Ibrahim, 2002; Simon et al., 1978; De Boer et al., 1995; Burke et al., 1990; Haskett & Tartera, 1965). Haskett and Tartera (1965) reported that crude oils with low asphaltene percentage could experience asphaltene precipitation/deposition due to pressure reduction in early stage recovery, as well as reservoir fluid composition variations during enhanced recovery by gas/chemical injection.

Different models have been reported in the literature to describe the behaviour of asphaltene deposition using different approaches. Hirschberg (1984) described a method based on solubility model using the Flory-Huggins theory with thermodynamic model considering temperature and pressure effects on asphaltene precipitation. Kawanaka et al. (1991) extended Hirschberg et al. (1984) approach considering asphaltene is a large nonhomogeneous polymers providing better fitting though increasing, the number of parameters to be adjusted. Thomas et al. (1992) derived an empirical correlation including the precipitated asphaltene as a multicomponent system using liquid-solid wax theory. Yang et al. (1992) described a modified Hirschberg solubility model and pointed that the oil phase should be modelled as a multicomponent system. Nghiem (1999) documented a thermodynamic solid model to see the dynamic description of asphaltene precipitation/deposited using a compositional simulator during miscible CO$_2$ injection. Paricaud et al. (2002) used statistical association fluid (SAFT) theory of the thermodynamics of large chain polymers used to model the onset of stability of polymer-colloid mixtures. Updated Statistical association fluid theory (SAFT) Equation of State derived by Chapman et al. (2004) was described providing the influence of polymer shape, Van der Waals interaction and aggregation of molecules. Kirangkrai et al. (2007) showed an empirical correlation between the solubility parameter limit and the molar volume of precipitants to observe the effect of dissolved gas on the onset solubility parameter of live oils. Gonzalez et al. (2008) demonstrated that CO$_2$ can be an inhibitor or a promoter of asphaltene precipitation depending on temperature, pressure, and composition studied. They have shown that at specific pressure and for specific live oil, CO_2 addition increases the asphaltene stability less than the crossover temperature, while above this point, the asphaltene is more unstable when the CO_2 concentration is increased. Hamouda et al. (2009) derived a modified solubility equation to take to account the effect of dissolved CO_2 fraction on asphaltene precipitation. The model considers the volume occupied by one mole of CO_2 and fluid as a function of the solubility parameter of the fluid. They have proved that below an onset CO_2 fraction (mol%) in fluid, asphaltene is stabilized, while asphaltene precipitates based on the model and observed experimental asphaltene solubility as a function of temperature and pressure. Verdier et al. (2006) considered the effect of pressure and temperature on asphaltene instability in presence of gas components. They observed experimentally and verified with the thermodynamic model, the solubility of asphaltene in oil increases with increasing pressure and decreasing temperature.

Compositional models considering oil as multicomponents system and asphaltene molecular shape and size can have a more robust estimation of asphaltene solubility/precipitation. Hamouda et al. (2009) reported higher oil recovery due to better sweep efficiency in cores with higher deposited asphaltene. Among proposed models, none of them can predict asphaltene precipitation which could results to divergence of CO_2 streamlines and higher sweep efficiency.

In spite of the fact that there are many literatures describing asphaltene solubility in oil and associated proposed thermodynamical models, few works can link the measured experimental asphaltene deposition in rocks and estimated values predicted by model. Within them, the influence of light components in oil (e.g., methane and propane) on asphaltene precipitation during CO_2 miscible flooding is even not addressed individually. The main objective in this paper is to address asphaltene precipitation due to CO_2 miscible flooding of oil containing methane and propane at different miscibility conditions. A comparison is made in the paper between asphaltene precipitated and estimated.

2. Experiments

2.1 Cores

Outcrop chalks and outcrop Benthiemer sandstones are used in this study. The chalks have porosity of forty up to forty eight percentages and absolute permeability of two to six mili Darcy. Sandstone cores have porosity near twenty to twenty five percentage and permeability range between six to nine hundreds mili Darcy. Both types of cores are in macro scale homogenous while sandstone cores are in micro scale heterogeneous.

2.1.1 Preparation of Fluids Containing Asphaltene and Light Components

The synthetic oil system is composed of asphaltene from crude oil in addition of $n-C_7$ (with the ratio of 1 to 40). Mixture of crude oil and heptane was shaken for at least two times a day and left for two days to reach equilibrium conditions, and then the solution was centrifuged and filtered through a 0.22 micrometre filter and dried for 1 day using a vacuum at room temperature. 0.25 g asphaltene was dissolved in toluene (22 ml) and mixed with 0.01 M stearic acid ($CH_3(CH_2)_{16}CO_2H$) dissolved in n-decane (42 ml) and then filtered for chalk cores, and 0.01 M N, N-dimethyldodecylamine ($CH_3(CH_2)_{11}N(CH_3)_2$) dissolved in n-decane (42 ml) and after filtration is used for sandstone cores. The live oils are made by mixing of methane and propane with the dead oil at a constant gas oil solubility ratio (GOR = 280 ft^3/bbl at standard conditions). The methane or propane was supported from cylinder at P_b to the dead oil cell containing dead oil by pump to make the live oil, and then the live oil cell was rotated by the rotating mixer with the speed of 50 rpm. The pressure of the cell was observed twice per hour. Then, the pressure was set as required using a pump, to keep a pressure higher than the calculated P_b of the fluid. Stable pressure was reached by mixing process for at least one day.

2.2 Experimental Procedure

Experimental apparatus is demonstrated in Figure 1. The CO_2 is injected after aging for cores by dead oil using core holder that consists of Hassler core holder and rubber/nylon sleeve to prevent from any corrosion by Carbone dioxide. For CO_2 injection of saturated cores with live oil, first the dead oil is removed by live oil near 1 PV with injection pressure of at least 20 bar higher than the P_b live oil. The oil saturated core (with live or dead oil) was flooded with CO_2 at pressure equal to MMP. CO_2 was flooded into the core at constant pressures of 90 ± 0.2, 120 ± 0.2 and 140 ± 0.2 bar for corresponding temperatures of 50, 70 and 80 °C. The pressure of gas is providing by injection pump. Mass flow meter one measures the inlet flow properties of Carbone dioxide (mass flow rate, density and total mass). A back pressure regulator is attached to the core to control the pressure during CO_2 flooding. The outlet flow properties (mass flow rate, density and total mass) of the produced gas/effluent were determined using flow meter two connected to the separator. The CO_2 injection continued for at least 4 PV until steady state condition that no extra oil is produced. Finally core was dried using vacuum at a temperature of

120 °C until a constant weight was reached. Flooded cores by CO_2 were crushed as powder and dried in oven under vacuum. The asphaltene precipitated is obtained from the weight of the dried core before the experiments and after the flooding from powdered cores at high temperature and under vacuum. The experimental errors of selected repeated crush cores are within ten percent. It is worth mentioning that the variations of observed asphaltene precipitation weights are more than this limit.

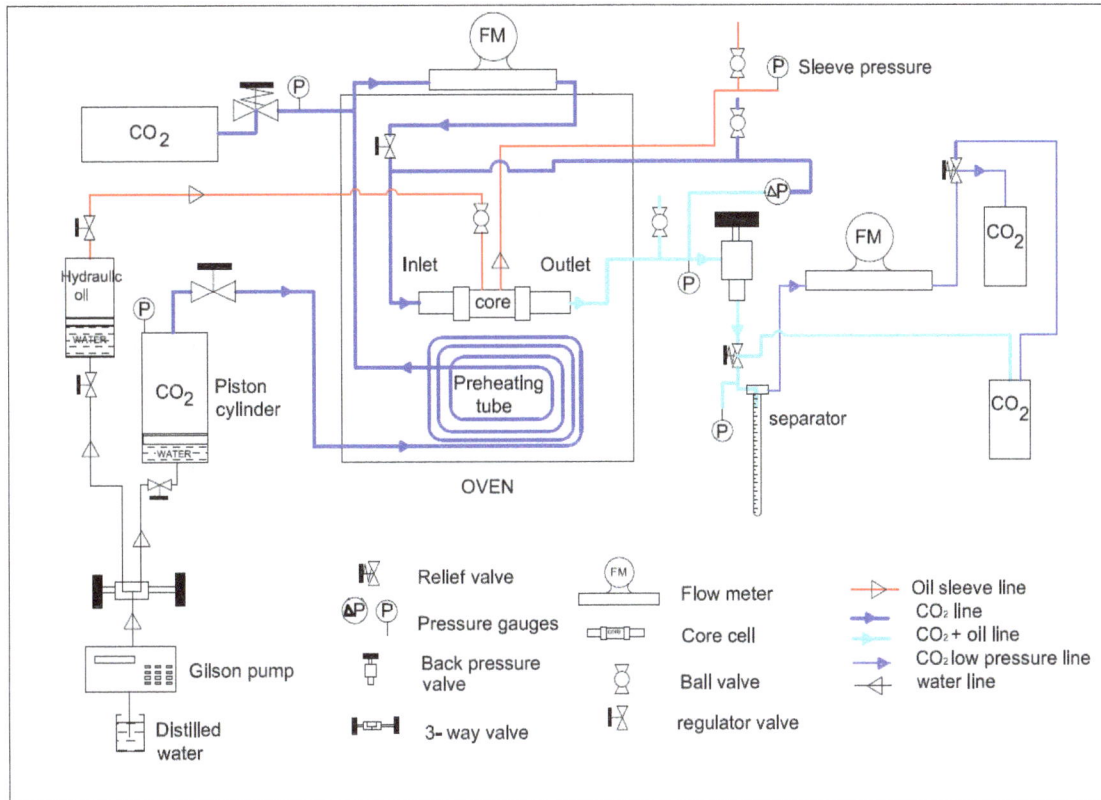

Figure 1. Experimental set up (Vahid Alipour Tabrizy, 2012)

3. Main Results and Analysis

3.1 Effect of Miscible CO_2 Flooding on Asphaltene Deposition

As CO_2 dissolves in the oil, the oil expands and the interfacial tension between oil and CO_2 reduces. Light hydrocarbon components can also be vaporized/extracted by Carbone dioxide. Oil composition gradients are developed from the injection inlet until production outlet depending on the mole percentage of injected Carbone dioxide and flooding conditions. To address the effect of the miscible CO_2 flooding on the asphaltene precipitation, it is important to determine the solubility parameter for different oil compositions (live and dead oils) and different flooding conditions. There are a wide spread definition of the solubility parameter (Verdirer & Anderson, 2005). The definition where the cohesive energy is equal to the residual internal energy is used by applying the equation below for liquid solubility parameter. This is adapted in this work to account for pressure and temperature effects.

$$\delta_L = \left(\frac{U_{vap}(T, P=0) - U_{liq}(T,P)}{V_{liq}(T,P)} \right)^{1/2} \tag{1}$$

Where U_{vap} and U_{Liq} are internal energy of vapour and liquid phases respectively and V_{Liq} demonstrate the volume of the one mole of mixture and are obtained from Redlich–Kwong equation of state -Peneloux Equation of State, using PVT simulator version Seventeen (2007).

The definition of asphaltene solubility parameter (δ_A) as a function of temperature, by Hirschberg et al. (1984) is used in this work.

$$\delta_A = 20.04 * \left(1 - 1.07 * 10^{-3} * T(C)\right) \tag{2}$$

At miscible flooding condition of 90 bar and 50 °C, a comparison of liquid solubility parameters for dead and live oils is shown in Figure 2 for the different mol% of CO_2 in the liquid. At this flooding condition, asphaltene precipitation is probable. The probability increases with the CO_2 content for both dead and live oil. Live oil #1 (methane with dead oil), shows slightly less difference between δ_A and δ_L at all CO_2 contents compared to live oil #2 (recombined methane and propane with dead oil recombined at same gas oil ratio), indicating asphaltene instability. Kirangkrai et al. (2007) observed similar trend in their study of asphaltene instability in live crude oils. Verdier et al. (2006) reported same observation, where solubility parameter of the liquids decreases with dissolved gas in the liquid. The effect of temperature and pressure is shown for model oil and live oils. The combined temperatures and pressures for miscible flooding of model oil are 50 °C/90 bar, 70 °C/120 bar and 80 °C/140 bar, however, for the live oil, the temperature is kept at 50 °C and the pressure changed similar to that for model oil. This is done in order to be within minimum miscibility pressure. The results obtained for the model oil at 50, 70 and 80 °C and pressures of 90, 120 and 140 bar are shown in Figure 3 to Figure 5, respectively. For a same CO_2 content, it is shown that the higher the pressure, the higher δ_L is. As CO_2 content less than 10% mole, the asphaltene is stable for combined miscible conditions of 50 °C and pressures of 140 and 120 bar (Figure 3). When the temperature is then increased, for CO_2 content less than 10% mole, to 70 and 80 °C with pressures 120 and 140 bar, the asphaltene became in the unstable region (Figure 4). However, Figure 5 indicates that asphaltene is in the unstable region for the combined miscibility conditions of 50 °C/90 bar, 70 °C/120 bar and 80 °C/140 bar. On the other hand for the live oil #1 and #2 as shown in Figure 6a and Figure 6b, respectively, in general, same trend as for the model oil is observed. However at all temperature/ pressure combinations, the oil lay in the unstable region. It can also be seen that in all cases, increasing pressure increases the asphaltene stability while increasing the temperature reduces the asphaltene stability. The asphaltene stability is related qualitatively to the solubility parameter. This is in agreement with the findings by Verdier et al. (2006).

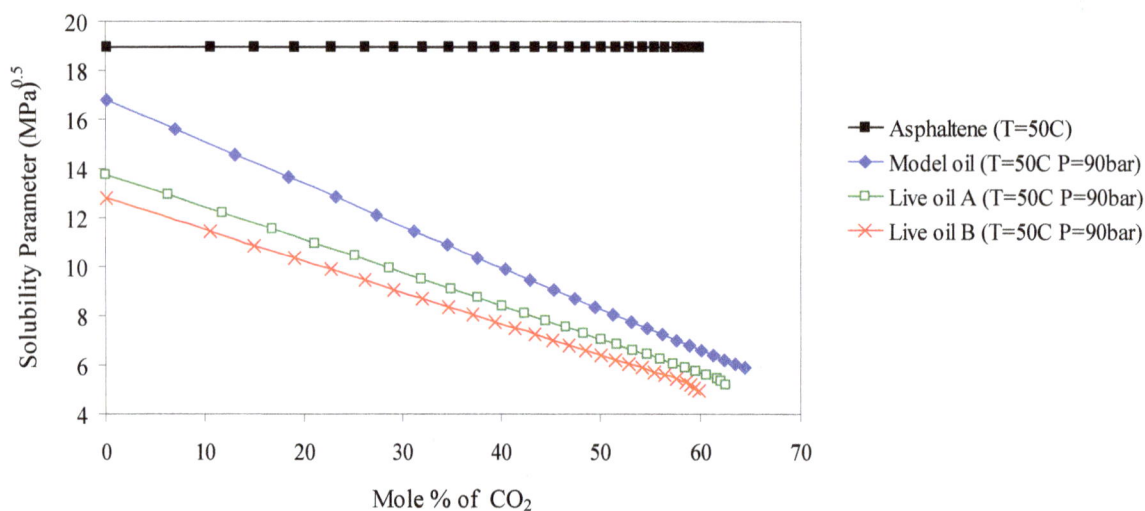

Figure 2. Comparison between liquid solubility parameter and asphaltene solubility parameter for different types of oil at flooding temperature of 50 °C and flooding pressure of 90 bar

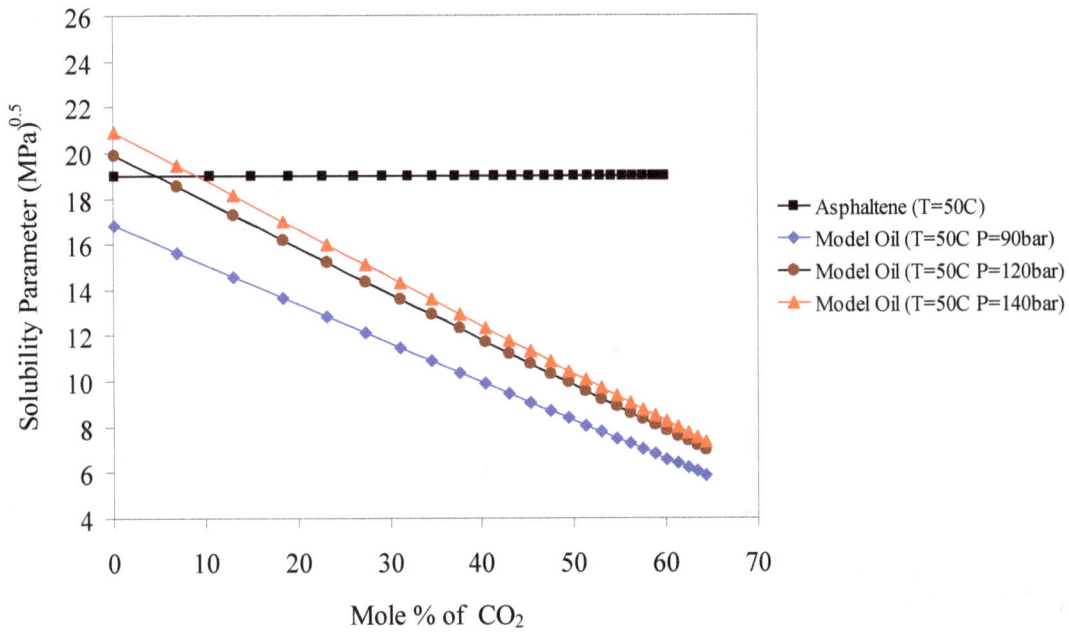

Figure 3. Comparison between liquid solubility parameters and asphaltene solubility parameter for model oil at flooding temperature of 50 $^{\circ}$C and different flooding pressures

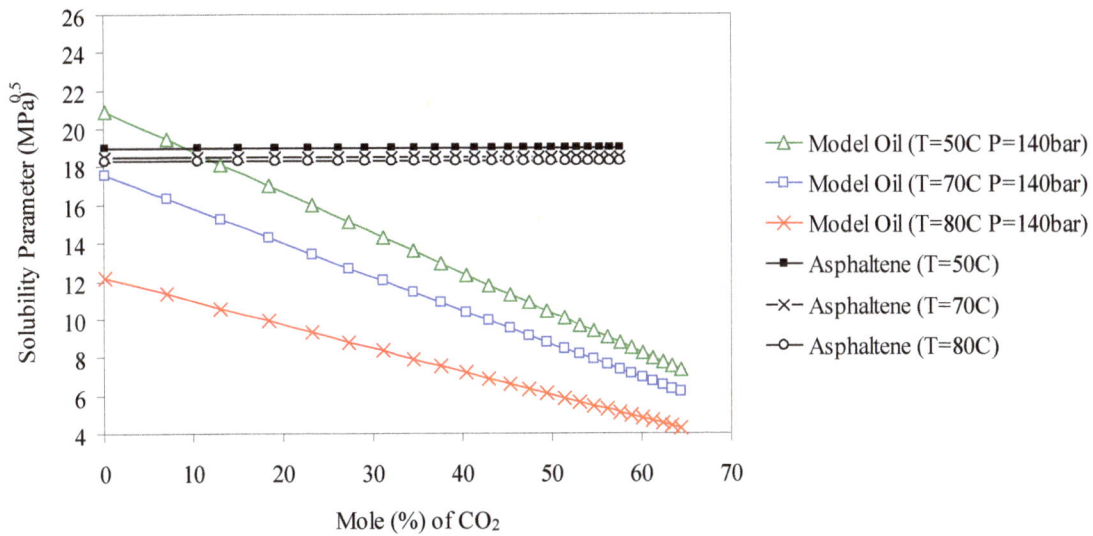

Figure 4. Comparison between liquid solubility parameters and asphaltene solubility parameter for model oil at flooding pressure of 140 bar and different flooding temperatures

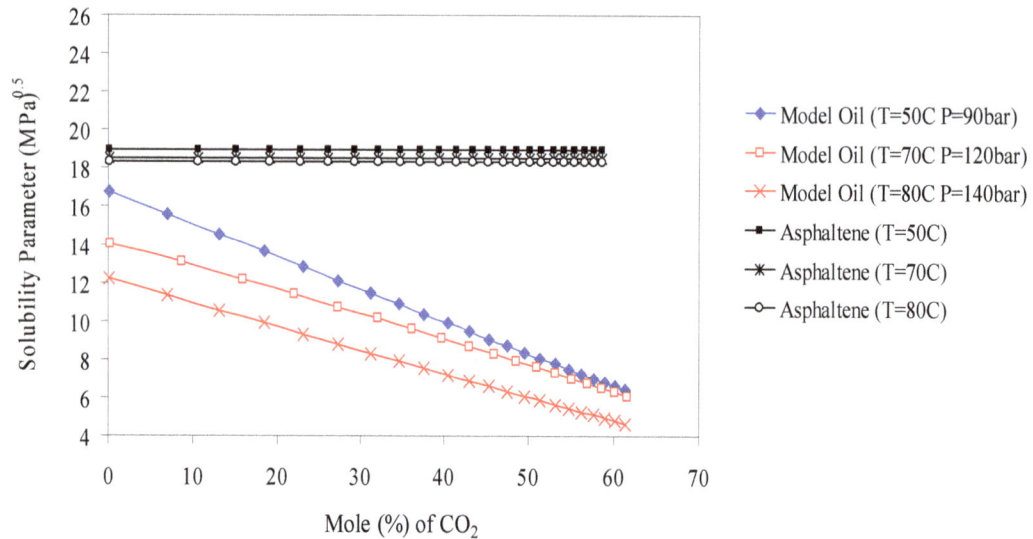

Figure 5. Comparison between liquid solubility parameters and asphaltene solubility parameter for model oil at different miscibility conditions (T=50 °C/P=90 bar, T=70 °C/P=120 bar, T=80 °C/P=140 bar)

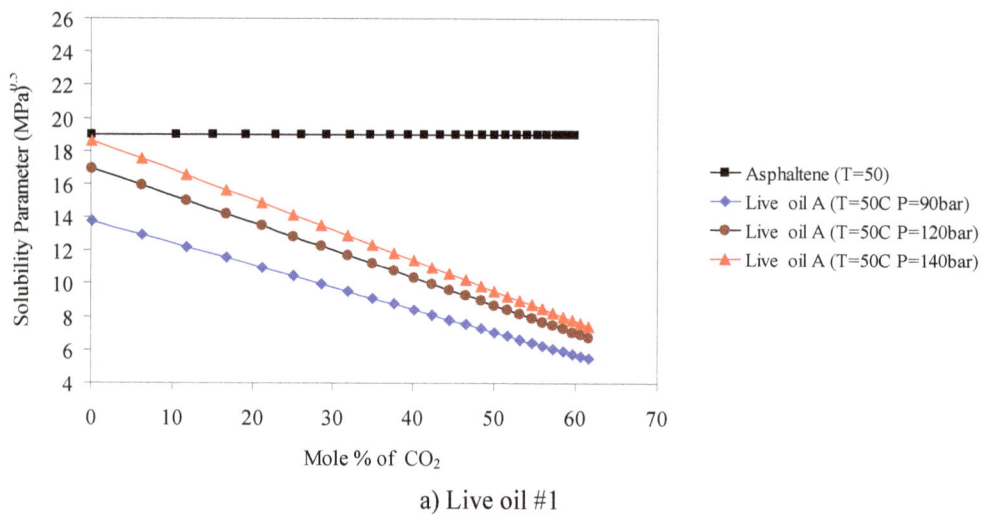

a) Live oil #1

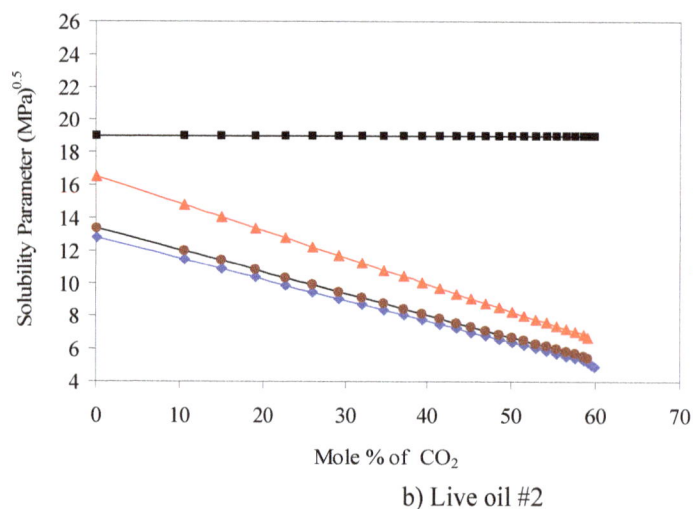

b) Live oil #2

Figure 6. Comparison of liquid solubility parameter and asphaltene solubility parameter at flooding temperature of 50 °C and different flooding pressures for a) live oil #1 and b) live oil #2

3.2. Comparison Between Precipitated and Estimated Asphaltene

Asphaltene deposition is estimated using the model developed by Hamouda at al. (2009) defined by Equation (3). Figure 7a and Figure 7b compare the estimated asphaltene from flooded cores, saturated with different dead and live oil types, at different miscibility conditions.

$$WA_{Model}(\%) = \frac{W_{TAL} - W_{AL}}{\left(V_{TL} - V_{TL} * exp\left[\frac{V_A}{V_L} - 1 - \frac{V_A}{RT}(\delta_A - \delta_L)^2\right]\right)\rho_A + W_{AL}} * 100 \tag{3}$$

Where, W_{TAL} shows the total asphaltene in the liquid (g), W_{AL} is the mass of asphaltene in the liquid phase (g), ρ_A and V_{TL} represents the asphaltene density(g/cm^3) and the total volume of solution (cm^3), A and L shows respectively to asphaltene and liquid phase, V(cm^3/mol) and T(K) shows volume of one mole and temperature, R (MPa.cm^3.mol^{-1} K^{-1}) is the gas constant and δ_A and δ_L (MPa$^{1/2}$) are solubility parameters for asphaltene and liquid, respectively. The experimental results for the different cases are shown in Figure 8a and Figure 8b, for sandstone and chalk, respectively. Higher asphaltene deposition corresponds to the higher miscible flooding conditions and presence of the light components (methane and propane), for both types of the cores. This is in line with the indicated results from the solubility parameter curves, where at the aforementioned conditions and presence of light components, the more the risk for precipitation as the difference between δ_A and δ_L increases. It can also be seen that the amount of asphaltene deposition is higher in the case of chalk compared to sandstone for the same oil type and flooding conditions. This may be due to the difference in the surface area for the two cores. Kozeny-Carman correlation demonstrated that chalk cores have tighter pore throats and larger surface areas compared to that for sandstone cores, where, the surface area changes between one and two m^2/g while for sandstone cores, the surface area varies between 0.02 and 0.04 m^2/g. It is attempted here to relate asphaltene deposition induced by miscible CO$_2$ flooding mechanism to the mixing zone index (I) defined by Equation (4).

$$I = 3.625\sqrt{\frac{K_L}{vx}} \tag{4}$$

In this equation, K_L is longitudinal dispersion coefficient (m^2/sec), v is the one dimensional velocity of the injected CO$_2$ through core and x is the length of core.

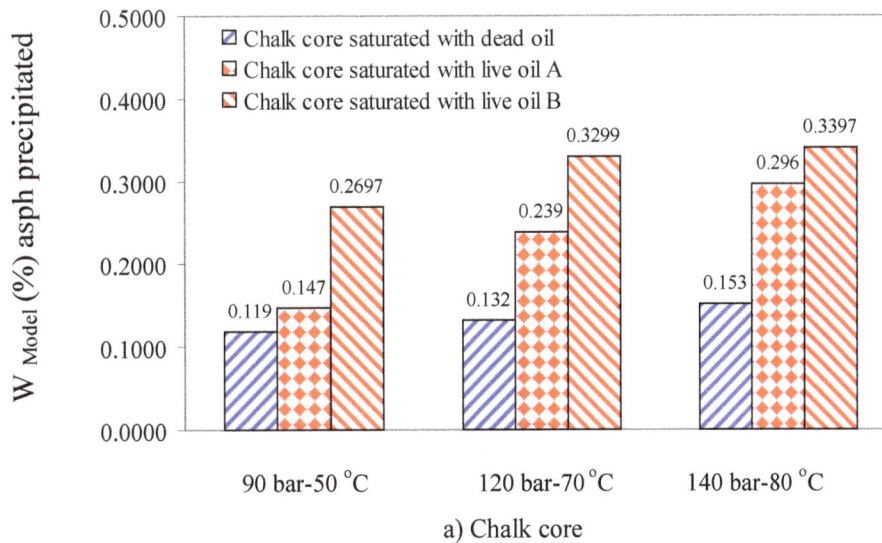

a) Chalk core

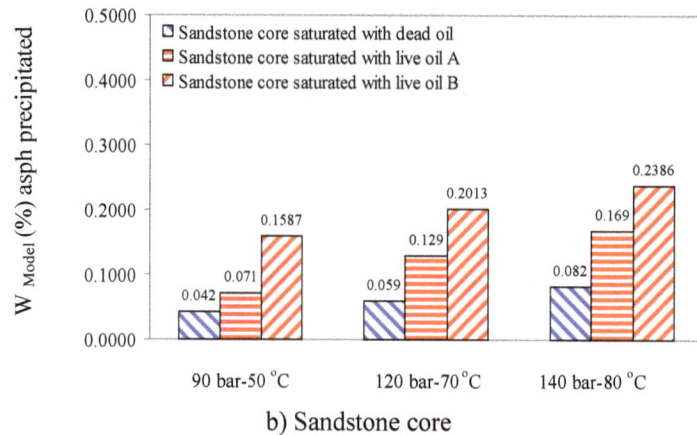

b) Sandstone core

Figure 7. Comparison between predicted amount of asphaltene deposition for different types of oil at different miscibility conditions (T=50 °C/P=90 bar, T=70 °C/P=120 bar, T=80 °C/P=140 bar) in a) chalk core and b) sandstone core

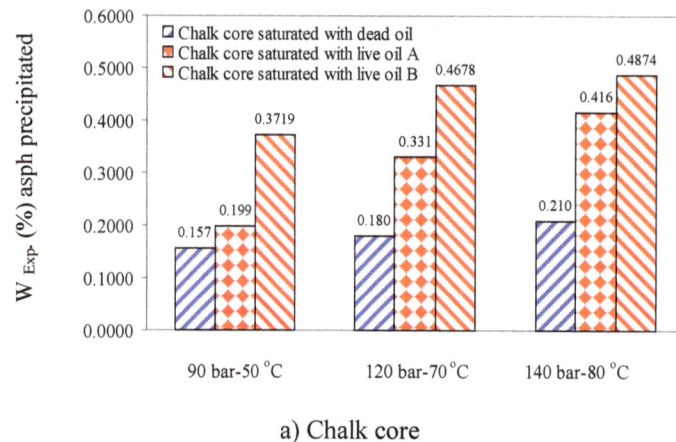

a) Chalk core

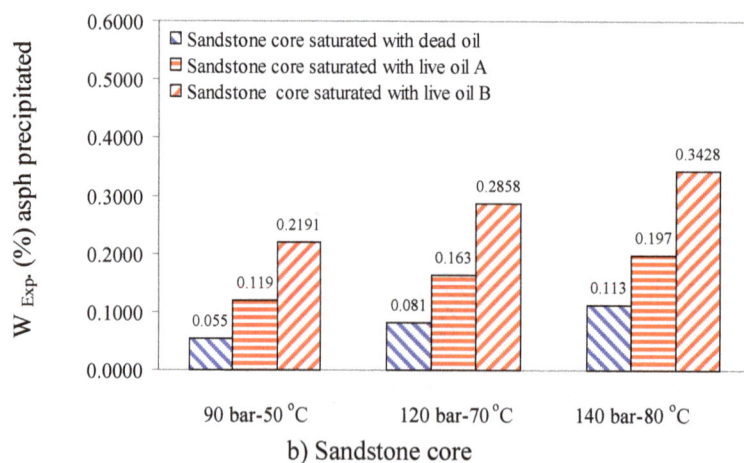

b) Sandstone core

Figure 8. Comparison between experimental measured asphaltene deposition for different types of oil at different miscibility conditions (T=50 °C/P=90 bar, T=70 °C/P=120 bar, T=80 °C/P=140 bar) in a) chalk core and b) sandstone core

Figure 9 compares the mixing zone index at three miscible flooding conditions for both sandstone and chalk cores. The figure shows that miscible bank zone increases as the miscible flooding conditions increase. In other words, surface area and the mixing zone enhance the contact efficiency between the flooding and flooded fluids.

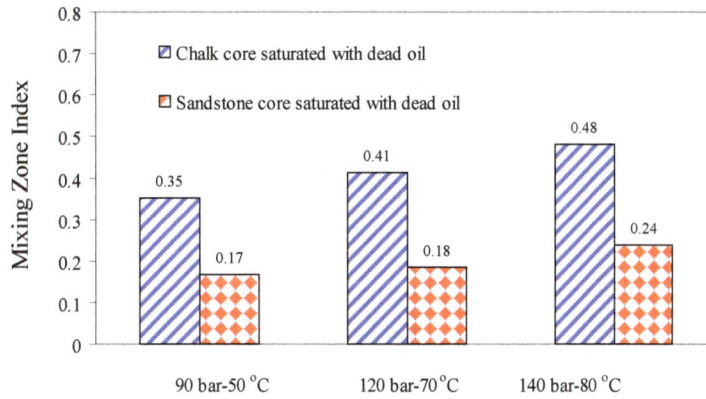

(a)

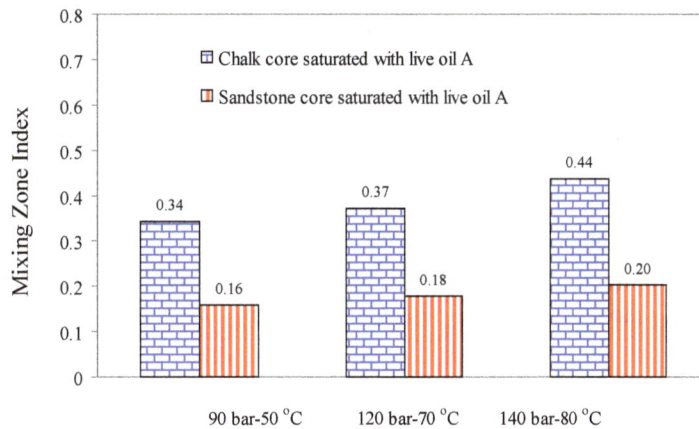

(b)

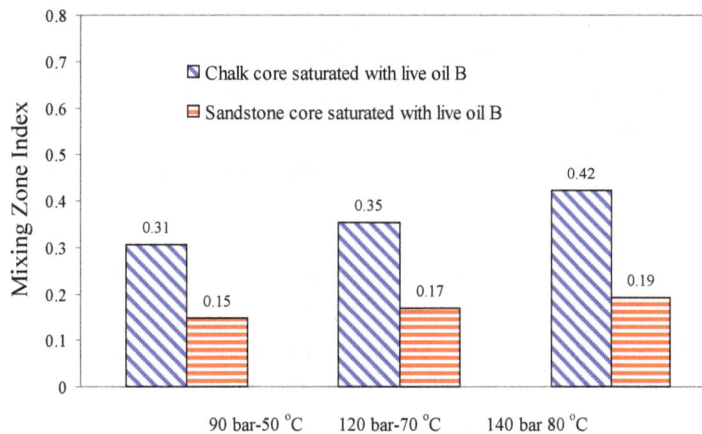

(c)

Figure 9. Comparison of mixing zone index for a) chalk and sandstone cores saturated with dead oil b) cores saturated with live oil #1 (with methane and gas oil ratio as 280 ft^3/bbl at standard conditions and c) cores saturated with live oil #2 with methane and propane and gas oil ratio as 280 ft^3/bbl at standard conditions)

4. Conclusions

An approach to assess the process for asphaltene deposition during the miscible CO_2 flooding is developed. The approach is based on solubility parameters/CO_2 fraction in the liquid. The used solubility parameter of the liquid is defined when cohesive energy is equal to the residual internal energy. The adapted equation accounts for pressure and temperature effects. This approach enables the determination of the critical CO_2 fraction that below it the asphaltene is stable.

Solubility parameters increases with pressure and decreases with temperature. In general, the presence of light components enhances the risk for the asphaltene instability. The comparison between predicted and experimental asphaltene deposition, is in agreement with the trend of the results obtained from the solubility parameters/CO_2 content, where the light components and the higher the miscible flooding conditions, induced asphaltene instability.

Acknowledgements

The authors wish to thank the University of Stavanger for the financial support, Krzysztof Piotr Dziadosz for his undeniable technical supports, Svein Myrhen for Lab view support and Inger Johanne for her positive attitude and great assistance in getting the chemicals used in this work.

References

Alipour, T. V. (2012). *Investigated miscible CO_2 flooding for enhancing oil recovery in wettability altered chalk and sandstone rock* (Doctoral dissertation). University of Stavanger, Norway.

Burke, N., Hobbs, R., & Kashou, S. (1990). Measurement and Modeling of Asphaltene Precipitation (includes associated paper 23831). *Journal of Petroleum Technology, 42*(11), 1440-1446.

Chapman, W. G., Sauer, S. G., Ting, D., & Ghosh, A. (2004). Phase behavior applications of SAFT based equations of state—from associating fluids to polydisperse, polar copolymers. *Fluid phase equilibria, 217*(2), 137-143.

Chukwudeme, E. A., & Hamouda, A. A. (2009). Enhanced oil recovery (EOR) by miscible CO_2 and water flooding of asphaltenic and non-asphaltenic oils. *Energies, 2*(3), 714-737.

De Boer, R. B., Leerlooyer, K., Eigner, M. R. P., & Van Bergen, A. R. D. (1995). Screening of crude oils for asphalt precipitation: theory, practice, and the selection of inhibitors. *Old Production & Facilities, 10*(1), 55-61.

Gonzalez, D. L., Vargas, F. M., Hirasaki, G. J., & Chapman, W. G. (2007). Modeling Study of CO_2-Induced Asphaltene Precipitation. *Energy & Fuels, 22*(2), 757-762.

Hamouda, A. A., Chukwudeme, E. A., & Mirza, D. (2009). Investigating the effect of CO_2 flooding on asphaltenic oil recovery and reservoir wettability. *Energy & Fuels, 23*(2), 1118-1127.

Haskett, C., & Tartera, M. (1965). A practical solution to the problem of asphaltene deposits-Hassi Messaoud Field, Algeria. *Journal of petroleum technology, 17*(4), 387-391.

Hirschberg, A., DeJong, L. N. J., Schipper, B. A., & Meijer, J. G. (1984). Influence of temperature and pressure on asphaltene flocculation. *Old SPE Journal, 24*(3), 283-293.

Idem, R. O., & Ibrahim, H. H. (2002). Kinetics of CO_2-induced asphaltene precipitation from various Saskatchewan crude oils during CO_2 miscible flooding. *Journal of petroleum science and engineering, 35*(3), 233-246.

Kawanaka, S., Park, S. J., & Mansoori, G. A. (1991). Organic deposition from reservoir fluids: a thermodynamic predictive technique. *SPE Reservoir Engineering, 6*(2), 185-192.

Kraiwattanawong, K., Fogler, H. S., Gharfeh, S. G., Singh, P., Thomason, W. H., & Chavadej, S. (2007). Thermodynamic solubility models to predict asphaltene instability in live crude oils. *Energy & fuels, 21*(3), 1248-1255.

Moritis, G. (2006). 2006 worldwide EOR survey. *Oil & Gas Journal*, 45-57.

Nghiem, L. X. (1999). *Phase Behaviour Modeling and Compositional Simulation of Asphaltene Deposition in Reservoirs* (Doctoral dissertation). University of Alberta, Department of Civil and Environmental Engineering, Edmonton, AB, Canada.

Paricaud, P., Galindo, A., & Jackson, G. (2002). Recent advances in the use of the SAFT approach in describing electrolytes, interfaces, liquid crystals and polymers. *Fluid phase equilibria, 194*, 87-96.

Simon, R., Rosman, A., & Zana, E. (1978). Phase-behavior properties of CO_2-reservoir oil systems. *Old SPE Journal, 18*(1), 20-26.

Verdier, S., & Andersen, S. I. (2005). Internal pressure and solubility parameter as a function of pressure. *Fluid phase equilibria, 231*(2), 125-137.

Verdier, S., Carrier, H., Andersen, S. I., & Daridon, J. L. (2006). Study of pressure and temperature effects on asphaltene stability in presence of CO_2. *Energy & fuels, 20*(4), 1584-1590.

Yang, Z., Ma, C. F., Lin, X. S., Yang, J. T., & Guo, T. M. (1999). Experimental and modeling studies on the asphaltene precipitation in degassed and gas-injected reservoir oils. *Fluid phase equilibria, 157*(1), 143-158.

3

Gas Permeation Study Using Porous Ceramic Membranes

Mohammed N. Kajama[1], Ngozi C. Nwogu[1] & Edward Gobina[1]

[1] Centre for Process Integration and Membrane Technology (CPIMT), IDEAS Research Institute, School of Engineering, The Robert Gordon University, Aberdeen, AB10 7GJ, United Kingdom

Correspondence: Professor Edward Gobina, Centre for Process Integration and Membrane Technology, School of Engineering, The Robert Gordon University, Riverside East, Garthdee Road, Aberdeen, AB10 7GJ, United Kingdom. E-mail: e.gobina@rgu.ac.uk

Abstract

A 6000 nm ceramic membrane was repaired with boehmite solution (ALOOH) through the repeat dip-coating technique. The permeance of hydrogen (H_2) and carbon dioxide (CO_2) were obtained through the membrane in relation to average pressure at room temperature for the support membrane and as cracked membrane. A repair process was carried out on the cracked membrane by same dip coating process and results obtained after first and second dips. The permeance of the support membrane obtained ranged between 1.50 to 3.04×10^{-7} mol m^{-2} s^{-1} Pa^{-1}. However, as a result of a crack that occurred during the removal of the membrane from the reactor, the permeance increased from 2.96 to 5.82 10^{-7} mol m^{-2} s^{-1} Pa^{-1}. Further application of boehmite solution on the membrane lead to an improvement on the surface of the membrane with some degree and surface cracks were reduced. This also decreased the permeance to $1.26 – 3.39 \times 10^{-8}$ mol m^{-2} s^{-1} Pa^{-1} after the second dip. Consequently, another silica based modified membrane was used for carbon dioxide and nitrogen (N_2) permeation. The plots show that carbon dioxide permeated faster than the other gases, indicating dominance of a more selective adsorptive transport mechanism. Accordingly, results obtained show an appreciable high carbon dioxide permeance of 3.42×10^{-6} mol m^{-2} s^{-1} Pa^{-1} at a relatively low pressure when compared to nitrogen confirming that the membrane has so far exhibited a high permeability, selectivity and high CO_2 gas recovery. The permselectivities of CO_2 over H_2 at room temperature was also obtained which were higher than the Knudsen selectivity.

Keywords: ceramic membranes, gas permeation, hydrogen selectivity, carbon dioxide selectivity, defect repair

1. Introduction

Carbon dioxide is one of the many greenhouse gases which contribute to global warming. Membrane technology has been suggested as a substitute to the use of conventional separation processes (e.g. dehydration, gas adsorption, distillation among others) due to their unique attributes such as thermal and mechanical stability, as well as harsh chemical resistance (Singh et al., 2004; Anwu et al., 1997). The employment of inorganic membranes in the industry has extended the application of membranes for hydrogen production, CO_2 recovery as well as H_2S removal from associated gas feed streams because of their simplicity and low energy requirement (Yildirim & Hughes, 2002).

There are several methods for porous membrane modification including; dip-coating, chemical vapour deposition (CVD), and pulsed layer deposition (Pejman et al., 2011; Benito et al., 2005; Koutsonikolas et al., 2010). Out of the mentioned modification methods, dip-coating method has many merits over the other methods such as its simplicity, uniform surface and the ability of controlling the pore structure of the membrane (Pejman et al., 2011). However, a lot of research is still needed to examine membrane modification through dip-coating method in order to elucidate the morphological effects of dip-coated membranes. Membranes defects are formed during preparation stages. It can be formed either during dipping, drying, calcination process and sealing (Koutsonikolas et al., 2010) or even during the process of inserting and removing it from the reactor. Any defect on a macroporous membrane (pore diameter > 50 nm) can be regarded as a crack. For example, a defect is considered in the presence of super-micropores (0.7 nm < pore diameter < 2 nm) instead of ultra-micropores (pore diameter < 0.7 nm) (Koutsonikolas et al., 2010). It is known that any amount of defect on the membrane can significantly lower the membrane selectivity.

Lambropoulos et al. (2007) repaired γ-alumina and silica membranes at 573 K by CVD process with a tetraethylorthosilicate (TEOS)/O_3 counter reactant configuration. The defect was characterised with a permeability technique and a novel mercury intrusion. However, Gopalakrishnan et al. (2007) applied a hybrid processing method for hydrogen-selective membrane preparation. They applied a primary sol-gel silica layer for the CVD zone thickness reduction, and a CVD modification with tetramethoxysilane (TMOS) and O_2 at 873 K. After which they only examined H_2/N_2 selectivity to be 2300 at 873 K. Pejman et al. (2011) modified the surface of ceramic supports to facilitate the deposition of defect-free overlying micro and mesoporous membrane. They investigated the effects of dipping time, heating rate, and number of coated layers on microstructure of the modified layers in their study. They have achieved a smoother surface and the cracks size was reduced dramatically after two dip-coating steps.

In this paper, we have repaired a crack on a ceramic membrane with boehmite solution (ALOOH), also carbon dioxide separation and the effects of permeation properties such as; permeation pressure was examined.

2. Experimental

The commercial ceramic membrane used was supplied by Ceramiques Techniques et Industrielles (CTI SA) France, consisted of (77% α-alumina + 23% TiO_2) with an average pore diameter of 6000 nm. The membrane has 19.8 mm and 25 mm internal and outer diameter respectively, and a permeable length of 318 mm. The feed pressure applied was between 0.05 to 5 bar at room temperature. Figures 1 and 2 show a SEM image of inside and outside surface of the porous ceramic membranes. The membrane's structure was defect-free before the 1st dip. An SEM of the cross section of the same membrane is also shown in Figure 3. During the removal of the membrane from the reactor; the membrane was cracked Figure 4 i.e. after gas permeation before modification.

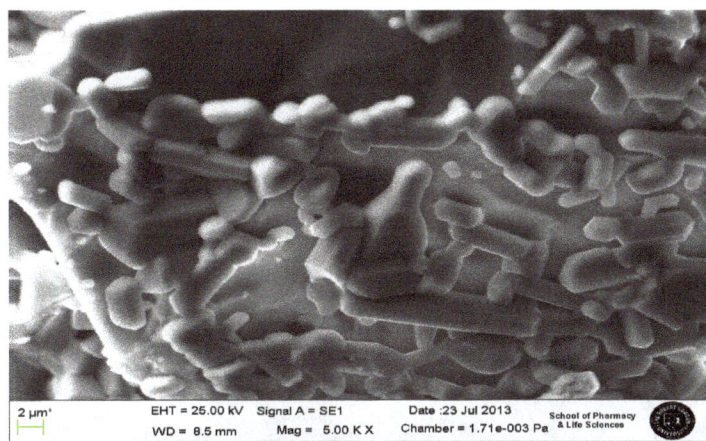

Figure 1. SEM image of the inside surface

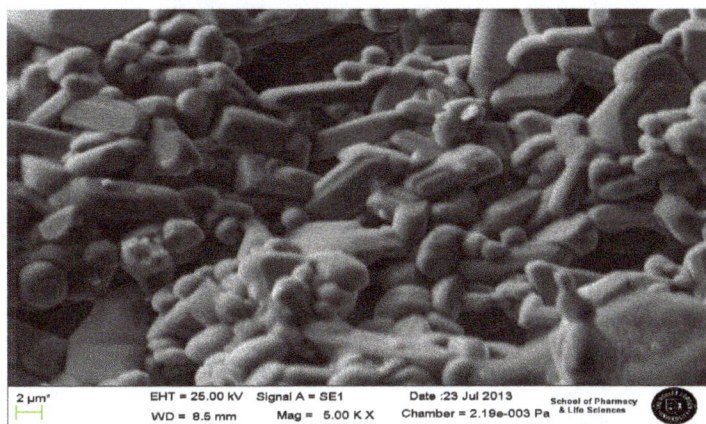

Figure 2. Outside surface of ceramic membrane

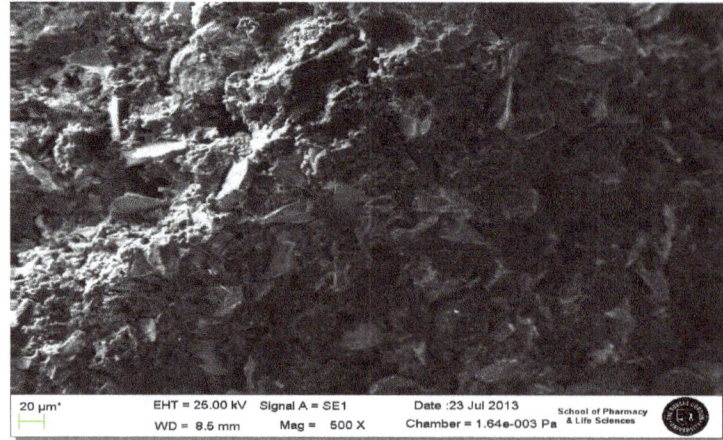

Figure 3. SEM image of the cross section of the membrane

In order to repair the defected surface of the substrate a 36 g/1000 L boehmite sol was used. A dip-coating method was applied to repair the defected membrane. The internal surface of the coarse alumina tube membrane was exposed to boehmite solution for 30 minutes. After this, the membrane was air-dried overnight and then heat-treated using the temperature profile shown in Figure 5. The dipping-drying firing procedure was repeated in order to achieve the required γ-alumina layer on the coarse support. The prepared membrane glazed at each end was sealed within the stainless steel reactor using graphite seals.

Figure 4. Pictorial view of the cracked membrane

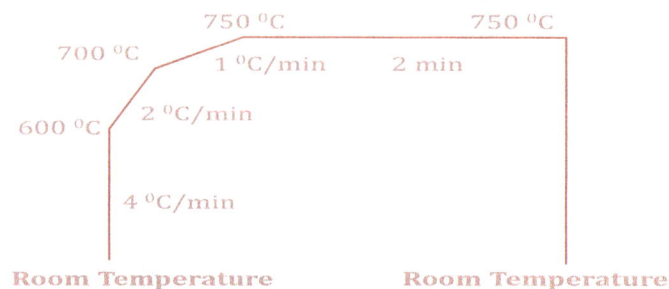

Figure 5. Membrane's heat-treatment profile

Single gas permeation measurements were carried out for gas components (H_2 and CO_2) at room temperature using the experimental setup shown in Figure 6 with the retentate valve fully open to the fume cupboard. The permeate flow tube was connected to the flowmeter to record gas flow rate. Gas permeance was obtained from the following expression;

$$Q = \left(\frac{q}{A.\Delta P} \right)$$ (1)

Where Q is the Permeance (mol m^{-2} s^{-1} Pa^{-1}); q is the molar flow (mol/sec); A is the membrane area (m^{2}); and ΔP is the pressure difference (Pa) across the membrane.

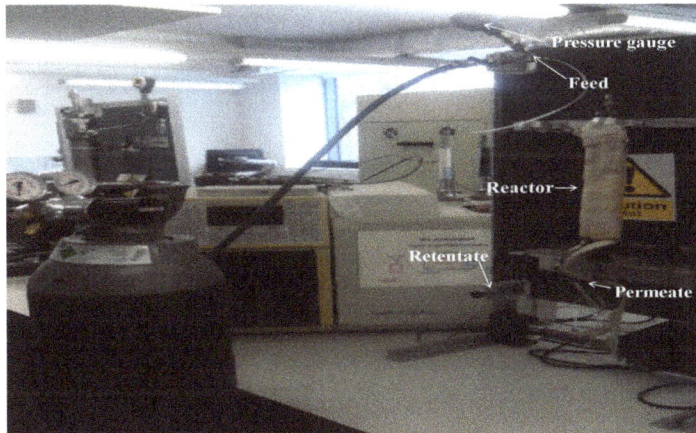

Figure 6. Experimental arrangement for the membrane reactor

The permselectivity of the membrane was also obtained using Equation (2);

$$S_{ij} = Q_i/Q_j$$ (2)

Where S_{ij} is the permselectivity of i to j; Q_i is the permeance of i (mol m^{-2} s^{-1} Pa^{-1}); Q_j is the permeance of j (mol m^{-2} s^{-1} Pa^{-1}).

3. Results and Discussion

Figure 7 shows the results of carbon dioxide permeance through the multilayered alumina membrane against average pressure at room temperature for the support membrane, cracked membrane, first and second dip repaired with boehmite solution stages. It can be seen that the permeance of the support is between 1.50 to 3.04 $\times$ 10^{-7} mol m^{-2} s^{-1} Pa^{-1}. However, due to the crack which occurred during the removal of the membrane from the reactor, the permeance increased to 2.96 – 5.82 10^{-7} mol m^{-2} s^{-1} Pa^{-1}. After exposing boehmite solution to the membrane, the surface was improved to some degree, surface cracks are also lowered, and the permeance decreased to 1.26 – 3.39 $\times$ 10^{-8} mol m^{-2} s^{-1} Pa^{-1} after the second dip which is almost parallel to the x-axis which indicates the applicability of Knudsen diffusion mechanism.

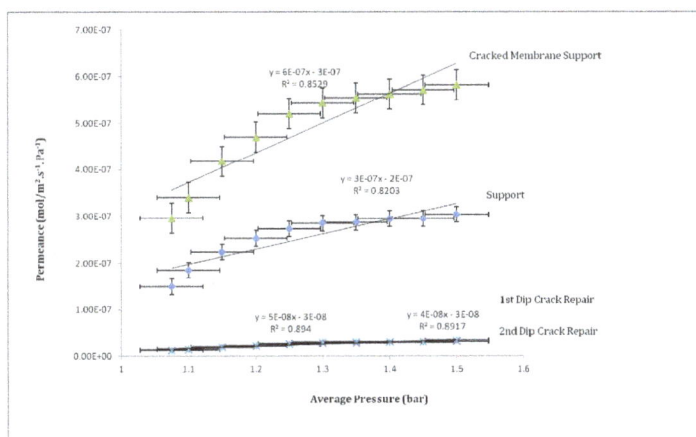

Figure 7. Permeance against average pressure for carbon dioxide at room temperature

Silica layer membrane was also prepared using the repeated dip-coating method (Gobina, 2006) which allows CO_2 to permeate faster from flue gas by maintaining a pressure drop across the membrane due to surface diffusion despite its higher molecular weight. It can be seen in Figure 8 that the graph showing the effect of feed pressure on CO_2 and N_2 permeance. From a feed pressure of about 1.4 bar, the permeance of CO_2 was quite higher than that of N_2, which indicates the presence of surface diffusion while that of N_2 reduced drastically which indicates the presence of Knudsen diffusion mechanism.

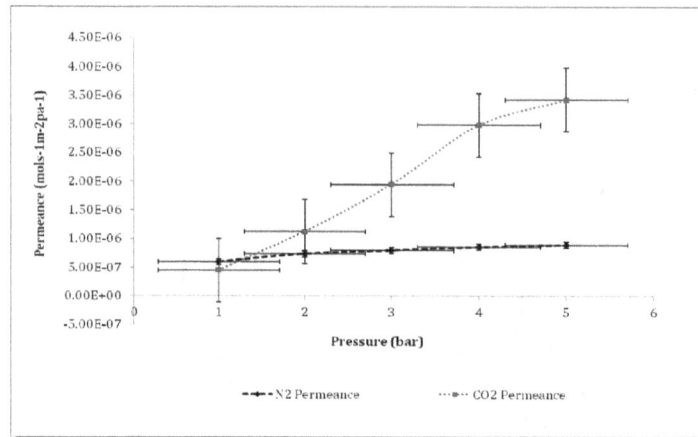

Figure 8. Effect of pressure on CO_2 and N_2 permeance

The permselectivities of CO_2 over H_2 at room temperature is shown in Figure 9. It can be seen that the selectivities obtained are higher than the Knudsen selectivity. After exposing the support to the boehmite solution, it was observed that a significant increase in the selectivity to CO_2 had occurred for the first dip crack repair as shown in Figure 9. Subsequent dips however reduced the CO_2 selectivity. This behaviour is related to the transport regime in the membrane. Transport owing to the combination of diffusion through the gas phase in the pores of the membrane and surface diffusion is a combination of mechanisms. The total transport in the absence of a pressure difference over the membrane of i owing to the combination of diffusion through the gas phase in the pores of the membrane and surface diffusion is described by Equation (3);

$$J_{i,\,tot} = J_{i,\,gas} + (4/d_p)\, J_{i,\,surf} \tag{3}$$

Where $J_{i,\,tot}$ = total molar flux owing to both transport mechanisms (mol-m^{-2}-s^{-1}), $J_{i,\,gas}$ = molar flux owing to transport in the gas phase in the pores (mol-m^{-2}-s^{-1}), and d_p = average pore diameter (m).

In the present study, it was assumed that the pores could be regarded as ideally cylindrically shaped although this assumption is questionable for alumina membranes. Another complicating factor is the homogeneity of the distribution of the γ-alumina over the α-alumina surface inside the membrane. As the main objective of the present study was to demonstrate the occurrence of surface diffusion effects in ceramic membrane reactors no special attention has been paid to the actual pore configuration. Moreover, pore shape only affects the value of the constants of the denominator of the term for $J_{i,\,surf}$ in Equation (3), e.g. rectangular pores have a value of 4. The geometric factor $4/d_p$ in Equation (3) arises from the fact that transport owing to diffusion through the gas phase in the pores is proportional to the cross-sectional area of the pores, therefore proportional to d_p^2 whereas transport owing to surface diffusion is proportional to the circumference of the pores, therefore proportional to d_p. The molar flux of component i through the gas phase in the pores of the membrane is described by Fick's law using the Bosanquet formula for the combination of Knudsen and continuum diffusion by the principle of resistances in series.

It is demonstrated experimentally that during the 1st dip repair surface diffusion of CO_2 in an alumina membrane impregnated with γ-Al_2O_3, can contribute substantially to the transport rate. Moreover, for this system it is of almost the same order of magnitude as the transport caused by ordinary diffusion through the gas phase in the pores. Subsequent dipping result in continuum diffusion and the molar flux is therefore independent of pressure. Surface diffusion however is linearly proportional to the pressure provided the fraction of the adsorption sites covered is very low. It is demonstrated that the resulting difference in pressure dependence of both transport mechanisms can be used to distinguish between diffusion through the gas phase in the pores and surface diffusion.

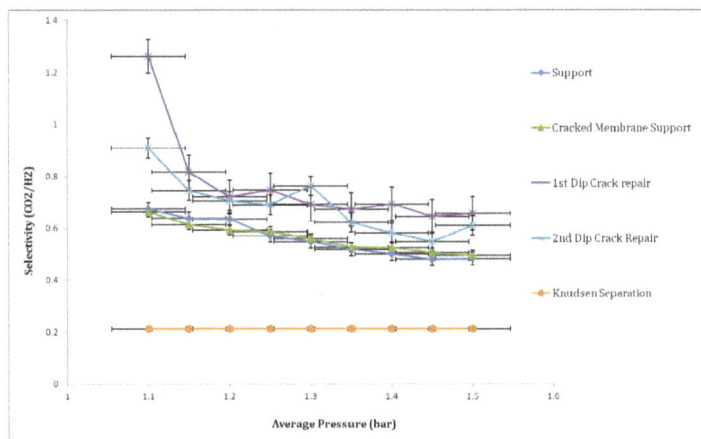

Figure 9. Selectivity of CO_2/H_2 against average pressure at room temperature

4. Conclusions

A simple but effective technique to modify nanostructure ceramic membranes which involves dip-coating method was studied and involved the use of boehmite solution to modify the membrane. Several parameters such as dipping time, number of coating, heating rate on the nanostructure were investigated. A high temperature heat-treatment was applied in order to repair defects on the commercial membranes, and to improve the permselectivity of these membranes above Knudsen regime selectivity. Single gas permeation tests were used to examine the permeance of the membrane. After the first and second modifications with boehmite solution, permeance decreased significantly which indicates the presence of Knudsen diffusion mechanism.

The silica membrane is thought to exhibit higher CO_2 permeance from flue gas N_2 by maintaining a pressure drop across the membrane due to surface diffusion mechanism. In general, inorganic membranes would exhibits an improved performance on carbon dioxide separation from flue gases.

Acknowledgements

The authors gratefully acknowledge Petroleum Technology Development Fund (PTDF) Nigeria for funding this research, and School of pharmacy & Life Sciences RGU for the SEM results.

References

Anwu, L. I., Hongbin, Z., Jinghua, G. U., & Guoxing, X. (1997). Preparation of γ-Al₂O₃ composite membrane and examination of membrane defects. *Science in China (Series B), 40*(1), 31-36. http://dx.doi.org/10.1007/BF02882185

Benito, J. M., Conesa, A., Rubio, F., & Rodriguez, M. A. (2005). Preparation and characterization of tubular ceramic membranes for treatment of oil emulsions. *Journal of the European Ceramic Society, 25*, 1895-1903. http://dx.doi.org/10.1016/j.jeurceramsoc.2004.06.016

Gobina, E. (2006). Apparatus and Methods for Separating Gases. United States Granted Patent No. US 7048778, May 23, 2006.

Gopalakrishnan, S., Yoshino, Y., Nomura, M., Nair, B. N., & Nakao, S. I. (2007). A hybrid processing method for high performance hydrogen-selective silica membranes. *Journal of Membrane Science, 297*, 5-9. http://dx.doi.org/10.1016/j.memsci.2007.03.034

Koutsonikolas, D., Kaldis, S., Sakellaropoulos, G. P., Loon, M. H. V., Dirrix, R. W. J., & Terpstra, R. A. (2010). Defects in microporous silica membranes: Analysis and repair. *Separation and Purification Technology, 73*, 20-24. http://dx.doi.org/10.1016/j.seppur.2009.07.027

Lambropoulos, A., Romanos, G., Steriotis, Th., Nolan, J., Katsaros, F., Kouvelos, E., ... Kanellopoulos, N. (2007). Application of an innovative mercury intrusion technique and relative permeability to examine the thin layer pores of sol–gel and CVD post-treated membranes. *Microporous and Mesoporous Materials, 99*, 206-215. http://dx.doi.org/10.1016/j.micromeso.2006.08.038

Pejman, A. N., Akbar, B. A., Elham, J., Majid, P., & Masoumeh, A. A. (2011). An optimum routine for surface modification of ceramic supports to facilitate deposition of defect-free overlaying micro and meso (nano) porous membrane. *Iran. J. Chem. Eng., 30*(3), 63-73.

Singh, R. P., Way, J. D., & McCarley, K. C. (2004). Development of a model surface flow membrane by modification of porous Vycor glass with a fluorosilane. *Ind. Eng. Chem. Res, 43*, 3033-3040. http://dx.doi.org/10.1021/ie030679q

Yildirim, Y., & Hughes, R. (2002). The efficient combustion of O-xylene in a Knudsen controlled catalytic membrane reactor. *Trans IChemE, 80*(Part B), 159-164. http://dx.doi.org/10.1205/095758202317576265

4

Influencing Carbon Behaviours: What Psychological and Demographic Factors Contribute to Individual Differences in Home Energy Use Reduction and Transportation Mode Decisions?

Clare Hall[1] & Fraser Allan[1]

[1] Land Economy, Environment and Society Research Group, SRUC, United Kingdom

Correspondence: Clare Hall, Environment and Society Research Group, Land Economy, SRUC, United Kingdom. E-mail: clare.hall@sruc.ac.uk

Abstract

As pressure mounts on countries to reduce carbon emissions, there is increasing interest in understanding what drives "carbon behaviours", in order to inform behavioural change policies. This study examined the impact of psychological and demographic variables, on "carbon behaviours". Secondary data analysis was carried out to investigate the antecedents of residential energy use reduction behaviours and choice of transportation mode for commuting and grocery shopping. Models explained 18.2% and 25.2% of variance in energy use and transport behaviours respectively. Being concerned about climate change and having an environmental identity increased household energy reduction behaviour but did not significantly affect travel mode choices. The antecedents of travel mode decisions were attitudes towards the travel mode itself, and demographic and structural variables such as income and distance travelled. Findings suggest that using "green" messaging will help encourage behavioural change in energy use, but contribute little to encouraging change in travel mode decisions.

Keywords: low carbon behaviours, energy use, travel mode, climate change, Scotland

1. Introduction

There is broad international acknowledgment that climate change, occurring as a result of human behaviours, is causing damage to the environment and is affecting ecological systems such as ocean currents and weather patterns (Parry et al., 2007). In order to stabilise global temperatures at levels which would limit the worst impacts, greenhouse gas (GHG) emissions must be reduced (Stern, 2006). Two sectors which make up a large proportion of global GHG emissions are household energy use and road transport. For example, emissions from household energy use are responsible for roughly 23% of GHG emissions in the UK (Palmer & Cooper, 2011), while emissions from passenger cars, buses, mopeds, and motorcycles account for around 13% (Department of Energy and Climate Change, 2012).

This means that reducing household energy use, and the use of GHG intensive transport, has an important part in lowering GHG emissions. These are both actions over which most people have direct control (Stead, 2007; Dijst et al., 2008), and thus offer a meaningful way for citizens to engage in GHG emission reduction. Governments recognise the important role that these kinds of voluntary behavioural changes have to play in reducing GHG emissions (see for example, Cabinet Office Behavioural Insights Team, 2011; Scottish Government, 2010). While this seeks to circumvent the need to impose regulatory restrictions on individuals, it still leaves a role for government in providing the necessary incentives and enablers. Thus governments need evidence of what influences peoples' behaviours. By way of illustration, at UK level, the Department of Energy and Climate Change began, in 2012, to collect data through a public attitudes tracker survey (DECC, 2013). This includes questions about attitudes to climate change.

While such data may help inform policy on how to meet emissions targets through behavioural change in the UK, the overall policy situation is somewhat complicated by devolution. The Scottish government passed the Climate Change (Scotland) Act 2009, setting targets for reducing GHG emissions by 42% by 2020 (higher than the rest of the UK) and 80% by 2050, from a baseline year of 1990 for carbon dioxide (CO_2) and 1990 or 1995 for other GHGs (Scottish Parliament, 2009) (Note 1). Thus Scotland presents an interesting case study through which to consider potential barriers to, and facilitators of, behavioural change relevant to achieving a lower carbon future.

Accordingly, this study uses data from a Scotland-wide survey to investigate how factors including attitudes to climate change, knowledge of climate change, and environmental identity relate to energy and travel behaviours. The remainder of this paper is structured as follows: the next section presents a review of models explaining how psychological and demographic factors affect behaviour, and an overview of empirical evidence from previous studies. Following that, the approach used in the present study is described, and results from the analysis are presented and discussed in relation to the existing theory and knowledge. The paper concludes with some recommendations and implications relating to how to motivate behavioural change in order to meet emission reduction targets.

1.1 Behavioural Models

The Theory of Planned Behaviour (TPB) (Ajzen, 1991) is a conceptual psychological framework for explaining why people undertake behaviour, which built on the Theory of Reasoned Action (TRA) proposed by Fishbein and Ajzen (1975). In its simplest form the TPB postulates that there are three determinants of a person's intention to perform an action: attitudes, subjective norms, and perceived behavioural control (PBC). Attitudes are the extent of favourable or unfavourable assessment of a particular behaviour (not the object of the behaviour) (Ajzen, 1991), while subjective norms are the degree to which a person believes "important others" would approve or disapprove of a given course of action (Fishbein & Ajzen, 1975). PBC is included as a determinant of intention because a perception of lack of control can form a real barrier (Dijst et al., 2008). PBC also acts directly on behaviour as a proxy for actual control (Fielding et al., 2008), based on a general assumption that perceived and actual control are highly correlated (Dijst et al., 2008).

It is now widely accepted that the TPB model has predictive limitations due to its generality, and in practice the majority of studies use additional determinants outside of the limitations originally envisaged (Armitage & Conner, 2001). Using additional context dependent variables can increase the explanatory power of the model (Fielding et al., 2008). Extensions have included *Knowledge*, typically added where it can be operationalised objectively (Fielding et al., 2008). *Identity* can be incorporated as a view of the self, where acting in a manner in accordance with one's identity gives a sense of validation, while acting contrary to it leads to inner tension (Fielding et al., 2008) (Note 2). *Prioritisation* of goals relates to the importance placed on different desired end results separate from the immediate outcome of the behaviour, and can also be important (Dijst et al., 2008; Sheppard et al., 1988).

Including both psychological and demographic factors typically explains a larger share of the variance, or spread in the data, than using one or the other alone (Fujii & Gärling, 2003; Dijst et al., 2008). Social scientists have tended to focus on how internal processes (e.g., values) affect behaviour, whereas economists tend to focus on external factors such as taxes or legislation (Guagano et al., 1995). Using demographic information as the sole source of prediction of behaviour implies that behaviour is constant as long as one's circumstances do not change, so that changing attitudes or emotions would have no effect (Dijst et al., 2008). To address this, Guagano et al. (1995) proposed their "ABC model" - which considers the effects of attitudes (A) and context (C) on behaviour (B). They suggested that the effects of different factors on behaviour are not necessarily constant, and that inclusion of both attitude and context variables, and the interaction between them, is needed to expand the predictive usefulness of behavioural models.

1.2 The Influence of Psychological Factors on Environmental Behaviours

This overview considers literature on energy use and transport behaviours, as well as environmental behaviours in general, to allow that environmentally driven behaviours may share underlying psychological determinants.

As discussed, attitudes are one of the fundamental determinants of intent and behaviour in many psychological models (Ajzen, 1988; Guagano et al., 1995). Environmental attitudes can have significant impacts on intention to engage in environmental activism (Fielding et al., 2008), and other psychological drivers such as identity can affect residential energy use decisions, particularly where personal and societal interests are in conflict (Poortinga et al., 2004).

Positive environmental attitudes have been associated with increases in support for government regulation of the environment (Poortinga et al., 2004), a readiness to take action against environmentally unfriendly companies, and willingness to pay taxes to protect forests (Stern et al., 1995).

Prioritising the environment relative to other concerns has been shown to have a positive effect on recycling behaviour, environmental stewardship and public activism (Huddart-Kennedy et al., 2009). Similarly, the importance a person places on climate change increases their propensity to engage in domestic energy conservation on a regular basis (Whitmarsh & O'Neill, 2010). Individuals who prioritise climate change are also more likely to engage in low carbon activities and lifestyles (Whitmarsh et al., 2011), and to engage in multiple types of environmental behaviours (Stern, 2000).

Identifying oneself with a particular group can lead to accentuating similarities (or differences) with those who are (or are not) members of the same group (Fielding et al., 2008). Having an *environmental* identity has been found to be a positive predictor of home energy conservation in the UK (Whitmarsh & O'Neill, 2010). Identity may also affect behaviour "in cases where our role identity dictates we behave in a certain way, irrespective of how we feel about that behaviour" (Whitmarsh & O'Neill, 2010). Social norms also impact on intent to engage in low-carbon purchasing and low-carbon energy use (Lingyun et al., 2011), and environmental activism (Fielding et al., 2008).

1.3 The Influence of Demographic Factors on Environmental Behaviours

Actual energy use tends to decrease with education level *ceteris paribus*, though education is highly correlated with income, which has a larger positive effect on energy use (Poortinga et al., 2004). Taken together, this suggests that when comparing two people on similar incomes, those with more education will, on average, use less energy. However, in other cases, education has not been found to be related to energy saving (Paço & Varejão, 2010; Neuman, 1986). Further, one survey of the UK public found that qualifications were a significant factor contributing to *reduced* energy conservation in the home (Whitmarsh & O'Neill, 2010). Thus the impact of education level on energy use behaviours is unclear.

Household income has been found to have a positive relationship with both environmental knowledge, and environmental behaviour (Jones et al., 1999). Income has been positively related to knowledge of climate change, but more strongly positively linked to self-reported knowledge of climate change (McCright, 2010). Similarly, though residential energy use increases with income, it is still possible to expect higher income to be linked to higher *reported* energy saving behaviour (McCright, 2010).

However, the expected correlation between income and education (Stead, 2007) can make it difficult to draw robust conclusions regarding the effect of one or the other depending on the form of the model.

Greater knowledge of the environment can increase uptake of actions to reduce a person's impact on the environment (Grob, 1995). For recycling behaviour, awareness of the environmental consequences of not recycling increases awareness of personal responsibility, which in turn has a measureable effect on actual behaviour, suggesting an indirect hierarchical effect (Guagano et al., 1995). Similarly, increased knowledge of climate change has been found to relate significantly to increased concern about the effects of climate change (McCright, 2010), which was found to affect behaviour.

Age has been found to be correlated with higher actual energy use (Poortinga et al., 2004), but negatively related to both actual and self-reported climate change knowledge (McCright, 2010). Willingness to pay more for environmentally friendly products increases slightly with age (Lynn & Longhi, 2011). Home energy use behaviours, such as always turning off lights when leaving the room, may be more prevalent as age increases (Lynn & Longhi, 2011). However some studies have found no impact of age on either knowledge, or environmental behaviour (Jones et al., 1999).

Women have been found to be more knowledgeable about climate change, and marginally more concerned about its effects (McCright, 2010), though not all studies found statistical significance. Women are also more likely to change their behaviour towards a company for environmental reasons (Stern et al., 1995), and exhibit slightly better energy saving behaviour in the home (Paço & Varejão, 2010). However men are more likely to report having increased their use of public transport and reduced their fuel use (Stead, 2007).

Energy use increases predictably with the number of people in the household (Poortinga et al., 2004), and those with dependent children are significantly less likely to be willing to pay more for environmental products (Lynn & Longhi, 2011). However, the number of adults in the household has been shown to have no effect on home energy conservation behaviours (Whitmarsh & O'Neill, 2010).

Urban residents may be more likely to be aware of environmental problems, to consider them important, and to take action to reduce their impact (Jones et al., 1999). However, emphasis on issues which are more significant in urban area (e.g., air and noise pollution), which assess services which may not be provided in areas with low population density (e.g., recycling or public transport), or which fail to address demographic differences can result in bias (Jones et al., 1999; Berenguer et al., 2005). Studies which balance rural and urban environmental issues have found that levels of environmentalism are similar in both (Huddart-Kennedy et al., 2009). However differences do exist , with findings that rural residents are more knowledgeable about the environment in general (Jones et al., 1999), more concerned with environmental conservation (Berenguer et al., 2005), and place more importance on stewardship of the environment (Huddart-Kennedy et al., 2009). On the other hand, urban residents have more pro-environmental attitudes, more frequently engage in pro-environmental behaviours (Berenguer et al., 2005), and are more likely to regularly engage in domestic energy conservation (Whitmarsh & O'Neill, 2010).

Demographic factors which affect behaviours directly have also been found in certain situations to act indirectly on behaviours, presenting hierarchical effects. Thus even where there is no direct effect, their influence may still be important. For example, education has been positively linked to climate change knowledge (Jones et al., 1999), and self-reported knowledge (McCright, 2010), which may in turn affect behaviour. It has also been found to relate to intention to reduce fuel use in the future and to increase the use of public transport (Stead, 2007), as well as being related to undertaking more environmental behaviour (Jones et al., 1999).

Overall, the evidence on the influence of demographic factors on environmental behaviours is largely inconclusive, supporting the need for additional consideration of psychological factors such as attitudes.

2. Methods

The research reported here is from secondary data analysis that investigated the antecedents of energy use reduction in the home and choice of transportation mode used for commuting and grocery shopping. The data used in this study comes from the Scottish Environmental Attitudes and Behaviours Survey (SEABS) (Davidson et al., 2009).

2.1 The Respondent Sample

The data was collected from interviews, generating cross-section data ($n = 3,054$) that is largely representative of the Scottish population. Respondents who did not fill in a self-completion part of the survey investigating attitudes, and those who did not reveal their income or their level of education were removed from the sample used for the present analysis (Note 3). The remaining sample ($n = 1,877$) had 944 males (50.3%) and 933 females (49.1%), aged between 16 and 92, with a mean age of 47.1 years ($S.D. = 17.3$). 40.6% of respondents were employed full time, 23.0% were permanently retired, and 4.6% were students (Table 1).

Table 1. Socio-demographic variables

Variable	Categories	Percentage of respondents
Age	16-24	10.2
	25-34	16.8
	35-54	41.0
	55+	31.8
	Refused	0.1
Sex	Male	50.3
	Female	49.7
Household income	Under £10,000	20.8
	£10,000-£19,999	24.5
	£20,000-£29,999	16.4
	£30,000-£39,999	13.6
	£40,000 - £49,999	8.8
	£50,000 or more	16.0
Employment Status	Employed full time	40.6
	Employed part time	9.0
	Self employed	5.1
	Unemployed	5.0
	School/further education	4.6
	Permanently retired	23.0
	Stay at home	7.2
	Other	5.5
Highest level of qualification obtained	Degree, professional	31.0
	HNC/HND or equivalent	9.3
	A, Higher, or equivalent	15.2
	O, Standard or equivalent	24.9
	None	19.6

How many children or babies are there in your household, aged under 16?	None	67.7
	One	15.1
	Two	12.6
	Three	3.3
	Four or more	1.3
Including yourself how many adults are there in your household, aged 16+?	One	28.9
	Two	53.3
	Three	12.8
	Four or more	5.0
Dwelling type	House or bungalow	70.1
	Flat, maisonette or apartment	29.9

An additional variable was included from the Scottish Household Survey (Scottish Government, 2011). This gave the percentage of the population in a datazone (a small region of between 500 and 1000 households) living near a high frequency bus service (one running five or more times an hour). The average percentage of the population of a datazone living near to a high frequency bus service was 32.2%. The minimum population percentage was 7.5% and the maximum was 59.3%.

2.2 Operationalisation of The Variables

2.2.1 The Behaviours

The SEABS dataset included the frequency of engagement with five energy use behaviours: turning down the heating when going out in winter, using energy-saving light bulbs, hanging washing out to dry in summer, turning lights off in unused rooms, and not overfilling the kettle. Responses ranged from "never" to "always".

To assess peoples' transport mode choices, respondents indicated their primary mode of transport to their place of work or study, and for grocery shopping. Responses were grouped into four categories (walk/cycle, public transport, multiple passenger, and single driver) to create an ordinal scale positive in more carbon intensive modes (adapted from Arnold, 2010).

Both energy use behaviour and choice of travel mode were tested as combined scales, and additionally as individual behaviours/travel mode to see if underlying factors differed between energy behaviours/travel modes, and to verify how well the scales captured the overall effects.

2.2.2 Attitudes, Importance and Identity

Attitudinal scales were derived from multiple Likert scale responses to related statements with responses ranging from 'strongly disagree' to 'strongly agree', which were recoded between "-2" and "+2".

A scale measuring attitudes to climate change was made up of nine statements which addressed climate change, environmental crisis, or household energy use, and one additional question which asked respondents which of four statements most closely reflected their views on climate change, with responses ranging from disbelief to urgency. Responses were recoded to be positive in increasing concern for climate change.

Attitudes towards travel modes were assessed from responses to statements regarding whether car use is necessary, if it is a right, and the reliability and convenience of public transport. These were recoded such that responses which were pro-car or negative towards public transport were positive.

Respondents were also asked how important it is that everyone in Scotland perform each of the household energy use reduction behaviours. Responses were scored on a five-point Likert scale ranging from "not at all important" to "very important" and were recoded and summed to create a scale measuring the perceived importance of household energy use reduction behaviour.

A variable was constructed to measure environmental identity based on the extent of respondents' discussions with others about the environment, promotion of behavioural change in others, and whether they are members of, or make regular donations to, environmental or land based charities.

Other factors included in the analysis were a mix of variables from SEABS and some specifically constructed for this analysis, designed to incorporate relevant psychological and demographic antecedents of behaviour. For example variables were included to test the effect of knowledge and residential location.

2.3 The Model

Based on the psychological literature and empirical findings examined, it was anticipated that hierarchical effects would exist, with psychological variables being influenced by demographic variables, and both affecting behaviour. A four-tiered model was hypothesised, with each tier affecting those below it (Figure 1). Secondary analysis was performed by testing the model by means of a multi-stage ordinary least squares (OLS) regression.

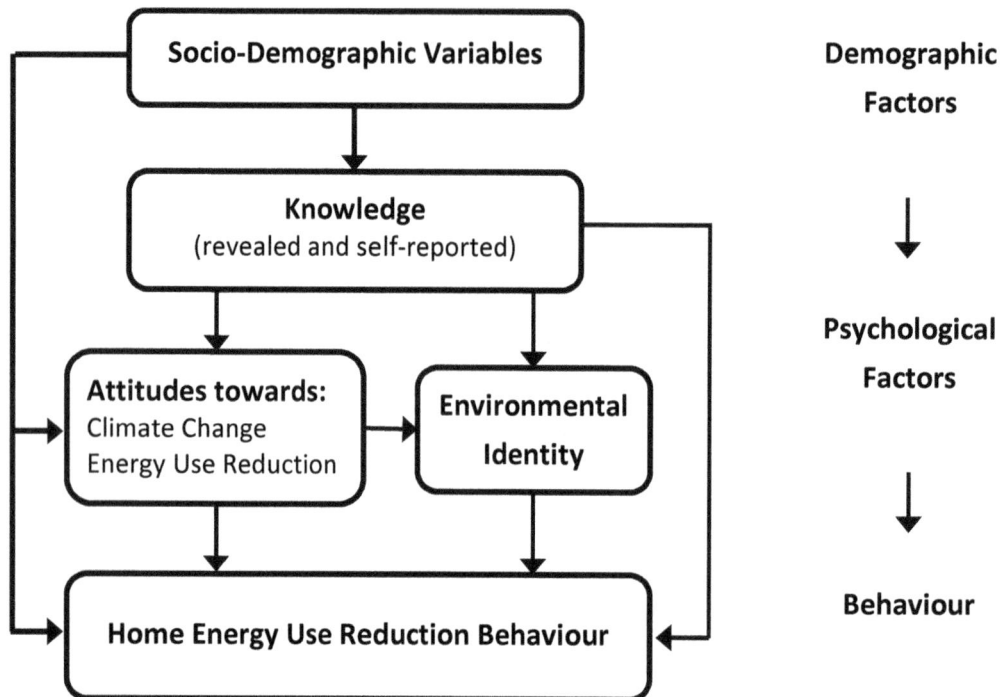

Figure 1. Model of hierarchical determinants of energy use reduction behaviour (A similar model was constructed for travel mode decisions for commuting and grocery shopping)

3. Results

3.1 Home Energy Use Reduction Behaviours

The model accounted for 18.2% of the variance in the home energy use reduction behaviours (Table 2, Adj R^2 = 0.182). Having an attitude of concern for climate change and having an environmental identity were positively related to energy use reduction behaviours. Greater revealed knowledge of climate change was found to have a significant impact on attitude to climate change and attitude to energy use reduction, but not behaviour directly. Self-reported knowledge was found to have a positive effect on attitudes, environmental identity, and energy use reduction behaviours. Prioritisation of climate change by respondents had no direct effect on behaviour, but had a significant positive effect on both environmental identity and concern for climate change.

Household income was found to have a significant *negative* effect on energy reduction behaviours, but no impact on attitude or identity. Contrastingly, educational achievement was non-significant for behaviour but had a positive coefficient impact on attitudes and increased the likelihood of having an environmental identity. Being female was found to have a direct positive effect on energy use reduction behaviours and attitudes to climate change. The number of children in the household was found to have a significant negative effect on the propensity to engage in energy use reduction behaviours. Other results were that students had lower performance of household energy reduction behaviours, and being in receipt of cold-weather benefit payments had a significant positive effect on household energy reduction behavior (Note 4).

Table 2. Home energy use results

Independent Variables (Range) [Units]	Dependent variable (n = 1,875)							
	Energy use reduction		Environ. identity		Attitude to climate change		Attitude to energy use reduction	
Environmental identity (0 : 10)	0.036	*						
Attitude to climate change (-1.9 : 2)	0.040		0.409	**				
Attitude to energy use reduction (-2 : 2)	0.030	*	0.043	*				
Revealed knowledge (-3 : 15)	0.013		0.009		0.068	**	0.029	*
Self-reported knowledge (-2 : 2)	0.057	*	0.265	**	0.165	**	0.151	**
Prioritisation of climate change (0 : 4)			0.104	**	0.132	**	0.056	*
Importance of energy behaviours (-2 : 2)	0.414	**						
Age (16 : 92)	0.002		0.016		0.020	**	0.030	**
Age squared+	0.000		0.000		0.000	**	0.000	**
Female (0 : 1)	0.073	*	-0.009		0.154	**	0.050	
Number of adults (1 : 4)	-0.018		-0.008		-0.010		0.009	
Number of children (0 : 4)	-0.042	*	-0.005		0.008		0.042	
Household income (0.125 : 12.5) [£ 10,000]	-0.052	*	-0.005		0.010		-0.015	
Household income squared+	0.001		0.000		-0.001		-0.001	
Level of education (0 : 4)	0.019		0.078	**	0.069	**	0.064	**
Living in an urban area (0 : 1)	0.034		-0.113	*	0.063		0.010	
Being a student (0 : 1)	-0.213	*	0.094		0.045		0.189	
Living in a house (0 : 1)	0.093	*	0.011		0.023		-0.081	
Cold weather benefit (0 : 1)	0.177	**						
Having a mortgage (0 : 1)	0.101	*						
Being a tenant (0 : 1)	0.044							
Adj. R^2	0.182		0.230		0.272		0.051	

Note:

+ Age and Income are included with their squares to allow for changes in their impact over the range of values without assuming a particular form of the relationship.

Table 3. Wald test statistics for location tests

Regional Division	Regression	χ^2 (d.o.f)	$Prob > \chi^2$
Urban	Energy use reduction behaviour	17.53 (14)	0.229
	Environmental identity	13.79 (19)	0.796
	Concern for climate change	12.12 (18)	0.841
Note:	(d.o.f.) indicate number of degrees of freedom.		

Tests of the location variables were performed by re-running the energy use reduction behaviour, environmental identity, and concern for climate change analysis separately for urban and non-urban respondents. In this study urban is defined as areas with a population greater than 10,000. Wald tests were conducted comparing the regression results, and all of the comparisons were outside of the critical region, implying that any differences between urban and non-urban residents are not statistically significant (Table 3) (StataCorp, 2009). However, it was found that concern for climate change was a significant driver of behaviour for urban residents (z = 2.40), while it was not for non-urban residents (z = 1.12).

Finally, the individual behaviours were each regressed using the same independent variables. However, the model explained much less variance when looking at the individual behaviours than when analysing the combined scale (Adj. R^2 of 0.072 → 0.105 compared with 0.182), which supports the notion that the behaviours are related and form a common scale.

3.2 Transport Mode Choice for Commuting to Work or College, or Going Grocery Shopping

A smaller sub-sample was used for the travel mode choice analysis, including only those working or studying. The model accounted for 25.2% of the variance in transport mode choice as measured by adjusted R^2 (Table 4). Positive attitudes towards cars were positively correlated with choosing more carbon intensive transport, as was living in a single dwelling house as opposed to a flat. People with higher incomes or further to travel chose more highly carbon-intense transport. Being a student tended to lower the use of carbon intense transport.

Similar effects were observed for travel to buy groceries, except that having greater education had a positive effect, and being a student was no longer significant. There was also an unexpected effect of greater concern for climate change being linked to *higher* carbon transport modes, though the effect was less than for attitudes to transport. Furthermore, greater knowledge and higher prioritisation of climate change, higher level of education, and being female all related to having less pro-car or more pro-public transport attitudes.

Table 4. Transport mode results

Independent Variables (Range) [Units]	Dependent variable (n = 1,016)		
	Transport mode: Work/school	Transport mode: Groceries	Attitude to transport modes
Environmental identity (0 : 10)	0.069	0.004	
Attitude to climate change (-1.9 : 2)	0.224	0.439 **	
Attitude to transport modes (-19 : 1.9)	1.348 **	1.708 **	
Revealed knowledge (-3 : 15)	-0.003	-0.005	-0.027 **
Self-reported knowledge (-2 : 2)	-0.089	-0.134	-0.048
Prioritisation of climate change (0 : 4)	-0.073	0.050	-0.107 **
Age (16 : 92)	-0.001	0.062	0.029 *
Age squared+	0.000	0.000	0.000 *
Female (0 : 1)	-0.043	-0.210	-0.085 *
Number of adults (1 : 4)	-0.183	-0.028	0.038
Number of children (0 : 4)	0.166	0.172	-0.004
Household income (0.125 : 12.5)	0.217 *	0.335 **	0.049
Household income squared+	-0.019 *	-0.023 *	-0.002
Level of education (0 : 4)	0.107	0.181 **	-0.050 **
Living in an urban area (0 : 1)	0.258	0.067	-0.117 **
Being a student (0 : 1)	-0.739 *	0.024	-0.212 *
Living in a house (0 : 1)	0.469 *	0.765 **	0.135 **
Inverse of miles to work/school/grocery store‡ (0.01 : 1)	-2.916 **	0.713 **	
Living in an area with high frequency bus service (7.5 : 59.3)	-0.006	-0.005	

Adj. R^2			0.146
Pseudo R^2	0.252	0.259	

Notes:

[+] Age and Income are included with their squares to allow for changes in their impact, e.g. consistent impact per year while of working age and no additional impact thereafter) without assuming a particular form of the relationship.

[‡] Distance is included as its inverse. This supposes that travelling twice as far away will always have the same impact, rather than assuming that the impact of travelling one mile further is constant, which is more consistent with travel patterns.

When the model was applied to individual transport modes, it was generally able to predict a higher proportion of the variance: 49% for cycling/walking, 25.7% for using public transport, 11.6% for getting a lift and 33% for driving (Table 5). Being more pro-car had a significant positive impact on choosing to drive, but lowered the use of both public transport and walking/cycling. However, attitudes towards cars had no significant correlation with whether or not someone got a lift. Unexpectedly, having attitudes of greater concern about climate change was positively correlated with driving and negatively with taking the bus, however the effect was smaller than for attitudes towards cars. The likelihood of choosing driving as the primary mode of transport also increased with rising income, while being a student was only linked to a positive impact on the likelihood of walking/cycling.

Table 5. Individual transport mode results

Mode	Positively significant variables		Negatively significant variables	
Walk/cycle	Age	(β=0.19, p<0.01);	Attitude to transport mode	(β=-0.80, p<0.01);
	Being a student	(β=1.38, p<0.01);	Age squared	(β=-0.00, p<0.01);
	Inverse of miles travelled	(β=4.85, p<0.01);	Number of children	(β=-0.44, p<0.01);
			House	(β=-0.82, p<0.01);
Public transport	Age squared	(β=0.00, p<0.01);	Attitude to climate change	(β=-0.43, p<0.05);
	Household income squared	(β=0.02, p<0.05);	Attitude to transport mode	(β=-1.79, p<0.01);
	High frequency buses	(β=0.03, p<0.01);	Age	(β=-0.23, p<0.01);
			Level of education	(β=-0.22, p<0.01);
			Inverse of miles travelled	(β=-2.86, p<0.01);
Multiple passenger			Attitude to climate change	(β=-0.62, p<0.05);
			Level of education	(β=-0.29, p<0.05);
Single driver	Climate change attitude	(β=0.41, p<0.01);	Household income squared	(β=-0.02, p<0.01);
	Attitude to transport	(β=1.66, p<0.01);	Inverse of miles travelled	(β=-1.99, p<0.01);
	Age	(β=0.11, p<0.05);	High frequency bus service	(β=0.01, p<0.05);
	Household income	(β=0.27, p<0.05);		
	Level of education	(β=0.17, p<0.05);		

The age of respondents was positively related to both walking/cycling and driving, but negatively related to using public transport. People living in an area with a more high frequency bus service were slightly more likely to take public transport, and less likely to drive. Distance travelled was highly correlated with both public transport and driving, while being highly negatively correlated with walking/cycling. Higher levels of education were significantly related to increased driving, and decreases in taking public transport and getting a lift. Finally, people living in a house, and those with more children in the household were less likely to walk or cycle.

4. Discussion

4.1 Energy Use Reduction

As expected, people who were more concerned about climate change were more likely to reduce their household energy use, which concurred with Poortinga et al. (2004) and Stern et al. (1995). People who had an environmental identity, were also more likely to participate in energy saving behaviours, also as anticipated (Sparks & Shepherd, 1992).

The prioritisation placed by respondents on climate change was expected to have a direct effect on behaviour (Huddart-Kennedy et al., 2009; Whitmarsh & O'Neill, 2010), however this study found that it only indirectly correlated with home energy use reduction through environmental identity and concern for climate change. The difference could be due to variable construction, as respondents were not directly asked to prioritise climate change relative to other issues. Prioritisation, as constructed here, could also be acting to mediate other determinants in a smaller number of cases where other factors conflict (Dijst et al., 2008), limiting its impact on the wider population.

Previous studies showed a positive impact of knowledge on behaviour (Grob, 1995), but no direct effect was found in this study. However, knowledge can also affect behaviour through identity and attitude (McCright, 2010), and this study found a link from knowledge to attitude, but not from knowledge to identity. Interestingly, self-reported knowledge of climate change did have a positive effect on behaviour, identity, and attitudes. This suggests that self-reported knowledge is a more important determinant than actual knowledge of climate change. This is possibly because knowledge in the absence of self-reported knowledge indicates uncertainty, which could make people less likely to act on what they know. However, it is important to note that the effects of revealed knowledge were still important and that the two variables were highly correlated (i.e. people who said they knew more generally did).

Environmental identity and concern for climate change were largely unimportant in choosing individual behaviours, despite being very important to the overall level of engagement. This suggests that how much you consider yourself to be environmental, or how concerned you are about the environment has little to say about which actions you will take, only that you are more likely to act.

Education had a positive effect on behaviour, which is consistent with some previous studies (Stead, 2007). The impact of household income was negative on behaviour, contrary to previous studies on *reported* behaviour, but in line with expected outcomes of *actual* energy use (Jones et al., 1999; Poortinga et al., 2004). These results further support the findings of Poortinga et al. (2004) that the level of education and level of income can have conflicting effects on behaviour while at the same time being linked. It was also interesting to see that income was not relevant to identity or attitudes, while education was. This suggests that though income and education are linked in their direct impact on behaviour, their effect on the psychological factors considered here are unrelated.

As anticipated, households with more children were less likely to reduce their energy use from everyday behaviours, particularly turning off the lights in unused rooms, or turning down the heat when going out (Poortinga et al., 2004). This could be because people want a warm and well lit home for their children to feel safe in, or because they have less time and/or other more pressing concerns. No effect was observed based on the number of adults in the house, in line with Whitmarsh and O'Neill (2010).

Tenants and people in flats were much less likely to hang their washing out to dry, likely for lack of space or not wanting to use communal areas. Women were also more likely to perform energy reduction behaviours, and exhibit concern about climate change, in line with previous studies (McCright, 2010; Paço & Varejão, 2010).

Overall levels of engagement in energy use reduction were found to be similar between urban and non-urban residents, in line with previous findings for other types of environmental engagement (Huddart-Kennedy et al., 2009). However, it was found that concern for climate change is a more important factor for urban residents than for rural residents, suggesting that further research into the effect of location should consider how attitudes affect behaviours in different areas, rather than just considering whether the attitudes are similar, as is typically done

(Huddart-Kennedy et al., 2009). To increase the information available, further studies could also follow the same population to generate panel data, which can yield more robust conclusions because it is possible to see what effect a change in a variable has on a person's behaviour (Tan et al., 2006). Panel data can thus offer further insight into what variables are most likely to *change* someone's behaviour, which may be a different, and perhaps more important, research question.

4.2 Travel Mode Choice

Contrary to energy use reduction behaviours, neither concern for climate change nor having an environmental identity were linked to the use of less carbon intensive modes of travel for commuting. Nor did they increase the likelihood of choosing one of the individual low carbon travel modes (such as walking or taking public transport). Murtagh et al. (2012) found that if people perceived a threat to their identity (e.g., as a motorist or a parent) there was reluctance to change behaviours. What their results demonstrate is that when it comes to travel choice behaviours in particular, there may be self-identities other than environmental identity that are important.

Further, Line et al. (2012) suggest that self-identity (image) is important in choosing to travel by car but these self-identities are not to do with environmental identity. This raises an interesting question of what happens when different self-identities compete. In this case, identities related to car ownership and motoring conflict with personal norms relating to climate change and environmental identity, as highlighted by the finding that prioritisation of climate change and being pro-car were opposed.

Interestingly, revealed knowledge of climate change was important in shifting attitudes towards public transport, while self-reported knowledge was not, in contrast to the importance of self-reported knowledge to attitudes to climate change. When considered with the impact of education on reducing pro-car attitudes, this suggests that education does have a role to play in changing travel mode decisions, though it is indirect. Other studies have also found knowledge plays a role in relation to intended transport behaviours in the context of climate change issues (Line et al., 2012).

The model offered the least explanation of why people get a lift to work, possibly because getting a lift is mostly related to whether a person is travelling the same direction as someone in their home or is on someone else's route, which are not captured in the study.

The relationship between age and transport mode did not follow the combined transport scale, rather people switched more to walking or driving as they got older. This may be due to a move from more communal to more solitary modes of transportation or a reflection of different living patterns at different ages. Previous work (Arnold, 2010) found a more direct relationship with older people more likely to drive than to take most other options transport options. Overall, there is scope for investigating the relationship between transport mode behaviours and life stage.

People were more likely to choose to use the bus if they lived near high frequency bus service routes, meaning that they do not need to plan what time they leave the house. Though proximity to high frequency bus service only had a small impact, it was an indirect measure of the datazone on average and not the respondent in particular. Its significance suggests this is an area which may have greater implications, and should be incorporated directly into future studies which include travel mode choices. Arnold (2010) concluded that policies aimed at structural change might be more effective than policies aimed at changing people's environmental values. The findings from the current analysis suggest that this may be true.

The underlying determinants of travel mode choice were attitudes towards travel mode directly (i.e. positive or negative opinions of the convenience and reliability of cars versus public transport), as well as demographic and structural variables such as income and distance travelled. This reinforces earlier findings (Line et al., 2012) that peoples' intention to use the car is at least in part related to positive attitudes towards cars and negative attitudes towards public transport. Arnold (2010) found a similar relationship between positive attitudes to cars and use of cars.

4.3 Policy Implications

This study has shown that energy use reduction behaviours and travel mode choices have both demographic and psychological determining factors, and suggests a hierarchical nature of some of the effects. Notably there are some important differences in influences on energy use behaviours as opposed to transport use behaviours. In particular having an environmental identity was found to be an important driver of household energy use reduction, as was concern for climate change indirectly, but neither affected choice of transportation mode. Knowledge also had a greater effect on household energy use reduction than on transport behaviours. Further, energy use reduction behaviours were best understood as a group of behaviours rather than individually while the opposite was true for the transport behaviours.

In terms of energy use behaviours, this means there is a role for information programmes, and they should seek to promote a range of behaviours for increased effect. This supports the current government strategy to engage individuals in energy use reduction through information and positive encouragement and not just regulation change (Scottish Government, 2010). Contrastingly transport mode decisions are best explained when looking at transport modes individually, and so promotion of less carbon intense modes of transportation should encourage particular travel modes explicitly rather than promoting "low carbon" travel in general. This means seeking to change attitudes directly related to the individual modes and their convenience rather than focusing on the implications for the environment. Policy attempting to change travel mode decisions also needs to affect structural change to make travel by public transport more desirable. One important finding is that appealing to an individual's environmental identity with messages about "green" behaviours may be more effective when targeting energy use behaviours in the home, than when focusing on travel behaviours.

The results of this new analysis showed that attitudinal factors contribute in an important manner to energy related behaviour. This study also supports recommendations that attempts at behavioural intervention through attitudes cannot be taken in isolation from other factors (House of Lords, 2011), and that information targeting campaigns are likely to be more effective when tailored to the target audience (Southerton et al., 2011). This analysis helps identify factors linked to individual behaviours, and thus could be used to help ensure the right people are targeted by the right type of information. It also offers further support to conclusions that attitudes lead to behaviours, and thus strengthens the case for seeking to influence attitudes as a viable means of affecting behaviour (Davidson et al., 2009). These findings add to existing literature on environmental behaviour and provide more rigorous analysis of available data in Scotland than previously existed. Greater information on the causal factors of behaviour can help policy makers be more effective at promoting behavioural change to encourage reduction of GHG emissions. The findings suggest that using 'green' messaging will likely help encourage behavioural change in home energy, but contribute little to encouraging change in travel mode decisions.

Acknowledgements

This research was undertaken within the Scottish Government Rural Affairs and the Environment Portfolio Strategic Research Programme 2011-2016, Theme Four 'A rural economy resilient to global and local change'. For more information please see: http://www.scotland.gov.uk/Topics/Research/About/EBAR/StrategicResearch/future-research-strategy/Themes/ThemesIntro

References

Ajzen, I. (1988). *Attitudes, Personality and Behavior*. Buckinham: Open University Press.

Ajzen, I. (1991). The theory of planned behavior. *Organizational behavior and human decision processes, 50*(2), 179-211. http://dx.doi.org/10.1016/0749-5978(91)90020-T

Armitage, C. J., & Conner, M. (2001). Efficacy of the theory of planned behaviour: A meta - analytic review. *British journal of social psychology, 40*(4), 471-499. http://dx.doi.org/10.1348/014466601164939

Arnold, S. (2010). *Environmental Decision Making and Behaviours: How do People Choose how to Travel to Work?* Working Paper. Bath, UK: Department of Economics, University of Bath.

Berenguer, J., Corraliza, J. A., & Martín, R. (2005). Rural-Urban Differences in Environmental Concern, Attitudes, and Actions. *European Journal of Psychological Assessment, 21*(2), 128. http://dx.doi.org/10.1027/1015-5759.21.2.128

Davidson, S., Martin, C., & Treanor, S. (2009). *Scottish Environmental Attitudes and Behaviours Survey 2008*. Edinburgh, UK: Scottish Government Social Research.

DECC. (2013). *DECC public attitudes tracker Wave 5*. Summary of key findings. Department of Energy and Climate Change, 2012. UK Emissions Statistics. DECC, London. Retrieved from http://www.decc.gov.uk/en/content/cms/statistics/climate_stats/gg_emissions/uk_emissions/uk_emissions.aspx

Fielding, K. S., McDonald, R., & Louis, W. R. (2008). Theory of Planned Behaviour, identity and intentions to engage in environmental activism. *Journal of Environmental Psychology, 28*, 318-326. http://dx.doi.org/10.1016/j.jenvp.2008.03.003

Fishbein, M., & Ajzen, I. (1975). *Belief, Attitude, Intention, and Behavior: An Introduction to Theory and Research*. Reading, MA: Addison-Wesley.

Fujii, S., & Gärling, T. (2003). Application of attitude theory for improved predictive accuracy of stated preference methods in travel demand analysis. *Transportation Research Part A, 37*, 389-402. http://dx.doi.org/10.1016/S0965-8564(02)00032-0

Grob, A. (1995). A structural model of environmental attitudes and behaviour. *Journal of Environmental Psychology, 15*, 209-220. http://dx.doi.org/10.1016/0272-4944(95)90004-7

Guagano, G. A., Stern, P. C., & Dietz, T. (1995). Influences on attitude-behavior relationships: A natural experiment with curbside recycling. *Environment and Behavior, 27*, 699-718. http://dx.doi.org/10.1177/0013916595275005

House of Lords. (2011). Behaviour Change-Science and Technology Select Committee 2nd Report of Session 2010–12. London, UK: Authority of the House of Lords.

Huddart-Kennedy, E., Beckley, T. M., McFarlane, B. L., & Nadeau, S. (2009). Rural-urban differences in environmental concern in Canada. *Rural Sociology, 74*, 309-329. http://dx.doi.org/10.1526/003601109789037268

Jones, R. E., Fly, J. M., & Cordell, H. K. (1999). How green is my valley? Tracking rural and urban environmentalism in the southern Appalachian ecoregion. *Rural Sociology, 64*, 482-499. http://dx.doi.org/10.1111/j.1549-0831.1999.tb00363.x

Line, T., Chatterjee, K., & Lyons, G. (2012). Applying behavioural theories to studying the influence of climate change on young people's future travel intentions. *Transportation Research Part D, 17*, 270-276. http://dx.doi.org/10.1016/j.trd.2011.12.004

Lingyun, M., Rui, N., Hualong, L., & Xiaohua, L. (2011). Empirical research of social norms affecting urban residents low carbon energy consumption behavior. *Energy Procedia, 5*, 229-234. http://dx.doi.org/10.1016/j.egypro.2011.03.041

Lynn, P., & Longhi, S. (2011). Environmental attitudes and behaviour: who cares about climate change? In S. L. Garrington (Ed.), *Understanding Society: early findings from the first wave of the UK's household longitudinal study* (pp. 109-116). Colchester, UK: Institute for Social and Economic Research: University of Essex.

McCright, A. M. (2010). The effects of gender on climate change knowledge and concern in the American public. *Population & Environment, 32*, 66-87. http://dx.doi.org/10.1007/s11111-010-0113-1

Murtagh, N., Gatersleben, B., & Uzzell, D. (2012). Self-identity threat and resistance to change: Evidence from regular travel behaviour. *Journal of Environmental Psychology, 32*, 318-326. http://dx.doi.org/10.1016/j.jenvp.2012.05.008

Neuman, K. (1986). Personal values and commitment to energy conservation. *Environment and Behavior, 18*, 53-74. http://dx.doi.org/10.1177/0013916586181003

Paço, A. D., & Varejão, L. (2010). Factors affecting energy saving behaviour: a prospective research. *Journal of Environmental Planning and Management, 53*, 963-976. http://dx.doi.org/10.1080/09640568.2010.495489

Palmer, J., & Cooper, I. (2011). *Great Britain's housing energy fact file*. London: Department of Energy and Climate Change (DECC).

Parry, M., Canziani, O., Palutikof, J., & Co-authors. (2007). Technical Summary. In IPCC, M. Parry, O. Canziani, J. Palutikof, P. v. Linden, & C. Hanson (Eds.), *Climate Change 2007: Impacts, Adaptation and Vulnerability*. Contribution of Working Group II to the Fourth Assessment Report of the Intergovernmental Panel on Climate Change (pp. 23-78). Cambridge, UK: Cambridge University Press.

Poortinga, W., Steg, L., & Vlek, C. (2004). Values, environmental concern, and environmental behavior: A study into household energy use. *Environment and Behavior, 36*, 70-93. http://dx.doi.org/10.1177/0013916503251466

Scottish Government (2010). *Low Carbon Scotland: Public Engagement Strategy*. Scottish Government, Edinburgh.

Scottish Government. (2011). *Scottish Household Survey-Survey Review 2010*. Retrieved November 2011, from http://www.scotland.gov.uk/Topics/Statistics/16002/SurveyReview2010

Scottish, Parliament. (2009). *Climate Change (Scotland) Act 2009 (asp 12)*. Edinburgh, UK: The Scottish Parliament.

Sheppard, B. H., Hartwick, J., & Warshaw, P. R. (1988). The Theory of Reasoned Action: A meta-analysis of past research with recommendations for modifications and future research. *Journal of Consumer Research, 15*, 325-343. http://dx.doi.org/10.1086/209170

Southerton, D., McMeekin, A., & Evans, D. (2011). *International Review of Behaviour Change Initiatives*. Edinburgh, UK: Scottish Government Social Research.

Sparks, P., & Shepherd, R. (1992). Self-identity and the Theory of Planned Behavior: Assessing the role of identification with "Green Consumerism". *Social Psychology Quarterly, 55*, 388-399. http://dx.doi.org/10.2307/2786955

StataCorp. (2009). *Stata Multivariate Statistics Reference Manual Release 11*. College Station, TX: StataCorp LP.

Stead, D. (2007). Transport energy efficiency in Europe: Temporal and geographical trends and prospects. *Journal of Transport Geography, 15*, 343-353. http://dx.doi.org/10.1016/j.jtrangeo.2006.11.009

Stern, N. (2006). *The Economics of Climate Change: The Stern Review*. Cambridge: Cambridge University Press.

Stern, P. C. (2000). Toward a coherent theory of environmentally significant behavior. *Journal of Social Issues, 56*, 407-424. http://dx.doi.org/10.1111/0022-4537.00175

Stern, P. C., Dietz, T., & Guagano, G. A. (1995). The New Ecological Paradigm in social-psychological context. *Environment and Behavior, 27*, 723-743. http://dx.doi.org/10.1177/0013916595276001

Tan, P. N., Steinbach, M., & Kumar, V. (2006). *Introduction to Data Mining*. Boston, Massachusetts: Addison-Wesley.

Whitmarsh, L., & O'Neill, S. (2010). Green identity, green living? The role of pro-environmental self-identity in determining consistency across diverse pro-environmental behaviours. *Journal of Environmental Psychology, 30*, 305-314. http://dx.doi.org/10.1016/j.jenvp.2010.01.003

Whitmarsh, L., Seyfang, G., & O'Neill, S. (2011). Public engagement with carbon and climate change: To what extent is the public 'carbon capable'? *Global Environmental Change, 21*, 56-65. http://dx.doi.org/10.1016/j.gloenvcha.2010.07.011

Notes

Note 1. Which include methane, nitrous oxide, and hydrofluorocarbons.

Note 2. Identity differs from TPB's use of subjective norms. The latter relates to the views of close relations, e.g. parents who may not be likeminded, whereas identity references a group holding shared values.

Note 3. Although there were demographic differences between respondents who were kept and those who were removed, probit analysis revealed only one significant variable for non-completion: education ($P > |z| = 0.000$), and the overall explanatory ability was low (Pseudo $R^2 = 0.0369$).

Note 4. Payable to certain welfare benefit recipients when local temperature is either recorded as, or forecast to be, an average of zero degrees Celsius or below over 7 consecutive days.

5

A Case Study of Monitoring Emission from CO_2 Enhanced Oil Recovery by Remote Sensing Data

Xiongwen Chen[1]

[1] Department of Biological & Environmental Sciences, Alabama A&M University, Normal, AL 35762, USA

Correspondence: Xiongwen Chen, Department of Biological & Environmental Sciences, Alabama A&M University, Normal, AL 35762, USA. E-mail: xiongwen.chen@aamu.edu

Abstract

Enhanced oil recovery with carbon dioxide (CO_2-EOR) is considered to be a cost effective way for carbon capture and storage. However, due to the complexity of geological structure in underground reservoirs, long-term leakage is possible. A case study of CO_2-EOR has been conducted at Citronelle, Alabama in the United State of America. A total of 8,036-ton of CO_2 were injected from November 2009 to September 2010 and some leakages via production were identified by isotopic analysis in May 2010. In this study, remote sensing data of CO_2 and methane (CH_4) concentrations, and aerosol optical depth (AOD) at a large scale were used to monitor emissions to atmosphere at the study site. Based on the observed monthly CO_2 and CH_4 concentrations in the atmosphere at the study site and surrounding areas, some abnormal values related to possible emission were identified at different time scales by correlation, variance and entropy analysis. The annual average of ratios between CO_2 concentration and CH_4 concentration, which might be due to CO_2 emission, reached the highest value in 2009. In comparison with surrounding areas, the monthly values of AOD at the study site were relatively higher, especially during the time periods of 2008, 2009 and part of 2010. Our results might confirm the isotopic analysis at the ground and may provide more detailed information. Therefore, through this approach remote sensing data could be used to monitor and evaluate emissions from areas involved in CO_2-EOR at a large scale and provide helpful information for ecological assessment of CO_2-EOR.

Keywords: carbon storage, CO_2 enhanced oil recovery, Citronelle in Alabama, emission, remote sensing

1. Introduction

Increasing atmospheric CO_2 and other greenhouse gases is considered to be a critical driving force of global climate change. Dramatic reduction in CO_2 emission is needed to significantly stabilize and decrease atmospheric greenhouse gas concentration while fossil fuels are continuously used. Carbon capture and geological storage, a means to bury CO_2 in geological reservoirs for a long time, is thus important to reduce the current CO_2 emission (Pacala & Socolow, 2004; Schrag, 2007). Carbon capture and storage are well-understood in oil and gas industries (Mertz, Davidson, de Coninck, Loos, & Meyer, 2005; Thomas & Benson, 2005). Based on the possible benefits for carbon storage, enhanced oil recovery with CO_2 (CO_2-EOR) is considered cost effective. CO_2-EOR has the potential to recover 30-60% more of the original oil in reservoirs based on Department of Energy Oil Recovery Program, and even without CO_2 credits or taxes, there are revenues from oil or gas production (Zweigel, Arts, Lothe, & Lindeberg, 2004; Thomas & Benson, 2005). CO_2-EOR practice has been applied for decades by the oil and gas industry in the USA and Canada (van Bergen, Gale, Damen, & Wildenborg, 2004; Thomas & Benson, 2005), such as those projects at Weyburn and Cranfield (Whittakers, White, Law, & Chalaturnyk, 2004; Meckel & Hovorka, 2009).

Although carbon capture and storage is very attractive in potential, this practice is still controversial. The main concern is its potential long-term leakage, which may be a result of fluid migration pathways from carbonate reservoirs in a tectonically complex region or new emerging pathways from reservoirs (e.g., Gilfillan et al., 2009; Jenkins et al., 2012). Roberts, Wood, and Haszeldine (2011) assessed the health risk of natural CO_2 seepage in Italy and indicated that human death is strongly influenced by seepage surface expression, local environmental condition, CO_2 flux and human behavior. Fear of surface leakage and a lack of local benefit are the main factors for negative public opinion of CO_2-EOR (Roberts et al., 2011).

The likelihood of surface leakage will depend on site specific geological structure and characteristics (Gilfillan et al., 2009; Roberts et al., 2011; Jenkins et al., 2012). Therefore, developing and implementing leakage monitoring and assessing procedures will help to evaluate the accuracy of current concerns (Roberts & Chen, 2012). One possible approach is the use of remote sensing data. The Atmospheric Infrared Sounder (AIRS: Pagano, Aumann, Hagan, & Overoye, 2003) and the Moderate Resolution Image Spectroradiometer (MODIS) data products might be used to monitor CO_2, CH_4 and aerosol optical depth (AOD) at a large area although there are limited applications for the study of industrial emission. With the emission of CO_2, then, the regional ratio of CO_2 and CH_4 should be changed. Also, aerosol concentration may increase with gas emission. As a case study, a CO_2-EOR experimental project has been conducted from 2007 to present at Citronelle, Alabama, USA. In May of 2010 CO_2 seepage relating to CO_2 injection activities was observed above ground and verified by carbon isotopic analysis. This pilot CO_2-EOR project in Citronelle, AL provides a case where known seepage occurred and thus provides an opportunity to assess the atmospheric changes relating to CO_2-EOR activities using remote sensing data. The objectives of this study are to (i) provide general information of monthly atmospheric CO_2, CH_4 and aerosol for evaluation; (ii) identify whether there are sudden increases of atmospheric CO_2, CH_4 and aerosol; and (ii) compare monthly atmospheric concentrations of CO_2, CH_4 and aerosol to characterize possible seepage.

2. Material and Methods

2.1 Study Site

Since 2007 a pilot CO_2-EOR project has been conducted at the Citronelle field (near 31°05'N, 88° 14'W) of the Mobile County, Alabama in USA. The Citronelle oil field, located on the crest of the dome, has produced millions of barrels of oil from sandstone in the Lower Cretaceous Rodessa Formation (Esposito et al., 2008). The local landscape is mainly composed of forest, agricultural land and town (urban) area. About 31– 34% of the original oil in this area has been collected by primary and secondary methods and CO_2-EOR was estimated to increase reserves by up to 20% (Esposito, Pashin, & Walsh, 2008). A total of 8,036-ton CO_2 injection was completed during the time period from November 2009 to September 2010. In May 2010 seepages were confirmed at the production well-fed tank battery and verified by carbon isotopic analysis. Higher CO_2 concentrations were also monitored several times at the study site and surrounding areas by routine monitoring work on ground. More information about the project may be found at the National Energy and Technology Laboratory website (www.netl.doe.gov).

2.2 Remote Sensing Data

The high-resolution Atmospheric Infrared Sounder (AIRS) was launched into Earth-orbit in May 2002, with the goal to support climate research and improve weather forecasting. AIRS is one of the six instruments on board the Aqua satellite, which is a part of the National Aeronautics and Space Administration (NASA) Earth Observing System. AIRS uses cutting-edge infrared technology and provides information related to air temperature, water vapor, trace gases and cloud property (e.g., Pagano et al., 2003; Chahine, Barnet, Olsen, Chen, & Maddy, 2006). AIRS CO_2 retrievals use an analytical method for the determination of carbon dioxide and other minor gases in the troposphere from AIRS spectra. The AIRS data have been shown to be accurate to within 1.20 ppm of simultaneous measurements by aircraft (Chahine et al., 2005). Moderate Resolution Imaging Spectroradiometer (MODIS) instruments onboard the Terra platform first and then onboard the Aqua platform are uniquely designed to monitor earth change. Terra MODIS and Aqua MODIS are viewing the entire Earth's surface every 1 to 2 days, acquiring data in 36 spectral bands, or groups of wavelengths (e.g., Salomonson, Barnes, Maymon, Montgomery, & Ostrow, 1989; Remer et al., 2005). Aerosol optical depth is negative natural logarithm of the fraction of radiation (e.g., light) that is not scattered or absorbed by aerosol and it is dimensionless. Wang & Christopher (2003) found a linear relationship between MODIS aerosol optical depth and $PM_{2.5}$ concentrations measured by tapered-element oscillating microbalances in Jefferson County, Alabama. Based on the knowledge in EOR practices, CO_2 emission may affect CH_4 concentration and AOD. Monthly data of CO_2, CH_4 from AIRS and AOD of 550 nm from MODIS at 1.0° × 1.0° near the study site (31°-32°N, 88°-89°W) and surrounding areas from January 2003 to September 2011 were used in this study. The extent of the surrounding areas includes the eastern area (31°-32°N, 87°-88°W), western area (31°-32°N, 89°-90°W), southern area (30°-31°N, 88°-89°W) and northern area (32°-33°N, 88°-89°W). Giovanni online data system developed and maintained by the NASA GES DISC were used (Acker & Leptoukh, 2007). The unit for CO_2 and CH_4 is ppm and there is no unit for AOD. More detailed information can be found at http://airs.nasa.gov and http://modis.gsfc.nasa.gov.

2.3 Data Analysis and Statistics

ANOVA was used to analyze monthly concentrations of atmospheric CO_2, CH_4 and values of AOD. Significant level is at $p < 0.05$. The variances of CO_2, CH_4 and AOD monthly values were calculated using the following:

$$V_i = \sum_{i=1}^{n} \left(x_i - \overline{x} \right)^2 \tag{1}$$

V_i is the variance at time scale j, x_i is the monthly concentration for atmospheric CO_2, CH_4 and AOD from remote sensing, $\overline{x}$ is the average value.

$$STD = \sqrt{V_i / n} \tag{2}$$

Where STD is the annual standard deviation.

$$T_j = \sum_{j=1}^{m} V_i \tag{3}$$

Where T_j is the sum of the variances at the scale j.

$$P_j = \frac{V_j}{T_j}\% \tag{4}$$

Where P_j is the percentage of variance in the total variance of this time scale.

$$S_j = P_j \log(P_j) \tag{5}$$

Where S_j is the entropy (similar as Shannon-Wienner index, no unit) at the time scale. The higher the values in entropy, the less variance there is or the more even for the concentration data. The high fluctuation of concentration data or local value of entropy might be related to seepages. In this study, the time scales for entropy calculation include 1, 3, 5, 7, 15 and 21 months. Similar method was used for studying plant spatial distribution (Chen, Li, & Scott, 2005).

3. Results and Discussion

3.1 Dynamics of CO_2 Concentration

The monthly atmospheric CO_2 concentration increased from 376 ppm to 394 ppm from 2003 to 2011 (Figure 1). The annual CO_2 concentration at the study site increased each year. There exists annual fluctuation of atmospheric CO_2 concentration at the study site. Usually the CO_2 concentrations reach the highest values near the start of summer (such as May or June) and then decline to the lowest values until fall time due to vegetation growth. If the monthly change of atmospheric CO_2 concentration should be smooth, then, the CO_2 concentrations during the following time periods in Figure 1 might be considered as abnormal: March – May 2005, January – April 2006, March – May 2008, January – March 2009, April –July 2010 and September – November 2010. The variances between the monthly CO_2 concentration at 2003-2006 and 2007-2010 were not significant ($p = 0.5218 > 0.05$). If the monthly CO_2 concentrations are substituted by the CO_2 concentration in January at each year and then these data are analyzed by ANOVA, the monthly CO_2 concentrations in 2010 and 2004 have more significant changes (Table 1). This also indicates that time scale is important for monitoring CO_2 emission.

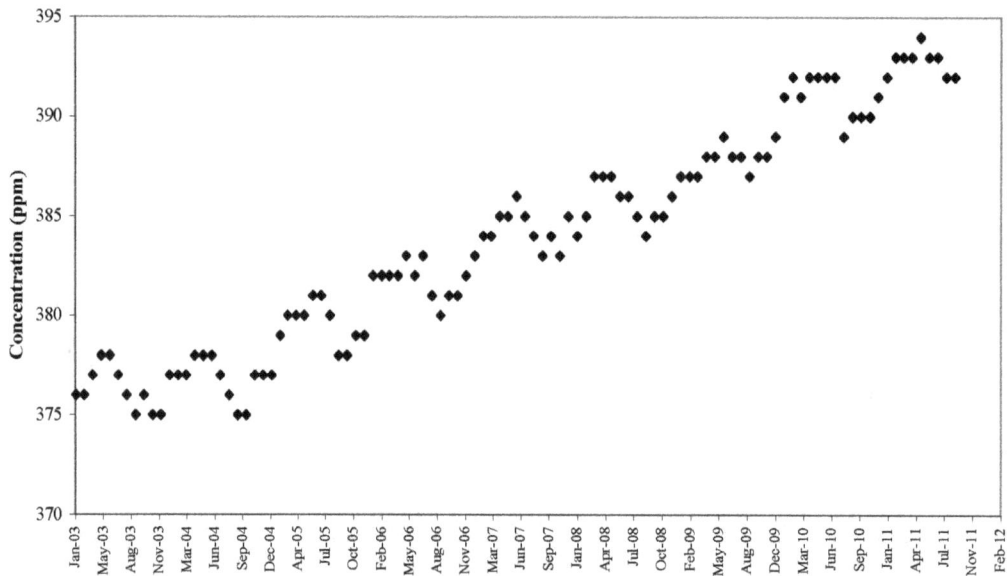

Figure 1. Dynamics of monthly CO_2 concentration in atmosphere at the study site

Table 1. The p values of ANOVA for monthly CO_2 concentration from 2003 to 2011 after the January CO_2 concentration in each year is substituted. $p < 0.05$ is statistically significant here

	Y04	Y05	Y06	Y07	Y08	Y09	Y10	Y11
Y03 (1.167 ±0.322)	0.006*	0.024*	0.002*	0.847	0.363	0.393	0.015*	0.333
Y04 (-0.167 ±0.297)		0.000*	0.832	0.002*	0.001*	0.012*	0.698	0.020*
Y05 (2.331 ±0.355)			0.000*	0.026*	0.128	0.002*	0.000*	0.002*
Y06 (-0.250 ±0.250)				0.001*	0.000*	0.003*	0.530	0.006*
Y07 (1.250 ±0.279)					0.435	0.244	0.006*	0.201
Y08 (1.583 ±0.313)						0.060	0.001*	0.049*
Y09 (0.833 ±0.207)							0.034*	0.857
Y10 (0.000 ±0.302)								0.052
Y11 (0.778 ±0.222)								

Note. Y03 represents year 2003 and others follow the same way. The numbers in ( ) of the first column at the left side are averages and standard errors of the above year.

The standard deviations of annual atmospheric CO_2 concentration declined from 2003 to September 2010, but there were fluctuations (Figure 2). There is a high correlation between the monthly atmospheric CO_2 concentrations in the same months among different years (Figure 3) ($R^2 = 0.9314$, $p < 0.05$). The following points were deviated from the regression line visually: (375, 375) September 2003 and 2004, (376, 375) October 2003 and 2004, (377, 382) January 2005 and 2006; (377, 381) July 2004 and 2005, (382, 386) June 2006 and

2007, (386, 386) June 2007 and 2008, (387, 387) March 2008 and 2009, (387, 392) February 2009 and 2010. The confidence interval is not given here since the X and Y values are not independent.

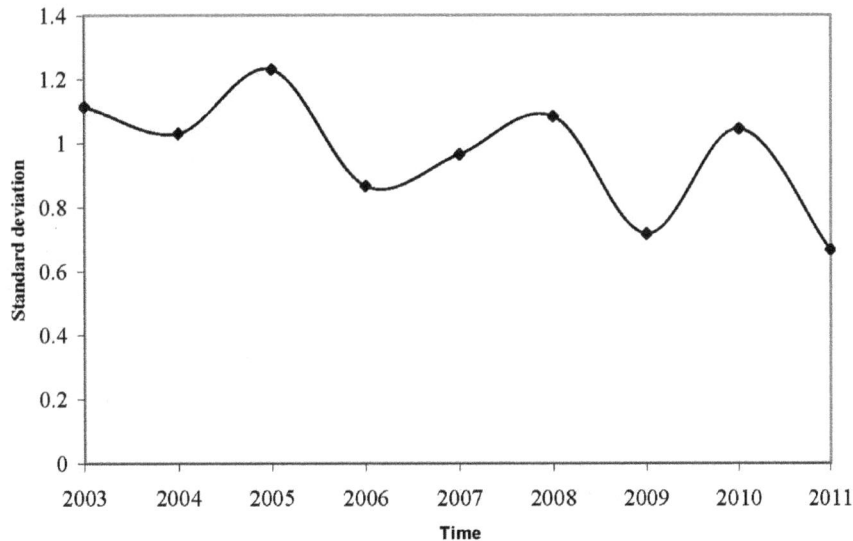

Figure 2. Change of annual standard deviations of monthly CO_2 concentration at the study site

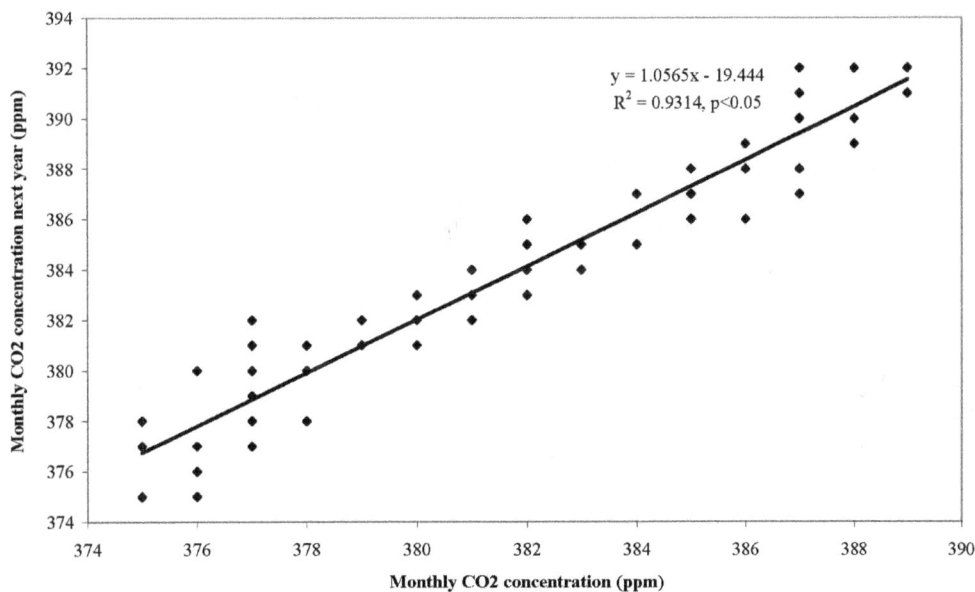

Figure 3. Correlation between monthly CO_2 concentrations at the same months but in different years

Sudden changes of atmospheric CO_2 concentration were obvious in the monthly data, but the linkage with CO_2 emission is not straightforward. The main problems are possibly related to (i) limited CO_2 emission into the atmosphere as CO_2 emission occurred on ground and the covered area is 1° by 1° in latitude and longitude and also the CO_2 concentration is measured from the mid-tropospheric layer; (ii) a total 8,036-ton of injected CO_2 might be too limited for CO_2-EOR activities to cause obvious change in atmospheric monthly CO_2 concentration. Any slight change in atmospheric CO_2 concentration might be related to a big change on ground; (iii) Citronelle is oil field with hundreds of oil wells. Emission from some wells was not monitored. Unidentified or missing emissions are possible. Remote sensing might provide more records of abnormal CO_2 concentrations than the discrete ground monitoring; and (iv) the rise of global atmospheric CO_2 concentration may hide local noises of CO_2 emission. Global CO_2 concentration increases about 2 ppm each year based on the NOAA observation data from Mauna Loa at Hawaii. All these factors might possibly work together to some extent, the original assumption about sudden increase of atmospheric CO_2 concentration related to emissions from CO_2-EOR became indirectly. However, from the decreasing annual standard deviations from the monthly CO_2 concentration, abnormal points in correlation map of CO_2 concentration at some months but in different years, and also the comparison of CO_2 concentrations at different time scales with CO_2 concentrations at surrounding areas, it still could be inferred that possible CO_2 emission occurred at the study site if there were no other sources (such as biomass burning).

The entropy of variance of CO_2 concentration at different time scales was similar with its entropy values in the eastern, western and southern areas, but lower than the value in the northern area (Table 2). This means the variances of atmospheric CO_2 concentrations at the study area were higher than the CO_2 concentrations in its northern area across all time scales.

Table 2. Comparison in the entropy (no unit) of atmospheric CO_2 concentration at the study site and surrounding areas across different time scales

Time scale (month)	Study site	East	West	South	North
1	17.77	17.69	17.76	17.76	19.6
3	64.29	64.32	64.31	65.23	65.97
5	85.36	85.18	85.36	85.36	86.88
7	99.35	99.05	99.35	99.35	100.84
15	129.96	129.88	129.96	129.96	131.39
21	144.68	144.54	144.68	144.68	146.51

The CO_2-EOR project, just like other geological carbon storage, was considered to operate under zero or very limited predictable leakage (Holloway, Pearce, Hards, Ohsumi, & Gale, 2007). Currently there is no standardized value for tolerable seepage. But the minimum retention for geological carbon storage is at least 99% stored CO_2 for 1,000 years. To ensure effective climate abatement, leakage rates of less than 0.1% y^{-1} are needed (Haugan & Joos, 2004). Roberts et al. (2011) estimated from Italian seepage that $0.1 — 1\%$ from storage of 3.6 Mt per year would be reasonable. The numerous models based on knowledge of fluid flows, usually turned out several orders lower in magnitude than the Italian gas seeps. In this study, it is still not clear how much atmospheric CO_2 concentration would increase if there is 1% emission of 8,036 tons injected CO_2.

3.2 Dynamics of CH_4 Concentration

In comparison to monthly CO_2 concentrations at the study site, monthly CH_4 concentrations were relatively stable from 2003 to 2011 and these values were close to 1.0 ppm (Figure 4). The correlation between monthly CH_4 concentrations at the same time among different years was not significant ($p > 0.05$).

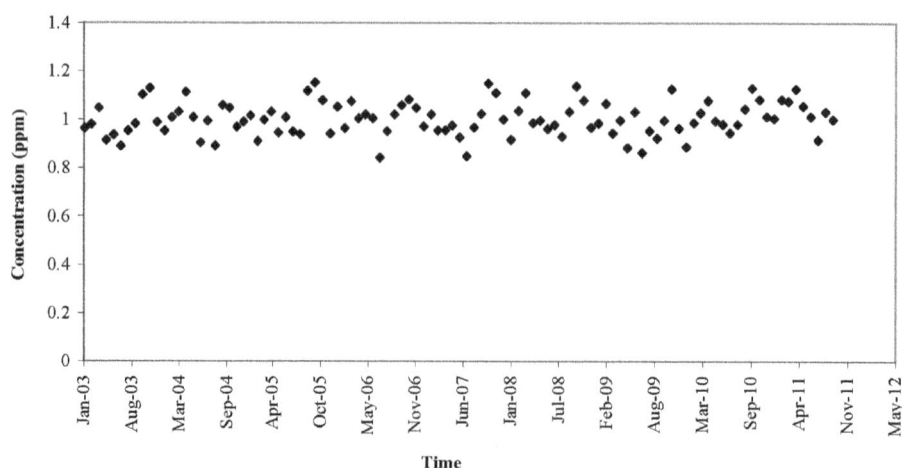

Figure 4. Monthly CH_4 concentration in atmosphere at study site

The entropy of CH_4 concentrations at the study area was lower than the values from the surrounding areas (Table 3), especially at the southern area, although the values were larger at some time scales. The annual average of ratios between CO_2 concentration and CH_4 concentration was relatively stable from 2003 to 2006, but it reached the highest value in 2009 (Figure 5).

Table 3. Comparison in the entropy (no unit) of atmospheric CH_4 concentration at the study site and surrounding areas across different time scales

Time scale (month)	Study site	East	West	South	North
1	28.26	31.71	27.86	28.41	28.38
3	55.51	58.01	55.17	57.71	55.36
5	73.01	74.07	72.21	75.33	74.32
7	85.17	85.45	84.11	86.79	86.57
15	116.14	117.31	116.51	116.67	116.80
21	130.80	130.78	130.31	130.89	131.46

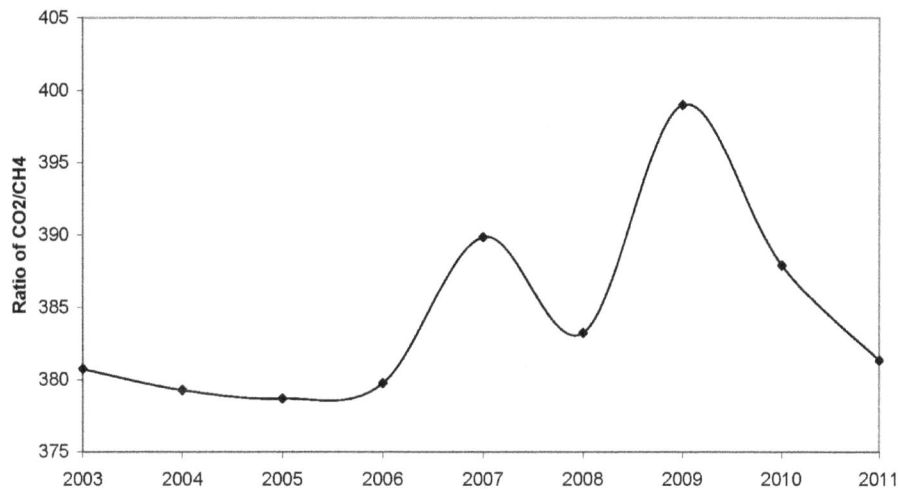

Figure 5. Annual average of ratios for atmospheric CO_2/CH_4 at study site

Although there were lots of small fluctuations in the monthly CH_4 concentration at the study site, the atmospheric CH_4 concentration at study site was within the range of 0.8 — 1.2 ppm, which is considered as relatively stable. However, there were large increases in the annual averages of the ratio of CO_2/CH_4 concentration in 2007 and 2009, which might be a helpful indicator for atmospheric CO_2 change, as this ratio might be a good indicator for some CO_2 sources.

3.3 AOD Change

The monthly AOD at the study site declined after 2007 (Figure 6). Usually AOD increases during spring and early summer time and reaches the highest value in July or August and then declines in the fall and winter time. The highest AOD in August 2009 was the lowest one of all these higher values. The annual average of monthly AOD is the lowest in 2009 for the study site and the surrounding areas (Figure 7). For the study site the annual average of monthly AOD was relatively lower than the AOD at the surrounding areas, but it was relatively higher after 2007.

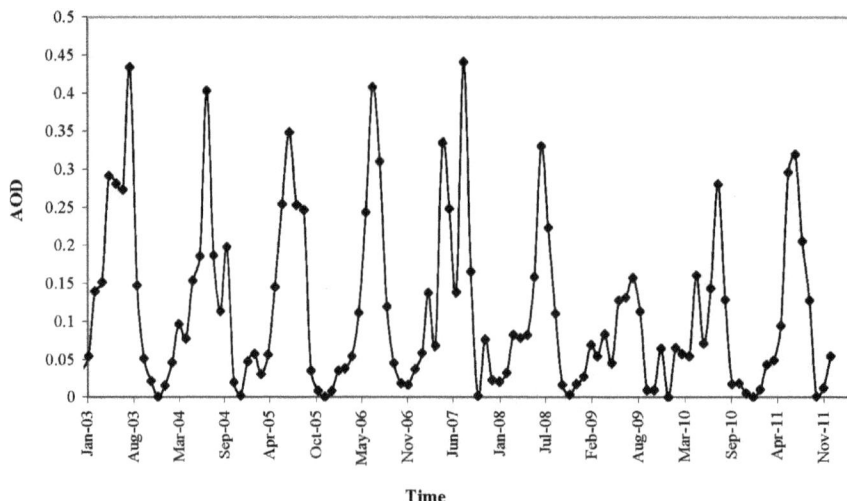

Figure 6. Monthly AOD (dimensionless) dynamics at study site

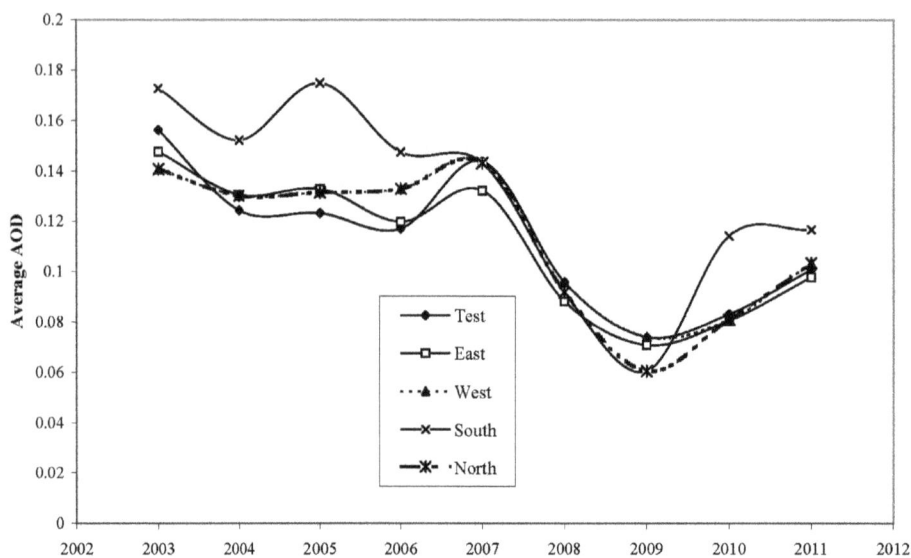

Figure 7. Comparison of annual averages of AOD (dimensionless) at the study site and surrounding areas

The annual averages of monthly fine AOD data at study site and surrounding areas decreased after 2008, which means there were less fine aerosols in the air. However, in comparison with surrounding areas, the AOD values at the study site were relative higher, especially during the time periods of 2008, 2009 and part of 2010. These relatively higher values of AOD might be related to emission from CO_2-EOR.

4. Conclusion

There was limited environmental monitoring for previous CO_2-EOR projects and this project is the first to implement some monitoring practices. However, due to the limitations of private land ownership at the study site, it is impossible to install monitoring instruments within the study site or have frequent visits. Thus, remote sensing is the best choice for environmental monitoring at the study site and surrounding areas at a large scale.

Through multiple-scale analysis of remote sensing data at the study site and surrounding areas, such as atmospheric CO_2 concentration, CH_4 concentration, the ratio of CO_2/CH_4 and AOD changes, it could be inferred that there were CO_2 emissions from the study site although the magnitude might be low. CO_2-EOR may not be operated with zero risk of CO_2 emission. However, it is necessary to conduct environmental monitoring and risk management procedures for CO_2-EOR projects. This method can be developed for monitoring emission from CO_2-EOR and other industrial operations at other locations.

Acknowledgments

This research was partially supported by US Department of Energy under Cooperative Agreement No. DE-FC26-06NT43029 and NSF CREST project (1036600). The author acknowledge the MODIS mission scientists and associated NASA personnel for the production of the data used in this study and the Giovanni online data system developed and maintained by the NASA GES DISC was used in this study. Thanks Mrs. K. A. Roberts for some editorial suggestions.

References

Acker, J. G., & Leptoukh, G. (2007). Online Analysis Enhances Use of NASA Earth Science Data. *Eos, Transactions American Geophysical Union, 88*, 14-17. http://dx.doi.org/10.1029/2007EO020003

Chahine, M. T., Barnet, C., Olsen, E. T., Chen, L., & Maddy, E. (2005). On the determination of atmospheric minor gases by the method of vanishing partial derivatives with application to CO_2. *Geophysical Research Letters, 32*, L22803. http://dx.doi.org/10.1029/2005GL024165

Chahine, M. T., Pagano, T. S., Aumann, H. H., Atlas, R., Barnet, C., Blaisdell, J., ... Zhou, L. (2006). AIRS: Improving Weather Forecasting and Providing New Data on Greenhouse Gases. *Bulletin of the American Meteorological Society, 87*(7), 911-926. http://dx.doi.org/10.1175/BAMS-87-7-911

Chen, X., Li, B. L., & Scott, S. (2005). Multiscale monitoring of a multispecies case study: two grass species at Sevilleta. *Plant Ecology, 179*(2), 149-154. http://dx.doi.org/10.1007/s11258-004-6802-z

Esposito, R. A., Pashin, J. C., & Walsh, P. M. (2008). Citronelle Dome: A giant opportunity for multizone carbon storage and enhanced oil recovery in the Mississippi Interior Salt Basin of Alabama. *Environmental Geoscience, 15*(2), 53-62. http://dx.doi.org/10.1306/eg.07250707012

Gilfillan, S. M., Lollar, B. S., Holland, G., Blagburn, D., Stevens, S., Schoell, M., ... Ballentine, C. J. (2009). Solubility trapping in formation water as dominant CO_2 sink in natural gas fields. *Nature, 458*(7238), 614-618. http://dx.doi.org/10.1038/nature07852

Haugan, P. M., & Joos, F. (2004). Metrics to assess the mitigation of global warming by carbon capture and storage in the ocean and in geological reservoirs. *Geophysical Research Letters, 31*, L18202. http://dx.doi.org/10.1029/2004GL020295

Holloway, S., Pearce, J. M., Hards, V. L., Ohsumi, T., & Gale, J. (2007). Natural emission of CO_2 from the geosphere and their bearing on the geological storage of carbon dioxide. *Energy, 32*(7), 1194-1201. http://dx.doi.org/10.1016/j.energy.2006.09.001

Jenkins, C. R., Cook, P. J., Ennis-King, J., Undershultz, J., Boreham, C., Dance, T., ... Urosevic, M. (2012). Safe storage and effective monitoring of CO_2 in depleted gas fields. *Proceedings of National Academy of Sciences USA, 109*(2), E35-E41. http://dx.doi.org/10.1073/pnas.1107255108

Meckel, T. A., & Hovorka, S. D. (2009). *Results from continuous downhole monitoring (PDG) at a field –scale CO_2 sequestration demonstration project, Cranfield, MS.* SPE International Conference on CO_2 Capture, Storage, and Utilization. Richardson, TX: Society of Petroleum Engineers.

Mertz, B., Davidson, O., de Coninck, H., Loos, M., & Meyer, L. (2005). *Carbon Dioxide Capture and Storage.* Cambridge, UK: Cambridge University Press.

Pacala, S., & Socolow, R. (2004). Stabilization wedges: solving the climate problem for the next 50 years with current technologies. *Science, 305*(5686), 968-972. http://dx.doi.org/10.1126/science.1100103

Pagano, T. S., Aumann, H. H., Hagan, D., & Overoye, K. (2003). Prelaunch and in-flight radiometric calibration of the atmospheric infrared sounder (AIRS). *IEEE Transactions Geoscience and Remote Sensing, 41*(2), 343-351. http://dx.doi.org /10.1109/TGRS.2002.808324

Remer, L. A., Kaufman, Y. J., Tanre, D., Mattoo, S., Chu, D. A., & Martins, J. V. (2005). The MODIS aerosol algorithm, products and validation. *Journal of Atmospheric Science, 62*(4), 947-973. http://dx.doi.org/10.1175/JAS3385.1

Roberts, K. A., & Chen, X. (2012). Considerations for the ecological monitoring of CO_2-mediated enhanced oil recovery. *International Journal of Ecological Economics and Statistics, 27*(4), 36-59.

Roberts, J. J., Wood, R. A., & Haszeldine, R. S. (2011). Assessing the health risks of natural CO_2 seeps in Italy. *Proceedings of National Academy of Sciences USA, 108*(40), 16545-16548. http://dx.doi.org/10.1073/pnas.1018590108

Schrag, D. P. (2007) Preparing to capture carbon. *Science, 315*(5813), 812-813. http://dx.doi.org/10.1126/science.1137632

Salomonson, V. V., Barnes, W. L., Maymon, W. P., Montgomery, H., & Ostrow, H. (1989). MODIS: Advanced facility instrument for studies of the Earth as a system. *IEEE Transactions Geoscience and Remote Sensing, 27*(2), 145-153. http://dx.doi.org/10.1109/36.20292

Thomas, D., & Benson, S. (2005). *Carbon Dioxide Capture for Storage in Deep Geologic Formations.* Amsterdam, Netherland: Elsevier Science.

van Bergen, F., Gale, J., Damen, K. J., & Wildenborg, A. F. B. (2004). Worldwide selection of early opportunities for CO_2-enhanced oil recovery and CO_2-enhanced coal bed methane production. *Energy, 29*(9-10), 1611-1621. http://dx.doi.org/10.1016/j.energy.2004.03.063

Whittaker, S., White, D., Law, D., & Chalaturnyk, R. (2004). *IEAGHG Weyburn CO_2 monitoring and storage project summary report 2000-2004.* Regina, SK, Canada: Petroleum Technology Research Center.

Wang, J., & Christopher, S. A. (2003). Intercomparison between satellite-derived aerosol optical thickness and PM2.5 mass: Implications for air quality studies. *Geophysical Research Letters, 30*(21), 2095. http://dx.doi.org/10.1029/2003GL018174

Zweigel, P., Arts, R., Lothe, A. E., & Lindeberg, E. B. G. (2004). Reservoir geology of the Utsira Formation at the first industrial- scale underground CO_2 storage site. In S. J. Baines & R. H. Worden (Eds.), *Geological Storage of Carbon Dioxide* (pp. 165-180). London, UK: Geological Society of London.

6

Change in the Annual Water Withdrawal-to-Availability Ratio and Its Major Causes: An Evaluation for Asian River Basins Under Socioeconomic Development and Climate Change Scenarios

Ayami Hayashi[1], Keigo Akimoto[1,2], Takashi Homma[1], Kenichi Wada[1] & Toshimasa Tomoda[1]

[1] Systems Analysis Group, Research Institute of Innovative Technology for the Earth, Kyoto, Japan

[2] Graduate School of Art and Science, The University of Tokyo, Tokyo, Japan

Correspondence: Ayami Hayashi, Systems Analysis Group, Research Institute of Innovative Technology for the Earth (RITE), 9-2 Kizugawadai, Kizugawa-shi, Kyoto 619-0292, Japan. E-mail: ayami@rite.or.jp

Abstract

More than half of the world's population lives in Asia, and ensuring a stable water supply is a critical issue. This study evaluates changes in the annual water withdrawal-to-availability ratio (WAR), and the major causes thereof, for each of Asian river basins under different socioeconomic development and climate change scenarios. According to our evaluation, the WAR will increase in 59%–61% of the Asian river basin areas by around 2030, as a result of population growth and the increase in per capita municipal and industrial water withdrawals. On the other hand, the WAR will decrease in 8%–16% of such areas, due to the increase in water availability associated with global warming and a decrease in per capita water agricultural withdrawal. After 2030, there will be a reduction of areas with increasing WAR because of a slowdown in the growth of both population and per capita municipal and industrial water withdrawals, while there will be an expansion of areas with decreasing WAR caused by continual decrease in per capita agricultural water withdrawal and intensified water availability. Significant measures to suppress WAR increase will differ by river basin, depending on the causes for the WAR increase. For instance, measures to deal with population increase and efforts to improve industrial and municipal water use by around 2030 will be important in the Huang He river basin. In the Indus river basin, coping with the decrease in water availability after around 2030 will be important. In addition, measures to handle population increase will be necessary.

Keywords: water resource management, water withdrawal-to-availability ratio, socioeconomic development, climate change

1. Introduction

More than half of the world's population lives in Asia, and the Asian population is projected to reach around 4.5 billion in the middle of this century (United Nations [UN], 2011). Furthermore, the number of people living in river basins threatened by severe water stress is estimated to increase, although the absolute value varies depending on assumptions on the criterion for severe water stress, scenarios on socioeconomic development and climate change and so on. For instance, according to an evaluation by Arnell (2004) based on a criterion, per capita annual water availability (PWA) < 1000 m^3, about 1 billion people in South Asia and East Asia were threatened by severe water stress in 1995. The 'water availability' includes renewable surface and subsurface runoff, and is also called as 'blue water resource'. Hayashi et al. (2013) estimated figures for the population under severe water stress based on the following criterion: $0.4 \leq$ a ratio of the annual water withdrawal to water availability (WAR) (Raskin et al., 1997), in which the annual water withdrawal is represented by the sum of annual water withdrawals of municipal, industrial, and agricultural sectors, and reported that the figure in Asia would increase from 1.4 billion in 2000 to 2.4 billion in 2050 and 2.1 billion in 2100. An evaluation based on the following criterion: cumulative withdrawal-to-demand ratio (CWD) ≤ 0.5 (Hanasaki et al., 2013) stated that the population in Asia under severe water stress would increase from 1.7 billion in 2000 to 2.0–3.2 billion by the middle of this century, and 1.6 to 3.1 billion by the end of this century. (The ranges mainly resulted from differences in assumptions related to population growth). These results suggest that a stable fresh water supply become a critical issue in Asia. Meanwhile, from the viewpoint of water resource management, important points include not only whether the indexes will exceed their criteria for 'severe water stress', but also whether they will become increasingly or decreasingly severe. If they change, an assessment of the causes would be meaningful.

Gosling and Arnell (2013) and Gerten at al. (2013) evaluated the impacts of climate change on the PWA and WAR, maintaining socioeconomic development scenario unchanged; and they stated that the severity of water stress will increase around the Mediterranean, in parts of Europe, North and South America, and southern Africa, while it will decrease in regions such as East Asia. Hanasaki et al. (2013) presented the CWD change for different scenarios on socioeconomic development and climate change. They stated that water stress conditions will be influenced by both of climate change and water use change, although the contributions by climate change and water use change will vary among regions and scenarios. Fung et al. (2011) evaluated the PWA for 112 large river basins throughout the world for two global mean temperature change scenarios, and stated that, assuming the 2060s population scenario (UN, 2007), the water stress condition would worsen in 71% and 74% of the river basins at scenarios of + 2 °C and + 4 °C (relative to the pre-industrial level), respectively. They described that the direction of change in the water stress condition was highly dependent on the population assumption for most of the basins, although climate change dominated the direction for a small number of basins. Alcamo et al. (2007) evaluated the WAR in the 2050s for the A2 and B2 scenarios of the Special Report on Emissions Scenarios (SRES) (Intergovernmental Panel on Climate Change [IPCC], 2000). They reported that the WAR would increase (i.e., the water stress condition would worsen) in over 62%–76% of the world river basin areas, and would decrease (i.e., the water stress condition would improve) in over 20%–29% of the world basin areas. The principal cause of the decreasing WAR is greater water availability due to climate change, while the principal cause of the increasing WAR is growing water withdrawals. Furthermore, for areas with increasing water withdrawals, they made estimations regarding the contributions of different sectors to increases in water withdrawals. They stated that the municipal sector would make the greatest contribution, since this sector represents the largest increases in 80%–86% of the aforementioned areas. These studies are interesting for understanding directions of change in water stress condition and the principal cause of the change. However, a limitation of these studies is a lack of quantitative comparisons of the impacts on the water stress change among major causes. For instance, the study by Alcamo et al. (2007) remains unclear whether the increase in the municipal water withdrawal or the decrease in water availability has the greatest impact on the WAR increase. In the study by Fung et al. (2011), the assumptions for water demand based on population figures are too simple to take into account impacts caused by changes in human activity.

To clarify the impacts on the water stress condition due to each of major causes, we decompose a change rate of the water stress index into the sum of change rates of major factors, and evaluate these change rates for different scenarios on socioeconomic development and climate change during this century. We adopted the WAR as a water stress index to take into account the human activity change in municipal, industrial, and agricultural sectors. A grid-based water supply-demand model is utilized to evaluate the WAR by river basin keeping a consistency with socioeconomic development and climate change scenarios. The evaluation is conducted focusing on Asian river basins, since a stable fresh water supply will probably become a critical issue in Asia as mentioned above, and information on the causes of the water stress change will be important for the water resource management. This paper describes the evaluations throughout Asian river basins. For some river basins that are currently threatened by high water stress, the details of the causes for the water stress change are described together with preliminary consideration for possible measures to alleviate the water stress.

2. Methodology

2.1 The Annual Water Withdrawal-to-Availability Ratio

The numerator of this ratio, the annual water withdrawals, are represented by the sum of annual water withdrawals of municipal, industrial, and agricultural sectors, and the denominator, the water availability, is represented by annual runoff (renewable freshwater). According to Raskin et al. (1997), the areas with WAR < 0.2 had low or no stress; areas with $0.2 \leq$ WAR < 0.4 had medium stress; and areas with $0.4 \leq$ WAR had high (severe) stress. This ratio is simple to apply to global-scale and long-term evaluations, and has been utilized in many studies (Vörösmarty et al., 2000; Oki & Kanae, 2006; Alcamo et al., 2007).

It also has the following disadvantages: 1) A gap between the subannual distribution of water availability and water demand is possibly overlooked; and 2) overestimation is possible if not only primary water withdrawal but also secondary water withdrawal (i.e., wastewater use) is counted. Hanasaki et al. (2008) proposed a sophisticated index, CWD, to cope with these disadvantages, in which daily water withdrawal from a river and daily potential water consumption demand is simulated by grid with a spatial resolution of one degree. According to their comparison between the WAR and the CWD, the WAR tends to underestimate the water stress in grids such as the Sahel, the Asian monsoon region, and southern Africa, and it has extremely large value (e.g., 100) in some grids. However, their results simultaneously show that there is a strong correlation between the WAR and the CWD for almost all grids. With an understanding of this situation, we adopted the WAR, which would not cause a serious problem in this study.

2.2 Change Rates of the WAR and Major Associated Factors

The *WAR* is denoted as Equation (1).

$$WAR = \frac{P \cdot D}{R} \tag{1}$$

Where *P*, *D*, and *R* are population, per capita annual water withdrawal, and annual water availability, respectively; and *D* is expressed by the sum of *M* (per capita annual municipal water withdrawal), *I* (per capita annual industrial water withdrawal), and *A* (per capita annual municipal water withdrawal).

Therefore, the change rate of WAR during a period from t_1 to t_2 is denoted as Equation (2).

$$\Delta WAR = \Delta P + \Delta D - \Delta R \tag{2}$$

Where ΔWAR, ΔP, ΔD, and ΔR are the change rates for *WAR*, *P*, *D*, and *R*, respectively; and the formula (ΔX) denoted by Equation (3) is applied to ΔWAR, ΔP, ΔD, and ΔR.

$$\Delta X = \frac{1}{t_2 - t_1} \cdot ln\left(\frac{X_2}{X1}\right) \tag{3}$$

Furthermore, we introduce the approximate expression of ΔD, as denoted by Equation (4), to understand the separate contributions of municipal, industrial, and agricultural water withdrawals on the per capita annual water withdrawal.

$$\Delta D \approx \Delta'M + \Delta'I + \Delta'A \tag{4}$$

Where the formulae applied to $\Delta'M$, $\Delta'I$, and $\Delta'A$ are:

$$\frac{1}{t_2 - t_1} \cdot ln\left(\frac{M_2 + I_1 + A_1}{M_1 + I_1 + A_1}\right), \quad \frac{1}{t_2 - t_1} \cdot ln\left(\frac{M_1 + I_2 + A_1}{M_1 + I_1 + A_1}\right), \quad \frac{1}{t_2 - t_1} \cdot ln\left(\frac{M_1 + I_1 + A_2}{M_1 + I_1 + A_1}\right), \text{ respectively.}$$

The introduction of this approximate expression makes it possible to quantitatively compare the contributions to the WAR change among sectoral water withdrawal changes, population changes, and water availability changes. The evaluation by sector based on Equation (4) was conducted for over 95% of river basins focused upon in this study, confirming that the approximation error for ΔD is small enough not to affect the evaluation result.

2.3 The Water Supply-Demand Model

The water supply-demand model is a grid-based model with basically 15 × 15 minute special resolution. As shown in Figure 1, it is integrated with an agro-land use model to have consistency with evaluations for agricultural land use. The details of the model are described in Hayashi et al. (2013). The outline is as follows: The grid-based annual water availability is estimated from the runoff data projected by atmosphere-ocean coupled general circulation models (AOGCMs) together with global mean temperature change scenario specified, based on a pattern-scaling method. The annual water withdrawal is estimated separately for municipal water, industrial water, and agricultural water. The grid-based municipal water withdrawal is calculated by the product of population and per capita municipal water withdrawal. The per capita municipal water withdrawal is calculated separately for urban and rural areas, taking into account the increases associated with economic growth by country. For the industrial water withdrawal, first, the country total is estimated based on scenarios for the production volumes for water-intensive manufacturing sectors (i.e., iron and steel, the chemical industry, and the pulp and paper sectors) and industrial water-use efficiency. The data on the production volume for each of these sectors and the efficiency are obtained from the Dynamic New Earth 21 plus (DNE21+) model (Akimoto et al., 2010). Then, the industrial water withdrawal is distributed to urban areas in proportion to the urban population in each grid. It is assumed that industrial water is used solely in urban areas. The data on grid-based urban and rural population is developed based on the data for the year 2000 (Netherlands Environmental Assessment Agency [PBL], 2009). It is assumed that the population will concentrate in urban areas and that urban areas will spread so as to be in close proximity to each other, according to a study by Grübler et al. (2007). For each of the municipal and industrial water withdrawals, the country total in 2000 agrees with the amount reported by the AQUASTAT database (Food and Agriculture Organization [FAO], 2011a).

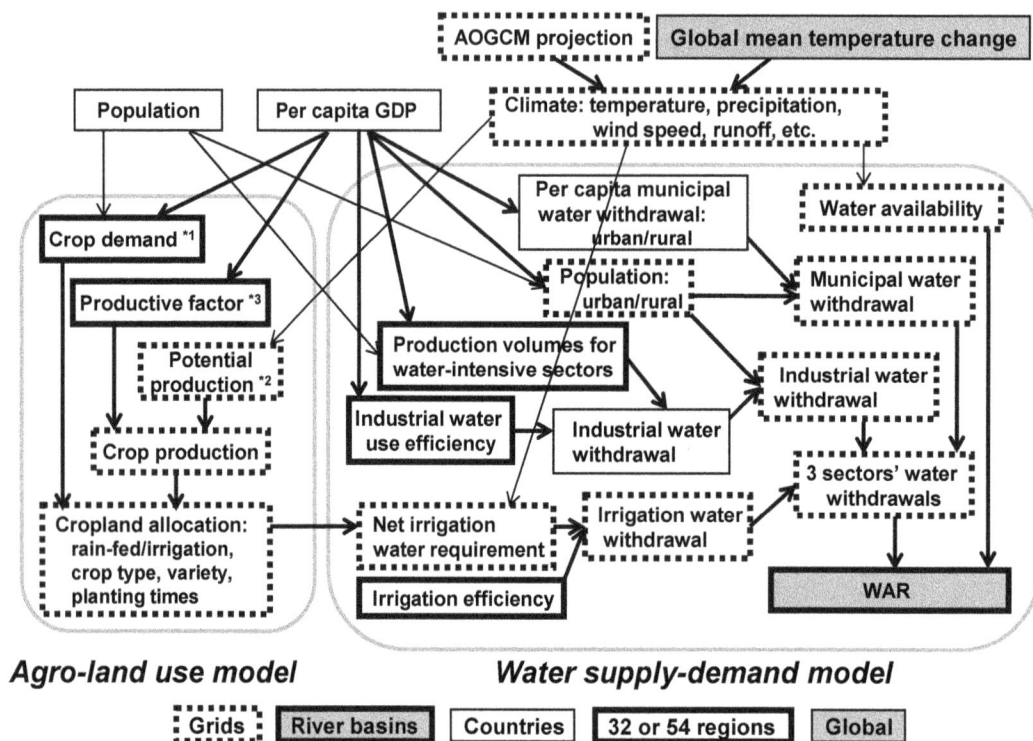

Figure 1. Framework of the water supply-demand model

*1: Crop demand (for wheat, rice, maize, sugar cane, soybeans, oil palm fruit, rapeseed, and others) is estimated to meet the predicted food demand in the future, as well as the biofuel demand (which is assumed to be same as the current level).

*2: The impact of climate change on crop production is estimated based on a framework of the GAEZ model. Adaptations by changing crop variety and planting times are taken into account.

*3: Improvement of productivity caused by technological progress associated with economic growth, land use constraints, etc. (which do not include impacts due to climate change) is assumed based on the historical trend during the period from 1961 to 2007 (FAO, 2011b). For all crops, greater growth is expected in developing regions than in developed regions. For instance, considering rice in scenarios A+4 and A+2 (for the definition of scenarios, refer to section 2.4), the improvement of this factor for Asian regions is assumed in a range from 0.1% p.a. (for Japan) to 0.8% p.a. (for Mongolia and the Democratic People's Republic of Korea). (These figures show mean values for the period from 2000 to 2030.)

As agricultural water withdrawal, we focus on irrigation water, which is the dominant form of agricultural water use. The grid-based irrigation water withdrawal is estimated based on the net irrigation water requirements and irrigation efficiency (i.e., the ratio of net irrigation water requirements to irrigation water withdrawals). The numerical values for the irrigation efficiency are derived from a study by Döll and Siebert (2002) (i.e., 0.35 for East and Southeast Asia, and 0.55 for South Asia; to be 0.1 lower for rice). Efficiency is assumed to be constant during this century. The net irrigation water requirement is estimated as the difference between potential evapotranspiration and actual evapotranspiration during the growing period for a crop planted in an irrigation grid. The potential and actual evapotranspiration values are calculated using the Penman-Monteith method and the water-balance method employed in the Global Agro-Ecological Zones (GAEZ) model (Fischer et al., 2002). Information on the irrigation grid, crop type, and planting times are obtained from the results using the agro-land use model. In this model, the grids available for irrigation are based on an irrigation map for the year 2000 (Siebert et al., 2007), and no expansion of the available irrigation grids is allowed in the future according to the assumptions of Alcamo et al. (2007). Our estimations for irrigated area and net irrigation water requirements in 2000 agree fairly well with the results obtained by Siebert and Döll (2010).

Data for the river basins are derived from the Total Runoff Integrating Pathways (TRIP) database (Oki, 2001). When we define Asian river basins by the criterion that more than half of a given river basin's area is included in an Asian region (i.e., East Asia, Southeast Asia, or South Asia excluding Iran), the TRIP data covers approximately 800 Asian river basins (20×10^6 km^2). Among them, approximately 700 river basins (18.5×10^6 km^2) are focused upon; other river basins are excluded mainly due to their very low population densities (i.e., < 1 person km^{-2} in 2000).

For grid-based climate projection by AOGCM, data from MIROC 3.2 (Medres) for the SRES-A1B emission scenario were obtained from a Program for Climate Model Diagnosis and Intercomparison (PCMDI) database (2004). Regarding the AOGCM projections, inter-model differences have been suggested (IPCC, 2007); therefore, sensitivity analysis is an ongoing issue.

The calculations are carried out for every decade from 2000 to 2050, and for specific time points for 2070 and 2100.

2.4 Scenarios

We set up three scenarios (A+4, A+2, and B+4), in relation to socioeconomic development and climate change, as shown in Table 1. The time series of socioeconomic development scenarios and climate change scenarios are presented in Tables 2 and 3, respectively. The socioeconomic development scenarios ALPS-A and ALPS-B were developed by Akimoto et al. (2013) giving consideration to historical trends and uncertainties in the future. In scenario ALPS-A, moderate growth of population and per capita GDP is assumed. On the other hand, low population growth and high per capita GDP growth is assumed in scenario ALPS-B.

The global mean temperature changes by 2100 for the climate change scenarios + 4 °C and + 2 °C are around + 4 °C and below + 2 °C, respectively. They were obtained from our previous study, in which the temperature change was estimated based on the greenhouse gases (GHGs) emission scenarios (Akimoto et al., 2012). The scenarios + 4 °C and + 2 °C correspond to a baseline GHGs emission scenario and a severe GHGs mitigation scenario, respectively. If the GDP losses due to climate mitigation measures and climate change impacts are taken into account, the GDP value is expected to be lower than that initially assumed without considering these effects. According to our previous study, a change of several percentage points can be expected for the world total in 2100 (Akimoto et al., 2012), and such figures are not large enough to seriously affect the evaluation of water; therefore, we adopted one set of GDP scenario for scenarios A+4 and A+2, regardless the level of global mean temperature change.

Table 1. Scenarios examined

	Socioeconomic development	Climate change [1]
A+4:	ALPS-A (Moderate growth of population and per capita GDP)	+ 4 °C
A+2:	ALPS-A (Moderate growth of population and per capita GDP)	+ 2 °C
B+4:	ALPS-B (Low population growth and high per capita GDP growth)	+ 4 °C

*1: Global mean temperature change in 2100 relative to the pre-industrial level (see Table 3).

Table 2. Time series of socioeconomic development scenarios

			2000	2030	2050	2070	2100
ALPS-A	Population [billion] [1]	:	3.3	4.3	4.6	4.5	4.3
	Per capita GDP [thousand 2000 US$] [2]	:	2.4	6.0	10	14	21
	GDP [trillion 2000 US$] [1]	:	8	26	46	62	91
ALPS-B	Population [billion] [1]	:	3.3	4.2	4.3	3.9	3.3
	Per capita GDP [thousand 2000 US$] [2]	:	2.4	6.4	12	19	39
	GDP [trillion 2000 US$] [1]	:	8	27	50	75	131

*1: Total for the Asian basins.

*2: Mean value for the Asian basins.

Table 3. Time series of climate change scenarios (global mean temperature change relative to the pre-industrial level) [°C]

	2000	2030	2050	2070	2100
+ 4 °C:	0.9	1.7	2.4	3.1	4.1
+ 2 °C:	0.9	1.6	1.8	1.9	1.9

3. Results and Discussion

3.1 WAR Change Under Scenario A+4

Figure 2 shows the WAR change rate estimated by river basin for scenario A+4. During the period from 2000 to 2030, an increase in the WAR is estimated for most of the Asian river basins. Large increases of more than + 2% p.a. are estimated for basins in island countries in Southeast Asia and coastal regions of the continent from China to Pakistan. In contrast, a decrease in the WAR is estimated for basins in Japan and part of the continent. During the period from 2030 to 2050, the increase in the WAR will continue in basins in Southeast Asia, Pakistan, Afghanistan, and northwest China, while the WAR will begin to decrease in large areas in China and India. After 2050, the WAR will gradually stabilize in most of the Asian regions, although the increase or decrease in the WAR will continue in some regions.

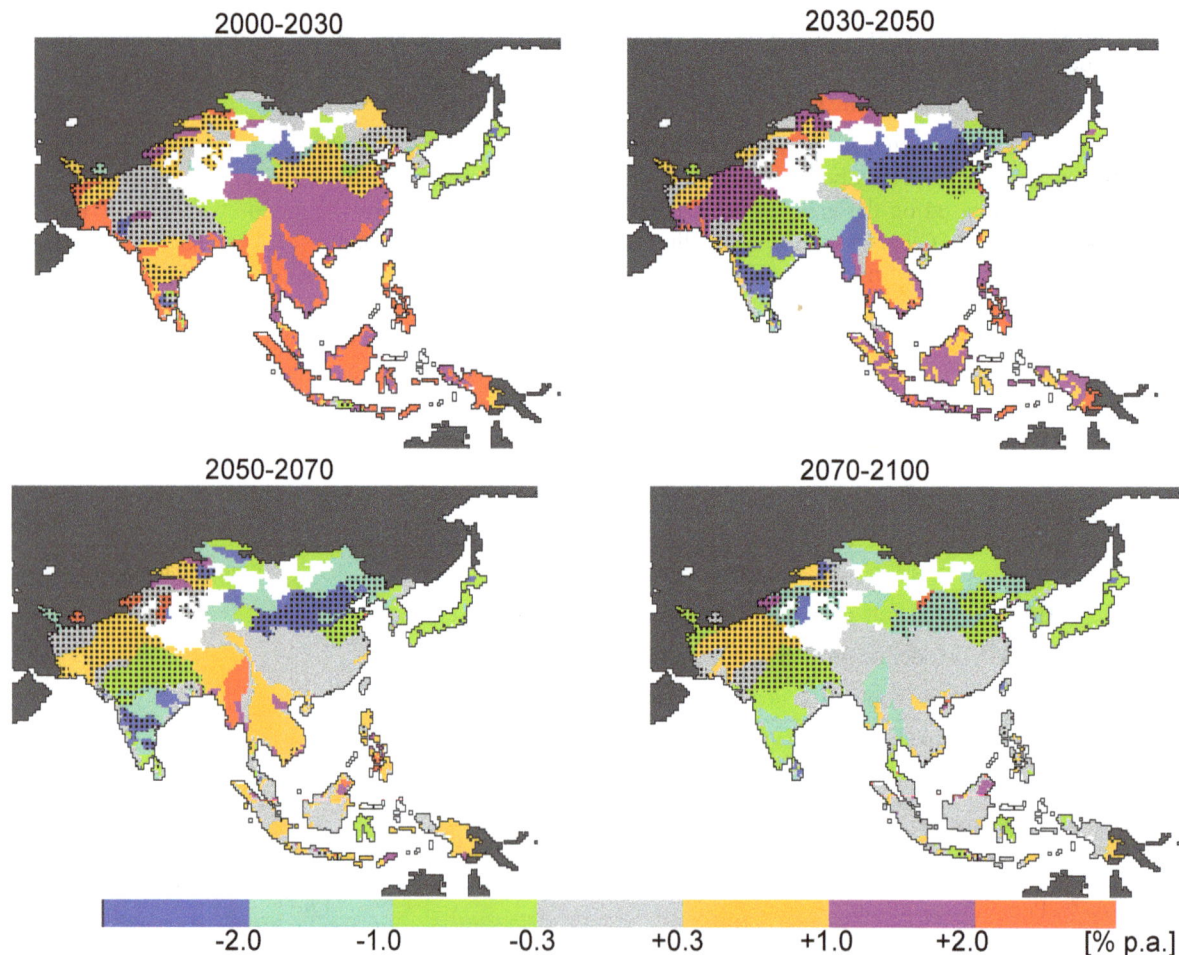

Figure 2. Change rate of the WAR for Asian river basins under scenario A+4

Dark gray indicates areas outside Asian river basins. White in Asian river basin areas signifies river basins excluded from this study, mainly due to their very low population densities (see the text in section 2.2). Dotted areas represent river basins estimated as 'river basins having high water stress' based on the criterion $0.4 \leq WAR_{t_1}$ for the period from t_1 to t_2.

To clarify the major causes of the changes mentioned above, the change rates of major factors constituting the WAR are evaluated. Figure 3 shows the results for the period from 2000 to 2030. From this figure, it can be noted that the increasing WAR estimated for most of the Asian river basins is mainly caused by population growth and increases in per capita municipal and industrial water withdrawals. For coastal regions of the continent from China to Pakistan, rapid increases in per capita municipal and industrial water withdrawals will significantly contribute to remarkable increases in the WAR. This will be associated with the progress of urbanization in these regions. In the urban areas, the per capita industrial water withdrawal will increase due to enhanced industrial activity, and the per capita municipal water withdrawal will increase due to the improvement of water accessibility and lifestyle changes (e.g., increased frequency of washing and cleaning) associated with economic growth.

The major causes of the decrease in WAR in some regions, such as the north of China and the northeast of India, will be a decrease in per capita agricultural water withdrawal and the increase in water availability. The estimated trend of the decrease in per capita agricultural water withdrawal agrees well with the historical trend since 1960, which was preliminary estimated by us based on the population and the total agricultural water withdrawal in Asia reported by Siklomanov (1999). According to our consideration based on the statistical data (FAO, 2011b) and a report (Alexandratos & Bruinsma, 2012), the decrease in per capita agricultural water withdrawal since 1960 was caused by the improvement of crop productivity and the decreasing per capita food consumption of rice. (Paddy rice accounts for large amounts of irrigation water withdrawal in Asia.) Both the trend of improvement of crop productivity and that of decreasing per capita food consumption of rice are taken into account in our future scenarios. The change in water availability associated with global warming will differ in different regions. It will increase in some basins in China, India, and Indonesia, leading to decreases in the WAR. However, it will decrease in some basins in Southeast Asia, leading to increases in the WAR.

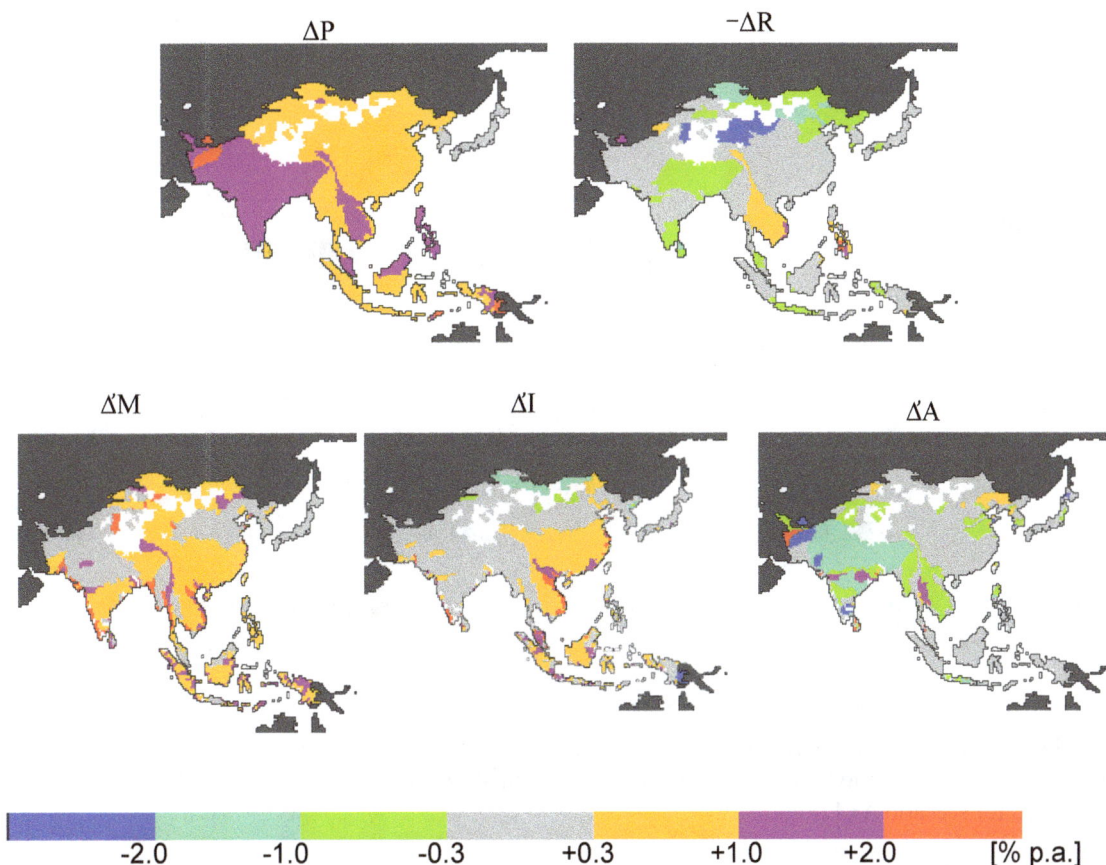

Figure 3. Change rates of the factors constituting the WAR (the period from 2000 to 2030, scenario A+4)

ΔP and ΔR represent change rates of population and annual water availability, respectively. ΔM, ΔI, and ΔA represent change rates caused by changes in the per capita annual withdrawals of municipal, industrial and agricultural water, respectively (see the definitions in section 2.2). White areas in Asian river basins show the river basins excluded from this study.

3.2 Comparisons of WAR Change Among Scenarios

The WAR change rate estimated by river basin for scenarios B+4 and A+2 is shown in Figure 4. By comparing scenarios B+4 (Figure 4 (a)) and A+4 (Figure 2), it is noted that the trend of the decreasing WAR will become remarkable after 2030 under scenario B+4. The per capita municipal, industrial, and agricultural water withdrawal changes are not so different between scenarios B+4 and A+4, and the water availability changes for these two scenarios are the same; therefore, the remarkable decrease in the WAR under scenario B+4 is mainly caused by a lower population growth in scenario B+4.

By comparing scenarios A+2 (Figure 4 (b)) and A+4 (Figure 2), it is clear that climate change impacts on the WAR will be suppressed under scenario A+2. For instance, the decrease in the WAR estimated in China and India for scenario A+2 is smaller than that for scenario A+4. The reason is that the effect due to the increase in the water availability and the decrease in per capita agricultural water withdrawal expected for these regions under scenario A+4 will be suppressed under scenario A+2; on the other hand, for some regions such as the northwest of Asia (e.g., Afghanistan and Pakistan), the increase in the WAR estimated under scenario A+4 is suppressed under scenario A+2 due to reduction of the decrease in the water availability and increase in the per capita agricultural water withdrawal caused by global warming.

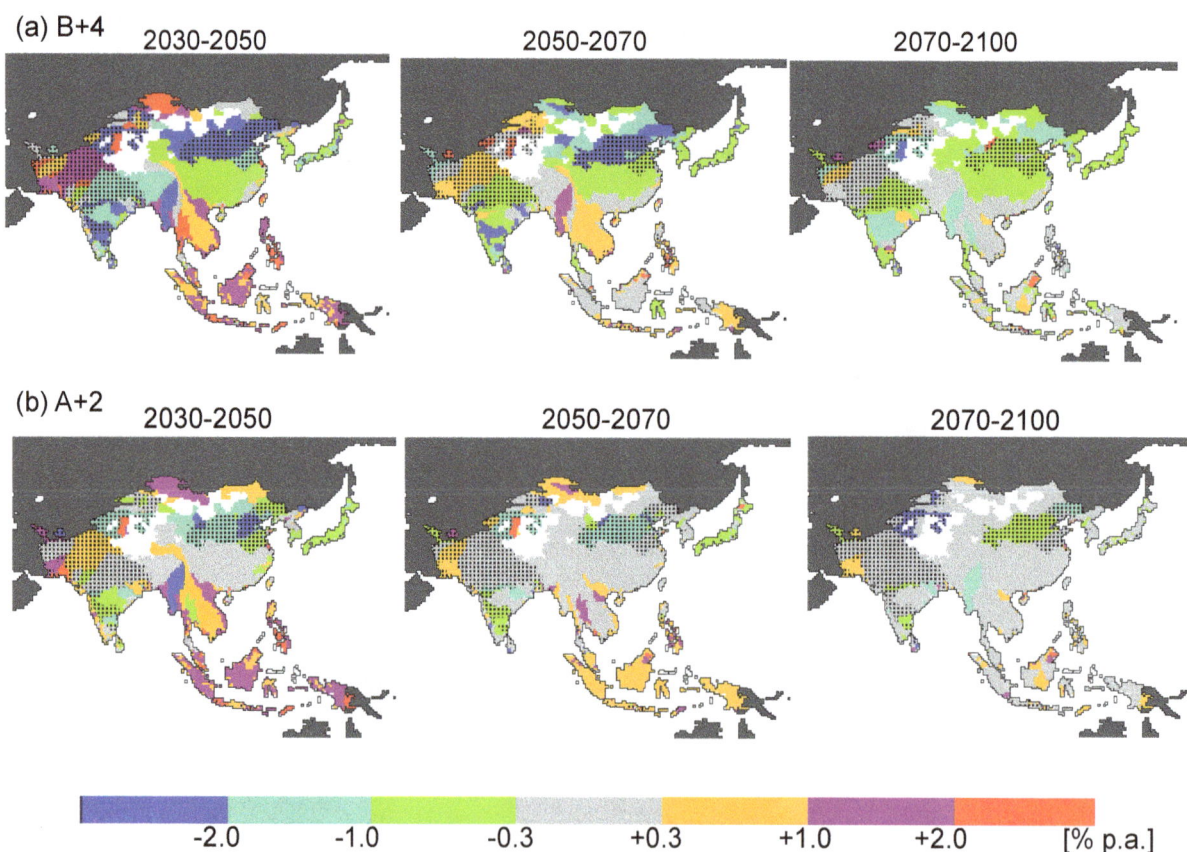

Figure 4. Change rate of the WAR under (a) scenario B+4, and (b) scenario A+2

For dotted areas, white areas, and dark gray areas, refer to the caption of Figure 2.
The figures for the period from 2000 to 2030 are omitted, since they are not so different from that estimated for scenario A+4 (Figure 2).

The percentage figures for areas with increasing and decreasing WAR under each of the scenarios are summarized in Table 4. Under scenario A+4, the figures for areas with increasing WAR, those with decreasing WAR, and those with little change are 61%, 13%, and 17%, respectively, for the period of 2000–2030. The remaining 9% corresponds to areas that are outside the scope of this evaluation, mainly for a reason of the very low population density. During the period of 2030–2050, areas with increasing WAR will be reduced because of a slowdown in the growth of population and per capita municipal and industrial water withdrawals; on the other hand, areas with decreasing WAR will be expanded owing to a decrease in per capita agricultural water withdrawal and the enhanced water availability associated with intensified global warming. After 2050, areas with little change will be expanded associated with the stabilization of the WAR.

The figures for scenario B+4 are close to those for scenario A+4, except for the larger values for areas with decreasing WAR after 2050. This is mainly caused by the remarkable population decrease under scenario B+4 as mentioned above. For scenario A+2, the figures for areas with increasing WAR are almost same as those for scenarios A+4 and B+4, while those for areas with decreasing WAR are smaller than those for scenarios A+4 and B+4. This is mainly caused by that the effect due to the increase in the water availability and the decrease in per capita agricultural water withdrawal expected in regions such as China and India under scenarios A+4 and B+4 will be suppressed under scenario A+2. Consequently, the figures for areas with little change for scenario A+2 are larger than those for scenarios A+4 and B+4.

The trend of areas with increasing WAR outnumbering areas with decreasing WAR during the period from 2000-2030, which is estimated for all three scenarios, is similar to the trend reported by Alcamo et al. (2007), although they evaluated only one period of 1995-2055. Our study evaluates the WAR changes for different periods, and shows that after 2030, changes in the WAR may differ remarkably depending on socioeconomic development and climate change.

Table 4. Percentages of river basin areas with increasing WAR, decreasing WAR, and little change for scenarios A+4, B+4, and A+2 [%]

		Areas with increasing WAR	Areas with decreasing WAR	Areas with little change [*1]
A+4	2000-2030:	61	13	17
	2030-2050:	36	45	10
	2050-2070:	30	34	28
	2070-2100:	11	40	41
B+4	2000-2030 :	61	16	14
	2030-2050 :	36	46	10
	2050-2070 :	25	47	20
	2070-2100 :	6	51	34
A+2	2000-2030 :	59	8	24
	2030-2050 :	37	27	28
	2050-2070 :	24	15	52
	2070-2100 :	6	14	71

[*1]: 'Little change' means the change rate is in the range from −0.3% to +0.3% p.a.

3.3 Details of WAR Change and Important Measures for River Basins Threatened by High Water Stress

The WAR increase in the future is a serious issue, particularly for river basins that are currently threatened by high water stress. In this section, we describe the details of the WAR change, and measures to suppress the WAR increase, taking the 'Huang He' and 'Indus' river basins as examples.

The Huang He river basin has an area of 0.8×10^6 km^2. Our study evaluates conditions there as follows: In 2000, 160 million people lived in this basin, and the annual water availability was 27 km^3 yr^{-1} (230 m^3 person^{-1} yr^{-1}). The WAR was over 3, and the basin was threatened by very high water stress. The annual water withdrawal figures were 5%, 15%, and 80% for municipal, industrial, and agricultural use, respectively. Paddy rice, wheat, and other crops were produced in the irrigated area.

Figure 5 shows estimations of the WAR and change rates of its factors in the future for this river basin. The WAR (Figure 5 (a)) is expected to increase by around 2030, and to decrease drastically thereafter. The expected major causes for the increase by around 2030 are population growth and increases in the per capita industrial and municipal water withdrawals, as shown in Figures 5 (b), (d), and (d). The major causes of the decrease after 2030 are expected to be an increase in water availability associated with climate change, and a decrease in per capita agricultural water withdrawal. The decrease in the per capita agricultural water withdrawal is expected as a result of climate change impacts, and the improvement of the crop productivity associated with the economic growth. The WAR decrease weakens in scenario A+2 compared to those in scenarios A+4 and B+4, since the effect due to climate change on the increase in water availability and the decrease in per capita agricultural water withdrawal will be suppressed in scenario A+2.

These results suggest that measures will be necessary in this river basin by around 2030. In particular, measures to handle the population increase and improvement of efficiency for industrial and municipal water use will be important.

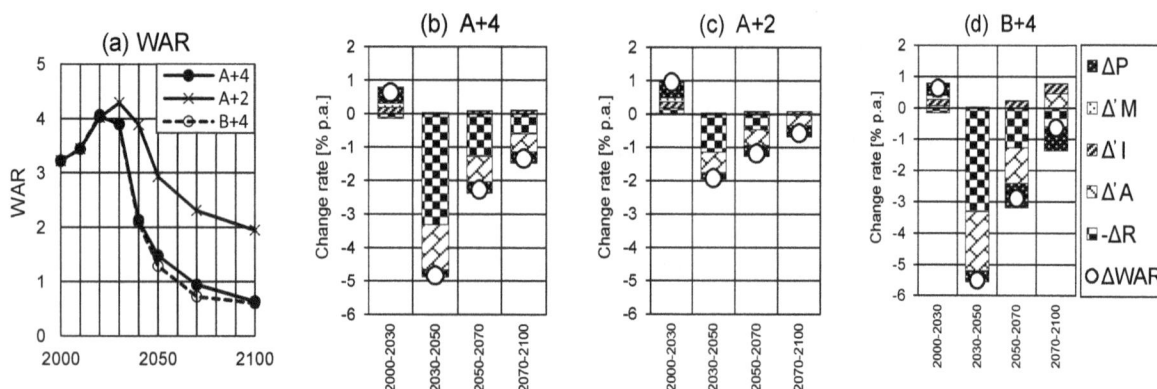

Figure 5. Time series for the Huang He river basin

(a) The WAR, and change rates for (b) scenario A+4, (c) scenario A+2, and (d) scenario B+4

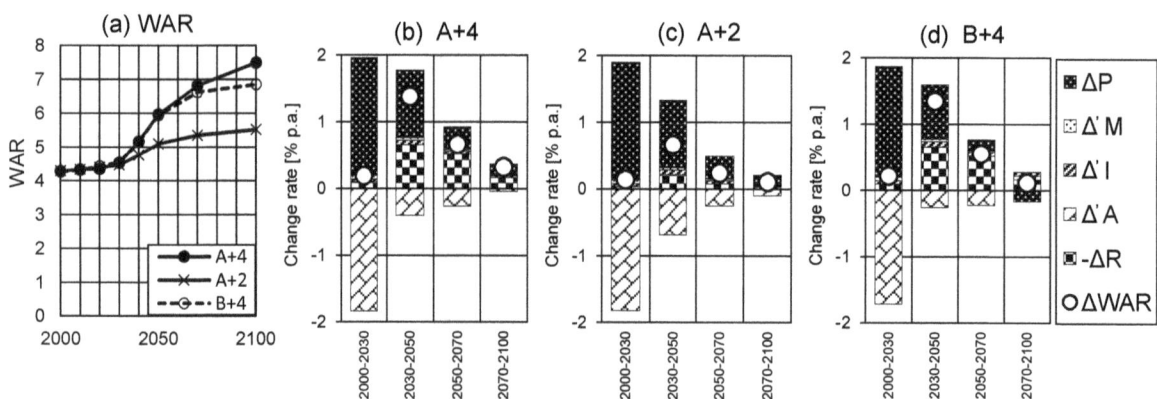

Figure 6. Time series for Indus river basin

(a) The WAR, and change rates for (b) scenario A+4, (c) scenario A+2, and (d) scenario B+4

The Indus river basin has an area of 1.0×10^6 km^2. According our estimates, in 2000, 190 million people lived in this basin, and the annual water availability was 70 km^3 yr^{-1} (370 m^3 person^{-1} yr^{-1}). The WAR was over 4, and the basin was threatened very high water stress. More than 95% of the annual water withdrawal was used for agriculture. Paddy rice, wheat, and other crops were produced in the irrigated area.

Figure 6 shows estimations for the WAR and change rates of its factors in the future for this river basin. The WAR is expected to be almost constant by around 2030, and to increase after that. For the period from 2000 to 2030, the population is expected to increase, and to contribute to the WAR increase. Meanwhile, the per capita agricultural water withdrawal is expected to decrease, and to contribute to the WAR decrease. Consequently, the WAR is expected to almost constant. The decrease in the per capita agricultural water withdrawal is mainly a result of improvement in crop productivity. The crop productivity in this region was very low (for instance, the for paddy rice in 2000 yield was 3.0, 3.9, and 6.3 ton ha^{-1} for Pakistan, the global mean, and China, respectively (FAO, 2011)); therefore, it is expected to steadily improve with economic growth.

After 2030, the decrease in the per capita agricultural water withdrawal dwindles, stemming from climate change impacts. The population growth also slows down; however, it is large enough to increase the WAR. In addition to these factors, the water availability decreases as a result of climate change, and becomes one of significant causes of the WAR increase.

These results suggest that coping with the decrease in the water availability, which is projected to become apparent after around 2030, will be important in this river basin. In addition, measures to handle the population increase may be necessary to suppress an increase of the WAR.

4. Conclusions

Ensuring a stable fresh water supply is a critical issue for Asia. To make progress with water management, understanding future changes in the ratio of the water demand and availability, as well as the major causes of such changes, are important. This study evaluates changes in the ratio by river basin in Asian regions based on the annual water withdrawal-to-availability ratio (WAR) under the three different future long-term scenarios for socioeconomic development and climate change. Furthermore, to clarify the major causes of the change, the contributions of major factors are evaluated. The findings of this study are as follows:

• The WAR will increase in 59%–61% of the Asian river basin areas by around 2030, as a result of population growth and the increase in per capita municipal and industrial water withdrawals. In some costal basins in which progress with urbanization is expected, increases in the per capita municipal and industrial water withdrawals may be more significant than population growth. On the other hand, the WAR will decrease in 8%–16% of the Asian river basin areas due to the increase in water availability associated with global warming and the decrease in per capita water agricultural withdrawal.

• After 2030, there will be a reduction of areas with increasing WAR because of a slowdown in the growth of both population and per capita municipal and industrial water withdrawals, while there will be an expansion of areas with decreasing WAR, mainly caused by continual decrease in per capita agricultural water withdrawal and intensified water availability. Differences in WAR changes among the three scenarios will become gradually clear. Under the low population growth scenario, the population decrease is expected to reduce the WAR in many Asian basins. Under the mitigated climate change scenario, changes in water availability and per capita agricultural water withdrawal are expected to be suppressed. Consequently, increases in the WAR estimated under the baseline climate change scenario will be suppressed in regions such as the northwest of Asia (e.g., Afghanistan and Pakistan); on the other hand, decreases in the WAR estimated under the baseline climate change scenario will be suppressed in regions such as China and India.

• Significant measures to suppress the WAR increase will be different in different river basins, depending on the causes for the WAR increase. For instance, measures to deal with the population increase and efforts to improve industrial and municipal water use by around 2030 will be important in the Huang He river basin. In the Indus river basin, the coping with a decrease in water availability after around 2030 will be important. In addition, measures to handle population increase will be necessary.

Our study clearly shows that the direction, the degree, the causes, and the period of the WAR change will differ among river basins, depending on the socioeconomic development and climate change. This implies that water resource management should be discussed by river basin, taking into account of uncertainties in major causes. This paper evaluates three highly consistent scenarios for socioeconomic development and climate change. However, uncertainties remain, particularly regarding regional climate change. We used the grid-based climate projection by an AOGCM (i.e., MIROC 3.2 (Medres)), and sensitivity analysis based on the projections by other AOGCMs remains to be addressed. In addition, the impacts of extreme weather events such as floods and droughts present uncertainties. Further research will be required.

Acknowledgements

The authors greatly appreciate the assistance of Professor Yoichi Kaya, President of RITE, and Professor Kenji Yamaji, Director General of RITE. We also acknowledge the assistance provided by the modeling groups in making their simulations available for analysis, the Program for Climate Model Diagnosis and Intercomparison (PCMDI) for collecting and archiving the CMIP3 model output, and the World Climate Research Programme's (WCRP's) Working Group on Coupled Modelling (WGCM) for organizing the model data analysis activity. The WCRP Coupled Model Intercomparison Project (CMIP3) multi-model dataset is supported by the Office of Science, U.S. Department of Energy.

References

Akimoto, K., Sano, F., Hayashi, A., Homma, T., Oda, J., Wada, K., & Tomoda, T. (2012). Consistent assessments of pathways toward sustainable development and climate stabilization. *Natural Resources Forum, 36*(4), 231-244. http://doi/10.1111/j.1477-8947.2012.01460.x/abstract

Akimoto, K., Sano, F., Homma, T., Oda, J., Nagashima, M., & Kii, M. (2010). Estimates of GHG emission reduction potential by country, sector, and cost, *Energy Policy, 38*(7), 3384-3393. http://dx.doi.org/10.1016/j.enpol.2010.02.012

Akimoto, K., Sano, F., Homma, T., Tokushige, K., Nagashima, M., & Tomoda, T. (2014). Assessment of the emission reduction target of halving CO_2 emissions by 2050: Macro-factors analysis and model analysis under newly developed socio-economic scenarios. *Energy Strategy Reviews, 2*(3-4), 246-256.

Alcamo, J., Flörke, M., & Märker, M, (2007). Future long-term changes in global water resources driven by sosio-economic and climate changes. *Hydrological Science, 52*(2), 247-275. http://dx.doi.org/10.1623/hysj.52.2.247

Alexandratos, N., & Bruinsma, J. (2012). *World agriculture towards 2030/2050: the 2012 revision.* ESA Working paper No. 12-03. Rome, FAO. Retrieved from http://www.fao.org/fileadmin/templates/esa/Global_persepctives/world_ag_2030_50_2012_rev.pdf

Arnel, N. W. (2004). Climate change and global water resources: SRES emissions and socio-economic scenarios. *Global Environmental Change, 14*, 31-52. http://dx.doi.org/10.1016/j.gloenvcha.2003.10.006

Döll, P., & Siebert, S. (2002). Global modeling of irrigation water requirements. *Water Resources Research, 38*(4), 8-1. http://dx.doi.org/10.1029/2001WR000355

Fischer, G., van Velthuizen, H., Shah, M., & Nachtergaele, F. (2002). *Global agro-ecological assessment for agriculture in the 21st century.* Retrieved from http://www.iiasa.ac.at/Research/LUC/SAEZ/index.html

Food and Agriculture Organization. (2011a). *AQUASTAT main country database.* Retrieved from http://www.fao.org/nr/water/aquastat/data/query/index.html?lang=en. Cited 10 Nov 2011

Food and Agriculture Organization. (2011b). *FAOSTAT database.* Retrieved from http://faostat.fao.org/site/291/default.aspx. Cited 11 Oct 2011

Fung, F., Lopez, A., & New, M. (2011). Water availability in+ 2 C and+ 4 C worlds. *Philosophical transactions of the Royal Society A: mathematical, physical and engineering sciences, 369*(1934), 99-116. http://dx.doi.org/10.1098/rsta.2010.0293

Gerten, D., Lucht, W., Ostberg, S., Heinke, J., Kowarsch, M., Kreft, H., & Schellnhuber, H. J. (2013). Asynchronous exposure to global warming: freshwater resources and terrestrial ecosystems. *Environmental Research Letters, 8*(3), 034032.

Gosling, S. N., & Arnell, N. W. (2013). A global assessment of the impact of climate change on water scarcity. *Climatic Change,* 1-15.

Grübler, A., O'Neill, B., Riahi, K., Chirkov, V., Goujon, A., Kolp, P., & Slentoe, E. (2007). Regional, national, and spatially explicit scenarios of demographic and economic change based on SRES. *Technological Forecasting & Social Change, 74*, 980-1029. http://dx.doi.org/10.1016/j.techfore.2006.05.023

Hanasaki, N., Fujimori, S., Yamamoto. T., Yoshikawa, S., Masaki, Y., Hijioka, Y., & Kanae, S. (2013). A global water scarcity assessment under shared socio-economic pathways – Part2: Water availability and scarcity. *Hydrology and Earth System Sciences, 17*, 2393-2413. http://dx.doi.org/10.5194/hess-17-2393-2013

Hanasaki, N., Kanae, S., Oki, T., Masuda, K., Motoya, K., Shirakawa, N., & Tanaka, K. (2008). An integrated model for the assessment of global water resources –Part 2: Applications and assessments. *Hydrology and Earth System Sciences, 12*, 1027-1037. http://www.hydrol-earth-syst-sci.net/12/1027/2008/hess-12-1027-2008.pdf

Hayashi, A., Akimoto, K., Tomoda, T., & Kii, M. (2013). Global evaluation of the effects of agriculture and water management adaptations on the water-stressed population, *Mitigation and Adaptation of Strategies for Global Change, 18* (5), 591-618. http://dx.doi.org/10.1007/s11027-012-9377-3

Intergovernmental Panel on Climate Change. (2000). *Special report on emissions scenarios.* Cambridge: Cambridge University Press.

Intergovernmental Panel on Climate Change. (2007). *Climate change 2007: the physical science basis.*

Cambridge: Cambridge University Press.

Netherlands Environmental Assessment Agency [PBL]. (2009). *History database of the global environment.* Retrieved from http://themasites.pbl.nl/en/themasites/hyde/index.html

Oki, T. (2001). *Total runoff integrating pathways (TRIP).* Retrieved from http://hydro.iis.u-tokyo.ac.jp/%7Etaikan/TRIPDATA/TRIPDATA.html

Oki, T., & Kanae, S. (2006). Global Hydrological Cycles and World Water Resources. *Science, 313*(5790), 1068-1072. http://dx.doi.org/10.1126/science.1128845

Program for Climate Model Diagnosis and Intercomparison. (2004). *WCRP CMIP3 multi-model database.* Retrieved from http://www-pcmdi.llnl.gov/ipcc/about_ipcc.php. Cited 7 May 2010

Raskin, P., Gleick, P., Kirshen, P., Pontius, G., & Strzepek, K. (1997). *Comprehensive assessment of the freshwater resources of the world.* Stockholm Environment Institute, Stockholm, Sweden.

Shiklomanov, I. A. (1999). *World water resources and their use a joint shi/unesco product.* Retrieved from http://webworld.unesco.org/water/ihp/db/shiklomanov/

Siebert, S., & Döll, P. (2010). Quantifying blue and green virtual water contents in global crop production as well as potential production losses without irrigation. *Journal of Hydrology, 384*(3-4), 198-217.

Siebert, S., Döll, P., Feick, S., Hoogeveen, J., & Frenken, K. (2007). *Global Map of Irrigation Areas version 4.0.1.* Johann Wolfgang Goethe University, Frankfurt am Main, Germany / Food and Agriculture Organization of the United Nations, Rome, Italy. Retrieved from https://www2.uni-frankfurt.de/45218039/Global_Irrigation_Map

United Nations. (2007). *World population prospects: the 2006 revision.* Retrieved from http://www.un.org/esa/population/publications/wpp2006/wpp2006.htm

United Nations. (2011). *World population prospects: the 2010 revision.* Retrieved from http://esa.un.org/wpp/Documentation/WPP%202010%20publications.htm

Vörösmarty, C. J., Green, P., Salisbury, J., & Lammers, R. B. (2000). Global water resources: vulnerability from climate change and population growth. *Science, 289,* 284-288. http://dx.doi.org/10.1126/science.289.5477.284

7

Natural Enemy Complex of Some Agroforestry Systems of Aizawl and Their Implications in Insect Pest Management

Rosy Lalnunsangi[1], Dibyendu Paul[2] & Lalit Kumar Jha[2]

[1] Department of Zoology, Mizoram University, Aizawl, Mizoram, India

[2] Department of Environmental Studies, North Eastern Hill University, Shillong, Meghalaya, India

Correspondence: Rosy Lalnunsangi, Department of Zoology, Mizoram University, Aizawl, Mizoram, India. E-mail: rose_khiangte@yahoo.com

Abstract

Agroforestry systems of the hilly regions of the north east are small scale, primitive land use systems, evolving through traditional knowledge requiring very little inputs in terms of irrigation, fertilizers and pesticides. The different components of the agroforestry systems were analyzed for their natural enemy complex through a monthly sampling programme. Maintenance and potential role of natural enemies' complex in various compartments of such systems are discussed in the light of land use practices and holding size.

Keywords: small scale, agroforestry, insect pest, natural enemy complex

1. Introduction

In Mizoram the practice of shifting cultivation also termed as "jhum" or "slash and burn" is still the predominant form of agriculture. This form of agriculture is subsistence level agriculture and is generally rainfed, often resulting in an enormous erosive soil loss during the heavy monsoons.

Among the various alternative to jhum, agroforestry is increasingly becoming popular because of its ameliorative potential and conservation of the biophysical characteristics of soil. Many of the benefits of agroforestry are derived from the increased diversity of these systems compared to traditional agricultural management practices (Holloway & Stork, 1991). In Mizoram traditional forms of agroforestry have been practiced by some farmers in their own way. It was first innovated by Jha and Lalnunmawia (2000, 2003) and named them as Tree-Greenhedge-Crop farming system and Bamboo based agroforestry system respectively. The various multipurpose tree species and predominant agroforestry systems prevalent in Mizoram have been reported by Lalramnghinglova and Jha (1996). Teak (*Tectona grandis*) is an indigenous timber species of the state occurring both naturally and in plantations. Traditional systems having *Tectona grandis* (Verbenaceae) and crops like *Zea mays* (Poaceae) and *Oryza sativa* (Poaceae) are the most common and successful agroforestry practices, particularly during the seedling, sapling/pole stages of growth of the tree component. Subsequently, depending on the existing spacing, cropping is abandoned if the canopy of the tree component closes; otherwise it is continued, simulating true agroforestry systems. Cultivation of paddy, maize or vegetables along with tung (*Aleurites* sp.) is another common combination in the state.

The insect fauna occurring on certain plant species, in any particular location, is more or less the same, whether that plant species is in monoculture or in a polycultural assemblage such as an agroforestry system. However, the activities of these insects are not likely to be identical in any two situations. There are several factors that influence the activities of insects in agroforestry and among these the role played by natural enemies is substantial.

The present investigation focuses on comparing the distribution of natural enemy complex in teak, tung and subabul based agroforestry system of Aizawl.

2. Materials and Methods

2.1 Study Site

The study was conducted in the outskirts and within Aizawl, which is the capital of Mizoram. It lies between 20°58'–24°35'N latitudes and 92°15'–93°29'E longitudes with an average annual rainfall of 2500 mm. Three experimental plots were selected, one each at Chanmari west, Zemabawk, and Sakawrtuichhun. For the purpose of sampling, each experimental plot was subdivided into three components, viz, tree, crop and fringe area component, each having an area of approximately 100 m × 100 m.

The tree components of these sites were repectively Teak (*Tectona grandis*), Tung *(Aleurites fordii)* and Subabul (*Leucaena leucocephala*). The crop component in all the sites were maize (*Zea mays*), *Phaseolus vulgaris* and/ or *Vigna sinensis*. Besides, the teak site also had *Clerodendrum colebrookianum* and *Solanum indicum* as other components. The fringe area comprises the natural vegetation immediately outside the cropping area and comprises weed species such as *Imperata cylindrica* L., *Drymaria chordate* L. (Caryophyllaceae), *Cyperus rotundus* L. (Cyperaceae), *Eupatorium sp*, *Ageratum conyzoides* L. (Asteraceae), etc. were common in the sites.

2.2 Sampling

A random sampling program (Southwood, 1978), replicated five times was undertaken for two years in each component of each site. The samplings were done by using sweep net or through handpicking. The fringe area and crop plots were sampled through sweeping with an insect collecting net. The sweep was adjusted such that each sweep covered an area of 1 m^2. Sampling through hand picking was undertaken for individual crop plants and tree components. All samples were replicated five times. The insects collected were identified from FRI Dehra Dun, ZSI, Shillong and ICAR, Shillong.

3. Results and Discussion

The natural enemy complex in order of importance encountered during the course of the study belongs to the following orders:

Mantoidea, Chilopoda, Hymenoptera, Araneida, Dermaptera, Hemiptera, Coleoptera.

Praying mantids were observed to frequent the cropping plots for preying. The undisturbed fringe area vegetation and the tree components provided refugia. The ability to fly also afforded a longer range of operation. The undisturbed (untilled) soils and litter around the tree component provided habitat conditions for the centipides (Chilopoda). Their ability to enter nooks and crevices made them excellent predators seeking out larvae. Ground foraging ants were ubiquitous in the cropping area. Wasps were also attracted to the cropping plots and the tree component and fringe area housed their nests. Hemiptera were observed to stay camouflaged on the bark of the tree component. Dermaptera and coleoptera were mostly encountered on the crop components foraging for larvae and aphids. Araneida (hunting spiders) were observed to actively seek prey on crop plants. Besides webs were also seen to trap small flying insects such as aphids, dipteral and coleopteran.

The percentage composition of the different arthropod orders in the *Aleurites* site of Aizawl is depicted in Figure 1. The tree component had 6 representative groups with araneida (15%), chilopoda (10%) and hymenoptera (21%), the three predatory groups constituting 46% of the total fauna while the predatory groups of the crop component were represented by dermaptera (11%), homoptera (14%), and hymenoptera (33%), i.e a total predatory fauna of 58%. The fringe area had 10 representative groups including larvae and the predatory fauna constituted 33% of the total fauna through araneida (11%), mantoidea (6%), hymenoptera (10%) and dermaptera (6%).

The percentage composition of the different arthropod orders in the Teak site of Aizawl is depicted in Figure 2. The tree component had 8 representative groups with araneida (10%), mantoidea (23%) and hymenoptera (18%) the three exclusively predatory groups constituting 51% of the total fauna. Among the crop component too, the predatory group constitute 33% of the total fauna through araneida (7%), chilopoda (5%) and hymenoptera (21%). The fringe area component had 8 representative groups, and predators constituted 47% of the total fauna, and consisting of mantoidea (13%), hymenoptera (22%) and araneida (12%).

The percentage composition of the different arthropod orders in the *Leucaena* site of Aizawl is depicted in Figure 3. The number of representative groups was 6, 5 and 10 in the tree, crop and fringe area components respectively and the predatory fauna constituted 26% through the orders dermaptera (11%), chilopoda (6%) and araneida (10%) in the tree area component and the predatory in the crop component is represented by 48% hymenoptera while 46% of the total predatory fauna in the fringe area components are constituted by hymenoptera (17%), araneida (9%), mantoidea (17%), dermaptera (7%).

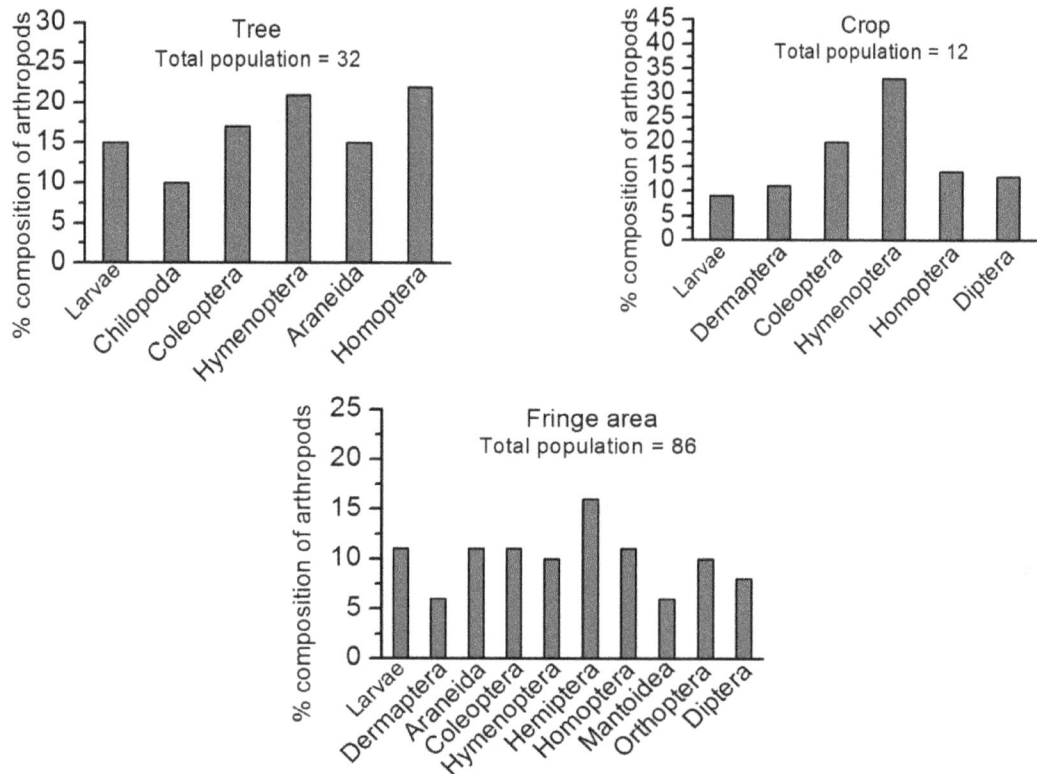

Figure 1. Percentage composition of arthropods in different plant habits (tree, crop and fringe) of Tung (Aizawl) site

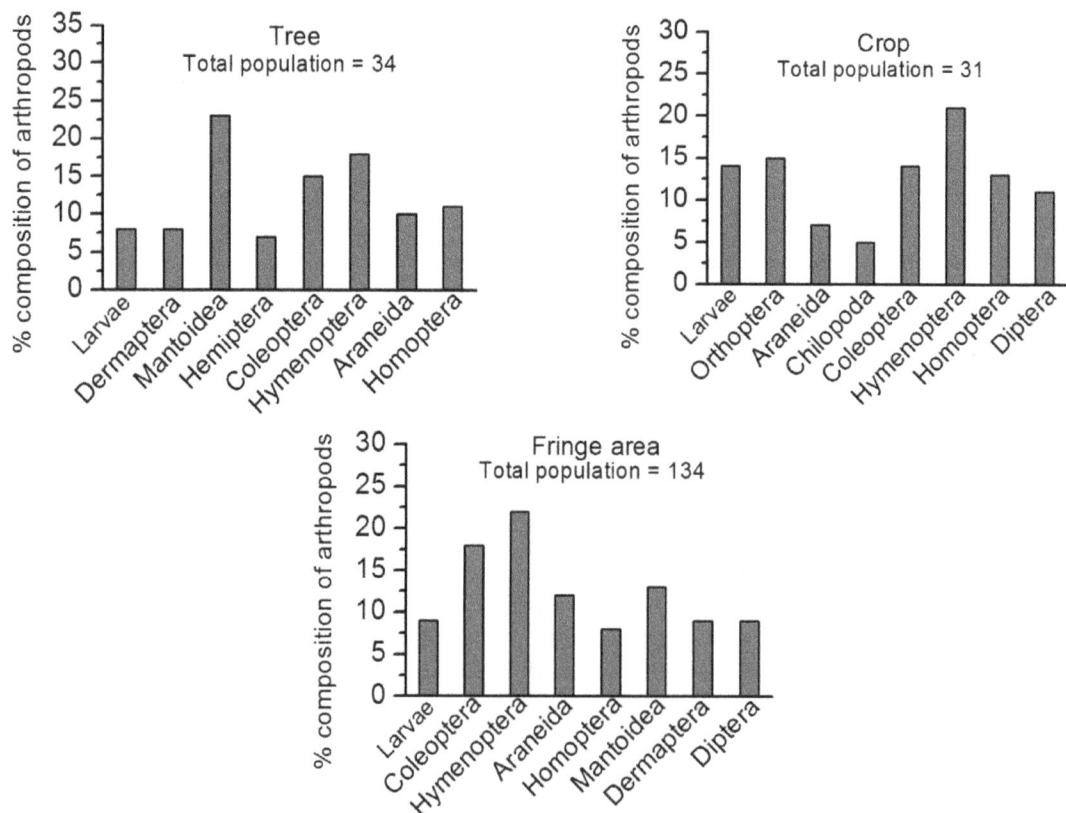

Figure 2. Percentage composition of arthropods in different plant habits (tree, crop and fringe) of Teak (Aizawl) site

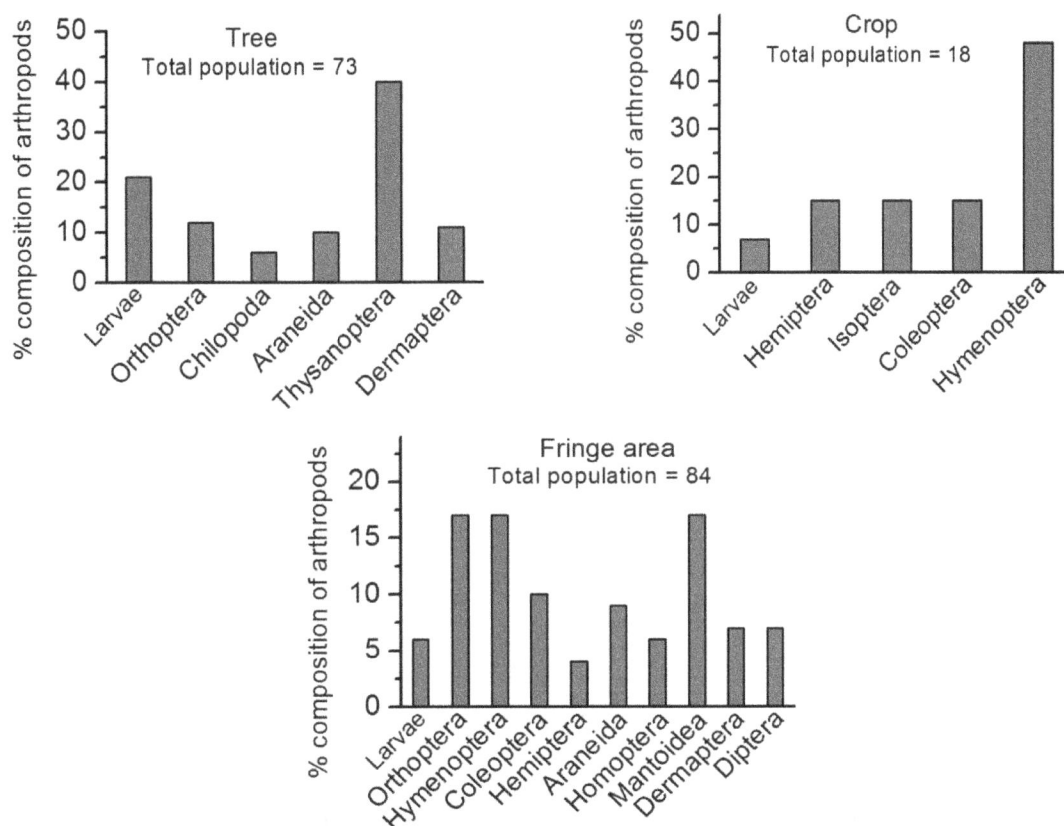

Figure 3. Percentage composition of arthropods in different plant habits (tree, crop and fringe) of Subabul (Aizawl) site

Agroforestry systems, particularly the complex ones, have a great potential for controlling pest populations through increasing the efficiency of biological control agents. Biological control by parasites/predators has been proved to be an excellent alternative to chemical pesticides in many ecosystems (Sithananthan et al., 1973; De Bach, 1974; Hagen et al., 1976; Kring et al., 1985; Yadav et al., 1985; Manjunath, 1988; Singh & Jalali, 1992).

The low level of infestation in the tree component is indicative of natural control through predator complex operating both from within and from the fringe area. The fringe area composing of dense, diverse vegetation, harbor different natural enemy population. Higher densities of predator species in dense vegetation have been reported by Sprenkel et al. (1979) and Horn (1981).

Diversified cropping systems based on agroforestry, cover cropping are considered more stable and more resource conserving (Vandermeer, 1995). Alteieri and Letourneau (1984) reported the effects of plant diversity on the regulation of insect herbivore populations by favouring the abundance and efficacy of associated natural enemies. Agroforestry, being perennial, along with the fringe area support the natural enemy complex within and between seasons, especially during the off season of the main crop by offering alternate prey, pollen and nectar sources and suitable micro-habits for parasitoids and predators (Jervis & Kidd, 1996; Landis et al., 2000; Begum et al., 2006)

The natural enemy complex operative in the systems is diverse and constitute a sizeable percentage of the total fauna. The presence of Hymenopterans like wasps and parasitoids are indicative of their potential in infesting the pest larval stages. Wasps have been observed to actively predate upon pest larvae. These larvae are enclosed in egg cases as food for the growing wasp larvae. Mantids and Chilopoda have been observed to predate upon a wide variety of insects such as larvae, orthopterans and cicada. Further, hymenoptera (Formicidae) have also been observed to forage and predate upon insect larvae. Dermaptera and coleopterans were seen to predate upon aphids.The presence of the fringe area of natural undisturbed vegetation is probably crucial in maintenance of such a healthy predatory complex. Further, the small size of the crop holdings (≈100 m²) ensures easy access of the whole plot to the natural enemy complex. The presence of untilled area around the tree component also provides undisturbed habitable conditions for predatory ground fauna like Chilopoda and predatory Coleoptera.

All these conditions contribute to a strong predatory component and is reflective of the low pest infestations.

References

Altieri, M. A., & Letourneau, D. K. (1984). Vegetation diversity and insect pests outbreaks. *CRC Critical Reviews in Plant Sciences, 2*, 131-169. http://dx.doi.org/10.1080/07352688409382193

Begum, M., Gurr, G. M., Wrattan, P. R., & Nicol, H. I. (2006). Using selective food plants to maximize biological control of vineyards pests. *Journal of Applied Ecology, 43*, 547-554. http://dx.doi.org/10.1111/j.1365-2664.2006.01168.x

De Bach, P. (1974). *Biological control by natural enemies* (p. 322). London: Cambridge Uinv., Press.

Hagen, K. S., Bombosch, S., & Mc Murtry, J. A. (1976). The biology and impact of predators. In C. B. Huffaker & P. S. Messenger (Eds.), *Theory and practice of biological control*. San Diego: Academic.

Holloway, J. D., & Stork, N. E. (1991). The dimensions of biodiversity: the use of invertebrates as indicators of human impact. In D. L. Hawksworth (Ed.), *The Biodiversity of Microorganisms and Invertebrates; Its Role in Sustainable Agriculture* (pp. 37-62). CAB International, Wallingford, UK.

Horn, D. J. (1981). Effect of weedy backgrounds on colonization of collards by green peach aphid, Myzus persicae, and its major predators. *Environmental Entomology, 10*(3), 285-289.

Jervis, M. A., & Kidd, N. A. C. (1996). Phytophagy. In M. A. Jervis, & N. A. C. Kidd (Eds.), *Insect Natural Enemies* (pp. 375-394). London: Chapman and Hall.

Jha, L. K., & Lalnunmawia, F. (2003). Agroforestry with bamboo and ginger to rehabilitate degraded areas in North East India. *J. Bamboo and Rattan, 2*, 103-109. http://dx.doi.org/10.1163/156915903322320739

Kring, T. J., Gilstrap, F. E., & Michels, Jr. G. J. (1985). Role of indigenous coccinellids in regulating greenbugs (Homoptera: Aphididae) on Texas grain sorghum. *J. Econ. Entomol, 78*, 269-273.

Lalramnghinglova, J. H., & Jha, L. K. (1996). Prominent agroforestry systems and important multipurpose trees in farming system of Mizoram. *The Indian Forester, 122*, 604-609.

Landis, D. A., Wratten, S. D., & Gurr, G. M. (2000). Habitat management to conserve natural enemies of arthropod pests in agriculture. *Annual Review of Entomology, 45*, 175-201. http://dx.doi.org/10.1146/annurev.ento.45.1.175

Manjunath, T. M. (1988). Mass production and utilization of *Trichogramma*. In *Biocontrol Technology for sugarcane pests: Proceedings of National Symposium* (pp. 249-253). Sugarcane Breeding Institute, ICAR, Combaitore.

Singh, S. P., & Jalali, S. K. (1992). Biological suppression of Chilo auricilius on sugarcane in India. *Trichogramma News, 6*, 25.

Sithanantham, S., Muthusamy, S., & Durai, R. (1973). Experiments on the inundative release of *Trichogramma australicum* (Gir.) in the biological control of sugarcane stem borer, *Chilo indicus* Kapur. *Madras Agricultural Journal, 60*, 457-461.

Sprenkel, R. K., Brooks, W. M., van Duyn J. W., & Deitz, L. L. (1979). The effect of three cultural variables on the incidence of Nomuraea rileye, phytophagous Lepidoptera, and their predators on soybeans. *Environmental Entomology, 8*, 334-339.

Vandermeer, J. (1995). The ecological basis of alternative agriculture. *Annual Review of Ecology Systems, 26*, 201-224. http://dx.doi.org/10.1146/annurev.es.26.110195.001221

Yadav, D. N., Patel, R. C., & Patel, D. S. (1985). Impact of inundative release of *Trichogramma chilonis* Ishii against *Heliothis armigera* (Hubn.) in Gujarat. *Journal of Entomological Research, 9*(2), 153-159.

8

An Experimental Analysis of a Nano Structured Inorganic Ceramic Membrane for Carbon Capture Applications in Energy Security Challenges

Ngozi C. Nwogu[1], Mohammed N. Kajama[1], Kennedy Dedekuma[1] & Edward Gobina[1]

[1] Centre for Process Integration and Membrane Technology, IDEAS Research Institute, Robert Gordon University, United Kingdom

Correspondence: Professor Edward Gobina, Centre for Process Integration and Membrane Technology, IDEAS Research Institute, The Robert Gordon University, Riverside East Garthdee Road, Aberdeen AB10 7GJ., United Kingdom. E-mail: e.gobina@rgu.ac.uk

Abstract

Nanostructured hybrid materials have the solution to facilitate renewable energy to cover up for anticipated energy gap and related ecological problems. In this work the design of a nano structured ceramic membrane is carried out using ceramic nanoparticles for application in energy security challenges. However the innovation is that a membrane porous network is modified through its immersion in silica based solution. This process helps to pull the gas of interest towards the membrane in this case CO_2 and allows the other gases to pass through. However the development of this hybrid ceramic gas separation membrane in this study elaborates on the recovery of hydrogen from fuel reforming unit for use in fuel cell applications. A detailed production and purification of hydrogen in a fuel processor using the advanced ceramic membrane is presented. A gaseous mixture of hydrogen and carbon dioxide is produced following fuel on-board reforming. To enhance the efficiency of the fuel cell, a clean hydrogen using membranes with a high permeability and selectivity for H_2 over N_2, CO_2 such that H_2 will permeate with high-purity. Accordingly, results obtained show an appreciable high flow rate of 5.045 l/min and 3.71 separation factor of hydrogen gas to CO_2 at relatively low pressure when compared to the other gases. Further confirmation of the dominance of Knudsen and surface flow mechanism in the entire experiments is also presented.

Keywords: ceramic membrane, energy security, renewable energy, carbon capture, hydrogen production, gas separation, fuel cell

1. Introduction

Carbon dioxide gas, a greenhouse gas has remained a contributor to global climate change. Recent reports from some researchers has shown a drastic increase in the concentration of atmospheric CO_2 from approximately 275 to 387 ppm with an annual average increment of 3 ppm in the last century which has resulted in temperature rise. This obviously is an indicator to future warming (Merkel et al., 2010). Therefore the protection and security of our climate is vital especially when there are efforts being put in place to reduce CO_2 emissions emanating from fossil fuel utilization which are major emitters. However various technologies are currently being developed to accomplish set goals to minimise global warming, a huge challenge resulting from the continual emission of these greenhouse gases. In this circumstance, identifying the method of capturing atmospheric CO_2 using low energy, high level performance and to meet global targets is pertinent (Favre, 2007). Energy security and curtailing energy involvement in global warming are two prevailing tasks being looked at in the power sector to attain high level of global sustainable energy security. Currently records show that about 1400 million people across the world do not have access to essential electricity. The people most affected by lack of electricity supply are those in the remote villages and lack good and hygienic cooking facilities. An example is the sub-Saharan Africa where about 15% of the world population live. This is a very pathetic and devastating situation to contend with and thus a huge impediment to economic and social growth due to unavailability of state-of-the-art facilities. Urgent attention is therefore needed to effect changes by putting together a combined effort and commitment to actualize and attain set objectives (Kaygusuz, 2012). Fossil fuel which is known globally as a major source of power generation as reported is not sustainable going by the ever growing global economy. In addition fossil fuel combustion through human related source has led to very hazardous and catastrophic effect on the earth's atmosphere. In that context, for fast developing and industrialised economies, energy is on high demand and this calls for an uninterrupted, available and affordable energy supply which will stand the test of time. Achieving this level of sustainable energy security and economic performance with current technologies has yielded little or no result in terms of balancing CO_2 emission abatement and the economic growth (Sartbaeva et al., 2008). Therefore the use of affordable technology which will be readily available and can provide cheaper and cleaner source of energy can be adapted to meet these challenges. For policies that have interest in fuel effectiveness, the likely option will be to mitigate CO_2 emissions. This can be realized from renewable as well as non-renewable energy sources (Favre, 2007). Hydrogen has been identified as an alternative to fossil fuel. As energy carrier, can be produced from renewable sources, namely: wind, solar, biomass and water. Natural gas and coal as non-renewable energy sources also have hydrogen and carbon dioxide as their combustion by-products. Hydrogen utilization has therefore paved way to the fabrication of very operational energy generated device including fuel cell (Edwards et al., 2008). By definition, fuel cell is a device that combines both electrical and chemical energy, whereby a reaction takes place in the presence of oxygen with further reforming process to produce electricity along with combustion by-products (Löffler, 2003). High-efficiency fuel cell operation from clean hydrogen can be utilised in all energy sectors, however its application in vehicular transportation and distributed power are more important. Accordingly fuel cell with clean hydrogen-rich fuel is top among solutions sort for in the transitional process to a CO_2-zero emission economy and a pathway to sustainable energy in the nearest future. Production of a cost efficient and sustainable clean hydrogen is among the top challenges which must be surmounted for the evolution from carbon based (fossil fuel) energy economy to hydrogen based economy (Edwards et al., 2008). Therefore an efficient and cost efficient technology for gas separation is highly required that which will have effect on the overall expenditure of the entire system. Membrane technology at the moment is being highly applied widely in hydrogen separation. Membrane applications have remained a more energy efficient substitute to conventional methods in gas separations. The outstanding difference in the flow rate and permeation features during the separation processes involving multi component gas mixtures makes the membrane a good candidate for gas separation (Murkowski, 2012). Generally two different types of membranes exist, organic and inorganic membrane (McCool et al., 2003).

2. Membrane as a Selective Barrier

The diagram in Figure 1 depicts a simple gas separation of two gas molecules across a membrane. The feed is a mixture of the two gases going into the upper level of the box and as expected separation takes place in this compartment.

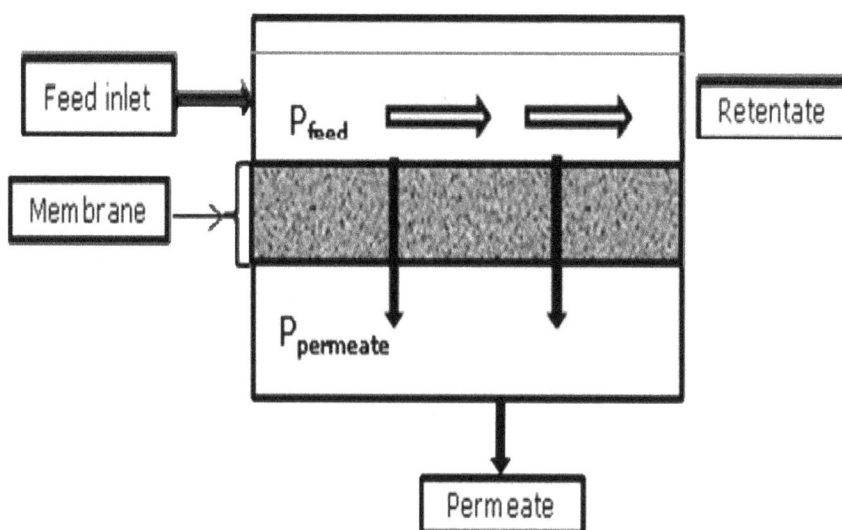

Figure 1. A simplified procedure of membrane separation process

3. Experimental

3.1 Membrane Material Characterization

The preparation and fabrication of inorganic ceramic membrane from different materials through various methods results in the development of membrane with distinct physical appearance which is tailored towards achieving a pore size, membrane area and pore size distribution which can be controlled resulting in a higher permeation rate of hydrogen gas with high separation factor. Membrane structure is one of the parameter in determination of membrane characteristics. This is done through scanning electron microscopy which produces micrographs and images. Nanoporous structural make up of the membrane is usually identified by the magnification of these images. Defects determination through cracks or pinholes of the membrane is also achieved; in addition the pore size distribution can be obtained through this means (Othman et al., 2004). The diagram shown below is an outer sectional area of the membrane support with ×2000 magnification.

Figure 2. SEM of Membrane support Images with ×2000 Magnification

Consequently, porous ceramic membrane consists of several layers of different materials namely: Aluminium oxide (Al_2O_3), Titanium Oxide (TiO_2), Zirconium Oxide (ZrO_2), Silicon dioxide (SiO_2), Silicon carbide, Zeolite or a mixture of two materials applied on an underlying porous stainless steel, α-alumina, γ-alumina, zirconium, zeolite supports (Shekhawat et al., 2003). Membrane separation of gases is a highly intricate process and therefore the material used for its preparation should demonstrate a long-lasting feature, stability and tailored in an advanced manner to be adapted to separate specific gases.

Membrane materials characterization can also be determined through the elemental composition of the sample by the analysis of the sample material which is identified using an Energy Dispersive x-ray analysis technique, (EDXA). The EDXA system is attached to the SEM. Data generated by EDXA analyses the true composition of the elements within the sample. Figure 3 shows the EDXA of the nano-structured ceramic membrane used in carrying out the gas permeation experiment. As observed, the elements present are Titanium oxide, silicon oxide and Aluminium Oxide.

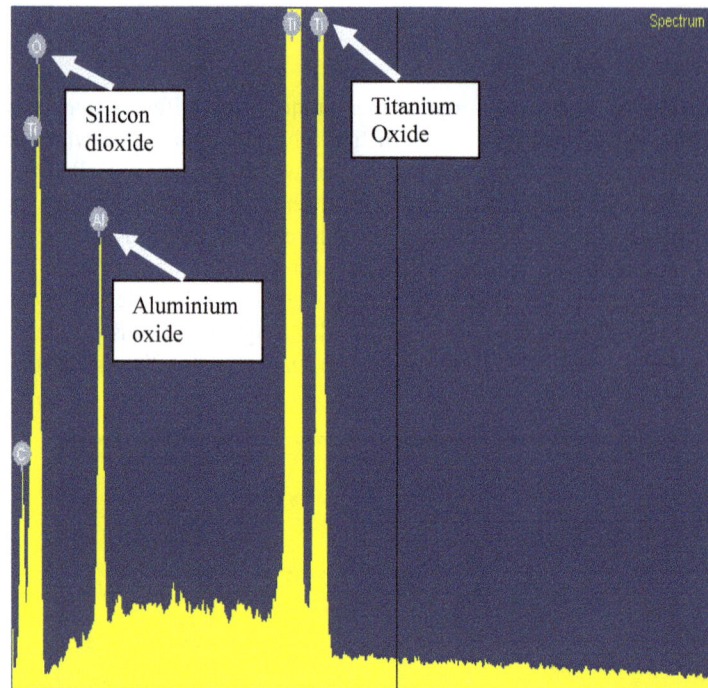

Figure 3. EDXA analysis of membrane sample elemental composition

3.2 Clean Hydrogen Production through a Fuel Processor

There are several ways to produce hydrogen gas since it does occur in the system in which it can be used. However in the thermo-chemical process, hydrogen production from fossil fuel or natural gas, its separation and purification are of great importance globally. As an illustration where water to gas shift reaction is involved for converting carbon monoxide to hydrogen, membrane technology show substantial promises for shifting the equilibrium to achieve the purpose. In general membranes are important for the refinement of Hydrogen (Lu et al., 2007). Consequently hydrogen can be produced on an economical scale by steam reforming (a reaction between steam and hydrocarbon) for instance

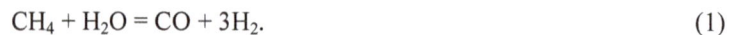

$$CH_4 + H_2O = CO + 3H_2. \tag{1}$$

Carbon monoxide further reacts with steam to form H_2 and CO_2 by the exothermic reaction, which is commonly referred to as the water–gas shift reaction:

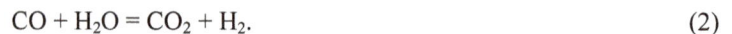

$$CO + H_2O = CO_2 + H_2. \tag{2}$$

Conversely if hydrogen is selectively recovered from the reaction system, the thermodynamic stability of the reaction is moved to the product side where higher conversion of CH_4 to H_2 and CO_2 can be attained at a relatively lower temperature. The characteristic features to achieve high level of CO_2 and H_2 separation include high separation selectivity, high permeability, the stability and durability of hybrid inorganic ceramic membranes (Xia et al., 2002).

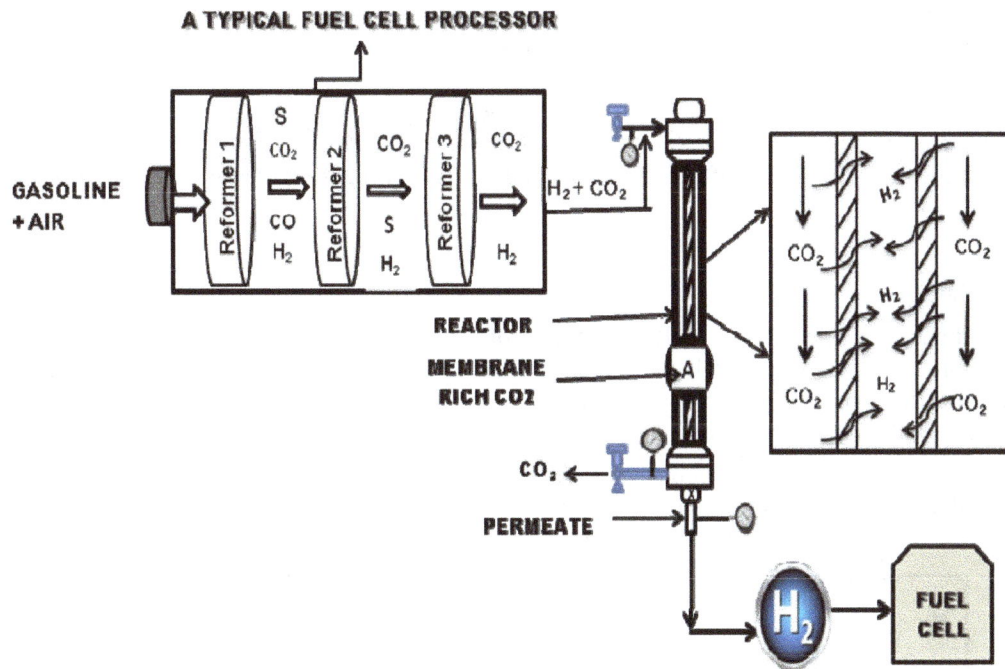

Figure 4. Hydrogen production and purification through a membrane

Figure 4 depicts a schematic diagram of clean hydrogen production through a membrane. It also shows the location of the membrane separation following fuel processing and fuel cell. In the fuel processing unit, the fuel comprises of gasoline in the presence of air and water). The first reformer takes in the syngas involving sulphur, carbondi oxide, carbon monoxide and hydrogen gases. Carbon monoxide is removed in the second reformer in the presence of a carbon monoxide to H_2 catalyst. The remaining products is channelled to a sulphur sop-up unit to get rid of the sulphur (Ahmed & Krumpelt, 2001). In order to achieve hydrogen of high purity from either the syngas or the products of the water–gas shift reaction, separation of H_2 from either CO or CO_2 is required. Competitive separation processes for hydrogen from such streams include amine absorption (CO_2 separation), pressure swing adsorption (PSA) and membrane separation. In the separation of hydrogen from other products in refineries, membrane systems are more economical than PSA in terms of both relative capital investment and unit recovery costs (Spillman, 1989). A gaseous mixture of hydrogen and carbon dioxide is produced following fossil fuel on-board reforming as by-products. At this point an advanced hybrid inorganic membrane is then installed which achieves high recovery of hydrogen over carbon oxide to produce clean hydrogen in a cost effective manner for use in various fuel cell applications. The efficiency of the fuel cell is thus greatly increased with further enhancement in the quest on a sustainable and affordable energy security to achieve a hydrogen economy.

3.3 Gas Permeation Experiment

The immersion of a membrane support in a silica based solution and repeated dip coating technique with intervals of hydrothermal treatment procedure has led to the manufacture of hybrid ceramic membrane for the production of clean hydrogen for fuel cell application. In this paper a 15 nm pore size nano structured modified commercial alumina ceramic membrane with an effective permeable length of 358 mm, outer and inner diameter of 10 mm and 7 mm respectively and membrane surface area of 0.0062 m^2 was placed in a membrane/reactor as housing with graphite rings as a seal to prevent gas leakages and at the same time enhance accuracy of data obtained. Pressure gauges were connected to the reactor at specified points on the flow line to measure feed, permeate and retentate pressures. This inorganic modified ceramic membrane is formed comprises of plurality of chemically discreet portion. The first part as the separating layer formed from silica oxide and the second the support made of aluminum oxide. The coated layer is used as a separating layer to allow faster permeation of one gas from a gas mixture and also due to its affinity for the gas (Nwogu et al., 2013; Gobina, 2006).

A single gas experiment comprising of N_2, H_2, CO_2 and O_2 gases was carried out to determine their individual flow rates with respect to pressure. Values obtained were calculated and recorded. Plot generated from the result is shown below.

4. Results and Discussion

Figure 5 shows the plot of the flow rate of H_2, CO_2, O_2 and N_2 gases versus pressure drop. It can be observed that hydrogen gas permeated faster with increase in feed pressure through the modified ceramic inorganic membrane more than the other three single gases, O_2, N_2 and CO_2. CO_2 which has a molecular weight of 44 had the lowest flow rate. The high flow rate of hydrogen however is attributed to its relatively smaller molecular weight in comparison to the other gases which affords its better mobility within the pore network of the membrane as illustrated below:

Flow Rates: $H_2 > N_2 > O_2 > CO_2$

Molecular Weights: $CO_2 > O_2 > N_2 > H_2$

Accordingly hydrogen gas which has a molecular weight of 2 had the highest flow rate of 5.045 l/min with increase in pressure drop above 4bars while carbon dioxide of a higher molecular weight 44 attained the lowest flow rate confirming the predominance of Knudsen flow mechanism in the separation process. This also shows that the flow is molecular weight dependent.

Figure 5 also confirms the membrane performance with respect to the gases that permeates faster than the others through the membrane. Generally the more permeable a membrane, the less selective it could be.

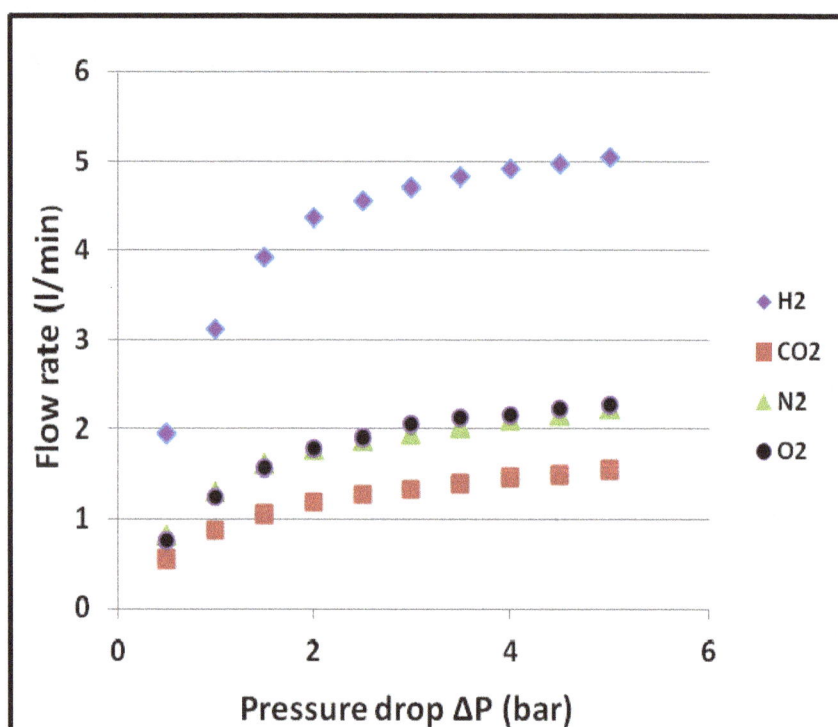

Figure 5. Effect of pressure drop on H_2, N_2, CO_2 and O_2 gases flow rate

The inclination of a membrane to allow or pass one gas and not another otherwise known as separation factor, α known as selectivity is also considered in this paper. This was determined through the membrane selectivity of H_2 over CO_2, O_2 and N_2 and given as the ratio of H_2 permeance to the other gases. For instance an estimation of H_2 selectivity to CO_2 can be calculated from the ratio of H_2 permeance to that of CO_2 given as,

$$\text{Selectivity,} \quad \alpha = \frac{\text{Permeance of } H_2}{\text{Permeance of } CO_2}$$

A selectivity factor of 1 is an evidence of no separation, the benchmark therefore in gas mixture selection should be geared towards achieving a selectivity higher than 1 because the more selective a membrane is to a particular gas, the higher the selectivity factor.

Figure 6 is a graphical representation of separation factor obtained in relation to pressure drop for H_2 gases over CO_2, N_2 and O_2 gases.

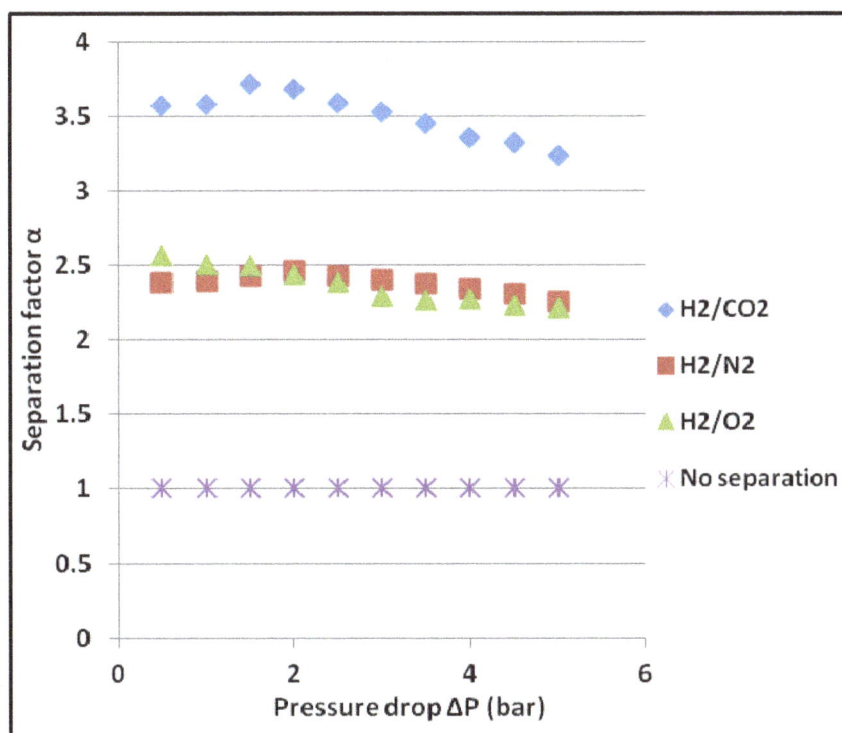

Figure 6. Effect of pressure drop on H_2 selectivity over CO_2, N_2 and O_2

Figure 6 presents separation factor results of H_2 over CO_2, N_2 and O_2. A high separation factor of 3.71 was obtained from the gas separation between H_2 and CO_2; this implies that more H_2 will be recovered from H_2 and CO_2 binary mixtures more than the other three gases.

5. Conclusions

Single gas permeation tests carried out have shown that the membrane exhibits good performance and higher permeability of hydrogen gas at low trans-membrane pressure and at room temperature. With 3.71 as the separation factor of H_2 to CO_2 at a low pressure of 1.5 bar, this shows that the membrane has a high level of hydrogen gas selectivity characteristics to CO_2 which is a good indication that it can be installed in a fuel processing unit for a high H_2 recovery from the rich CO_2 stream. Thus the membrane which has several inbuilt characteristic features over other membranes such as high mechanical strength, high chemical stability, easy cleaning, high resistance to acidic chemicals and the ability to withstand high pressure and temperature has the potential in the production of clean hydrogen gas for fuel cell applications for a sustainable and affordable energy security globally.

References

Ahmed, S., & Krumpelt, M. (2001). Hydrogen from hydrocarbon fuels for fuel cells. *International Journal of Hydrogen Energy, 26*(4), 291-301. http://dx.doi.org/10.1016/S0360-3199(00)00097-5

Edwards, P. P., Kuznetsov, V. L., David, W. I., & Brandon, N. P. (2008). Hydrogen and fuel cells: towards a sustainable energy future. *Energy Policy, 36*(12), 4356-4362. http://dx.doi.org/10.1016/j.enpol.2008.09.036

Favre, E. (2007). Carbon dioxide recovery from post-combustion processes: can gas permeation membranes compete with absorption? *Journal of Membrane Science, 294*(1), 50-59. http://dx.doi.org/10.1016/j.memsci.2007.02.007

Gobina, E. (2006). 'A membrane apparatus and method separating gases' (US Patent 7297184 B2), May 23, 2006.

Kaygusuz, K. (2012). Energy for sustainable development: A case of developing countries. *Renewable and Sustainable Energy Reviews, 16*(2), 1116-1126. http://dx.doi.org/10.1016/j.rser.2011.11.013

Löffler, D. G., Taylor, K., & Mason, D. (2003). A light hydrocarbon fuel processor producing high-purity hydrogen. *Journal of Power Sources, 117*(1), 84-91. http://dx.doi.org/10.1016/S0378-7753(03)00357-4

Lu, G. Q., Diniz da Costa, J. C., Duke, M., Giessler, S., Socolow, R., Williams, R. H., & Kreutz, T. (2007). Inorganic membranes for hydrogen production and purification: a critical review and perspective. *Journal of colloid and interface science, 314*(2), 589-603. http://dx.doi.org/10.1016/j.jcis.2007.05.067

McCool, B. A., Hill, N., DiCarlo, J., & DeSisto, W. J. (2003). Synthesis and characterization of mesoporous silica membranes via dip-coating and hydrothermal deposition techniques. *Journal of membrane science, 218*(1), 55. http://dx.doi.org/10.1016/S0376-7388(03)00136-4

Merikoski, R. (2012). *Flue gas processing in Amine-based capture* (pp. 15-16). Tampere University of Technology.

Merkel, T. C., Lin, H., Wei, X., & Baker, R. (2010). Power plant post-combustion carbon dioxide capture: An opportunity for membranes. *Journal of Membrane Science, 359*(1), 126-139. http://dx.doi.org/10.1016/j.memsci.2009.10.041

Nwogu, N. C., Gobina, E., & kajama, M. N. (2013). Improved Carbon dioxide Capture using Nanostructured Ceramic Membranes. *Low Carbon Economy Scientific Journal, 4*(3), 125-128.

Othman, M. R., Mukhtar, H., & Ahmad, A. L. (2004). Gas permeation characteristics across nano-porous inorganic membranes. *IIUM Eng. J, 5*(2), 17-35.

Sartbaeva, A., Kuznetsov, V. L., Wells, S. A., & Edwards, P. P. (2008). Hydrogen nexus in a sustainable energy future. *Energy & Environmental Science, 1*(1), 79-85. http://dx.doi.org/10.1039/b810104n

Shekhawat, D., Luebke, D. R., & Pennline, H. W. (2003). A review of carbon dioxide selective membranes. *US Department of Energy.* http://dx.doi.org/10.2172/819990

Spillman, R. W. (1989). Economics of gas separation membranes. *Chemical engineering progress, 85*(1), 41-62.

Xia, Y, Lu, Y., Kamata, K., Gates, B., & Yin, Y. (2002). Macroporous materials containing three-dimensionally periodic structures. *Macroeconomics for Business and Society: A Developed/developing Country Perspective on the "new Economy"*, 69.

9

Regional Environmental Quality and Cost Efficiency of International Tourist Hotels in Taiwan

Chin-Tsu Chen[1], Jin-Li Hu[2] & Shin-Lung Lin[2]

[1] Department of Commercial Design and Management, National Taipei University of Business, Taiwan

[2] Institute of Business and Management, National Chiao Tung University, Taiwan

Correspondence: Jin-Li Hu, Institute of Business and Management, National Chiao Tung University, Taiwan.
E-mail: jinlihu@gmail.com

Abstract

The purpose of this study is to compute the cost efficiency of international tourist hotels (ITHs) in Taiwan and analyze how regional environmental quality affects cost efficiency. The study collects a variety of operating data of international tourist hotels in Taiwan between 1998 and 2009 from the *Annual Tourist Hotels Operational Analysis Report*, as published by the Tourism Bureau. The data of regional environmental quality are collected from the Taiwan Environment Data Warehouse. This study first uses data envelopment analysis (DEA) for computing cost efficiency, and then applies the truncated regression to investigate the impact of environmental indicators on cost efficiency. Namely, it is a two-stage DEA-regression method. The empirical results indicate that the density of pungent air pollutants, such as O_3 and SO_2, have significantly negative impacts on the cost efficiency of ITHs in Taiwan. However, the high volume of waste disposal and low recycling rate reflect a higher business activity intensity in a region, which positively enhances ITHs' cost efficiency.

Keywords: international tourist hotels, cost efficiency, environmental quality

1. Introduction

In recent years, the rapid economic development and improved quality of material life have led to a relatively busy, fast, and oppressive pace of life. Therefore, the demand for leisure travel has increased dramatically, and the tourism industry has been regarded as one of star industries of the 21st century. In recent years, due to intense changes in natural environments, the world focuses on issues of environmental protection. Consequently, environmental issues have a far-reaching impact on the tourism industry. In addition to the economic development and employment opportunities brought by the tourism industry, countries in the world pay greater attention to the use of natural and human resources to develop and maintain natural ecological environments that facilitate sustainable development.

According to the statistical data of the Tourism Bureau, the Ministry of Transportation, 9.415 million tourists went sightseeing in Taiwan in 2010, growing from 8.142 million in 2009 (15.6%). Moreover, in the most recent 10 years, tourists to Taiwan and Taiwanese tourists going overseas have been increasing. The percentage of tourists for sightseeing purposes, against the total number of tourists, grew from 36% in 2001 to 58% in 2010; with mainland China the largest source of tourists, followed by Japan, Southeastern Asian countries, Hong Kong, Macao, and the United States. According to the data of World Tourism Organization (2011), the total number of tourists in the world in 2010 was 935 billion, an increase of 7% compared to that in 2009. The Asia-Pacific region is the first area of economic recovery, and attracted a total of 204 million international tourists in 2010, growing by about 13%, as compared to 181 billion people in 2009, as the second only to the Middle East. In addition, according to the statistical data of World Travel & Tourism Council (WTTC), in 2010, the global tourism industry (including tourism-related industries, input, and taxation) accounted for about 9% of the GDP (equivalent to 5.7 trillion USD), growing by about 1.7%, as compared to that in 2000. In 2010, the tourism industry gradually recovered.

Taiwan is surrounded by the sea and located in a maritime transport hub in East Asia. In history, Taiwan is a central location fought over by western and eastern powers and has become a rare cultural melting pot. In terms of geographic resources, as Taiwan is located in the Pacific seismic belt, geological activities are rich, which render Taiwan a rich geographic environment with great sightseeing values. Therefore, tourism has become one of the most promoted industries of the government, which launched various policies to attract international tourists; for example: in 2011, the Tourism Bureau shot Taiwan's first global sightseeing movie, marketing Taiwan to the world, signed free-visa agreements with 42 countries, and opened to mainland China tourists for sightseeing tours in Taiwan. Counties and cities in Taiwan host various festivals with local characteristics to attract tourists. According to the 2010 survey by the Taiwan Tourism Bureau on the consumption and movements of tourists to Taiwan, from 2001 to 2010, tourists to Taiwan increased from 2,831,035 to 5,567,277, growing by about 87%; and tourism foreign exchange earnings rose from 4,335 billion USD to 8,719 million USD, growing by about 101%.

The tourism industry will become an important industry with a direct impact on Taiwan's development, and the government has increased emphasis on tourism industry development. Studies on the tourism industry will contribute to industrial development and provide the Government with directions and reference for policy-making. According to the ITH summary of operations in 2010, as published by the Tourism Bureau, Ministry of Transportation, accommodation for tourists to Taiwan is mainly in hotels, including ITHs, general hotels, and tourist hotels. Therefore, when attracting more international tourists to Taiwan, the hotel industry is an important link for the development of the industry. Taiwan's ITHs rose from 44 hotels in 1985 to 70 hotels in 2010. With the increasingly fierce competition in the hotel industry, only those that can control costs and properly use resources can achieve the highest efficiency to survive in the tourism industry.

According to the 2010 survey and statistics on the tourism of Taiwanese people, conducted by the Tourism Bureau, Ministry of Transportation, regarding tourism activities, people are mainly engaged in "natural sightseeing activities", "cultural experience", "sports activities", "amusement park activities", "gourmet activities", and "leisure activities". "Natural sightseeing activities" accounts for 53.8% as the highest. It thus can be concluded that Taiwan's natural environment has a considerable impact on the tourism industry. The selection of hotels is an important activity in sightseeing. The natural environment quality of ITHs, such as air quality, waste recycling, and processing procedures, can be major factors affecting tourists when determining the location of accommodations. Therefore, the purpose of this study is to investigate the linkage between natural environmental indicators and the international tourist hotel's cost efficiency.

2. Literature Review

2.1 Environment Indicators

Many scholars have studied the relationship between the tourism industry and the environment. Studies have pointed out that the success of tourism and hotel industries depends largely on providing a clean natural environment. According to the survey, the European hotel industry and Scandinavian hotel guests believe that environment is the major key factor to the development and success of the tourism industry (Bohdanowicz 2003). According to the needs of tourists, the environmental quality of natural attractions is an important part of tourism quality. Therefore, to maintain a high quality overall tourism environment is very important to the competitiveness of most tourist destinations (Mihalic 2000).

Huybers and Bennett (2003) pointed out that, as the tourism industry depends on natural environment quality, tourism operators will have motivations for voluntary environmental protection activities. Some studies have suggested that environment is very important to the competitiveness of the tourism industry in tropical Northern Queensland, and undamaged environments have a significant impact on the needs of tourists. By questionnaire survey, Li (2004) pointed out that there are five types of tourist attractions: plants, animals, ecosystems and landscapes, history and religion, fresh air and water. The most attractive type is fresh air and water; 68.3% of tourists to the Nature Reserve of Tianmu Mountain in Zhejiang come because of the quality of air and water. The second mostly preferred is the ecosystems and landscapes type; 59.2% of tourists enjoy ecosystems and beautiful sceneries. The third most popular type is plants, with 59% of tourists claiming they are attracted by the primitive plants in the Nature Reserve of Tianmu Mountain in Zhejiang. In 2006, the significance of high environmental quality to tourism industry development was recognized by the World Trade Organization (WTO) (Bohdanowicz 2006). However, Kirk (1995) pointed out that the hotel industry also produces some undesirable emissions, such as carbon dioxide, CFCs, noise, smog, and odors, as well as wasted resources, such as energy, water, food, and packaging, which may result in natural appeal losses for beautiful scenery, natural hydrological structures, clean water, fresh air, and species diversity, due to pollution. Therefore, as the tourism industry depends on natural environment quality, it has a special position in the sustainable development of environments. Although the tourism industry may lead to environmental damage, it can positively contribute to natural environments; for example, advocacy of the value of natural environments to people, the creation of incentive mechanisms for improving environments, improvement of environmental protection awareness, and encouragement of tourists to exercise environmental protection actions (Blanke and Chiesa 2007).

In recent years, due to rising environmental protection awareness, many industries are faced with pressure to enhance environmental protection awareness, and the hotel industry is no exception. The hotel industry can contribute to the improvement of environmental quality by reducing its impact on the environment. However, for operators of the hotel industry, most are profit-making units, and whether implementations of environmental protection activities can produce sufficient benefits is a key issue. Although benefits from implementing environmental protection activities may not be sufficient, the hotel industry is actually faced with pressures from both consumers and law enforcement (Knowles et al. 1999).

Grove et al. (1996) pointed out that organization may protect the environment through recycling of packaging, recycling materials, and reducing data. For tourism and hotel industries, it is not difficult to make commitments for environmental protection, as recycling waste water is a means of environmental protection.

Fairweather et al. (2005) also pointed out that the most commonly seen types of information expected by tourists from environmental protection labels are detailed certified information, recycling information, and information regarding measures taken to protect the environment. Previous studies have also pointed out that the tourism industry should be concerned about the repeated use of resources in environmental management, rather than the reduced use of resources (Hunter and Green 1995).

Font and Tribe (2001) studied a variety of different awards and labels on environmental management. In some cases, only one standard, such as the presence or absence of chlorine, is evaluated; while in other cases, multiple indicators are evaluated. However, the greatest disadvantage of these awards is that they cannot be universally applied, that is, they are not applicable to all industries or hotels in all countries, as environmental problems vary in different countries' hotel and industrial environments.

2.2 Hotel Efficiency

Most previous studies on hotel operating efficiency conducted analysis using the inefficient frontier method. Wassenaar and Stafford (1991) proposed using hotel room usage as the measurement indicator (Wassenaar and Stafford 1991). However, due to the development of DEA and SFA in recent years, and the wide use of these two methods, more and more studies adopt DEA and SFA to measure hotel industry operating efficiency, in particular, DEA is the most widely applied. With many characteristics, DEA can convert multiple inputs into outputs to analyze efficiency; therefore, DEA is preferred in many studies (Charnes 1994). For example, Anderson, Fok, and Scott (2000) used DEA to compute the overall efficiency, technology efficiency, pure technology efficiency, and scale efficiency of 48 hotels in the United States by using five input items, including number of full-time staff, number of rooms, game-related total cost, total cost of food and beverage, and other cost, and the total revenue as the output items, and found that the overall average efficiency is about 42%, while the major cause of relatively low efficiency is low allocative efficiency.

By using the number of full-time employees, labor costs, number of rooms, the hotel area (square meters), the book value of assets, operating expenses, the external cost of input items and turnover, number of tourists, and accommodation revenue as output items, Barros (2005) used the CCR model of DEA to explore the efficiency of 43 hotels in the Pousadas de Portugal chain group of Portugal in 2001. According to the findings of the study, most hotels have production efficiency, and technical efficiency is 90.9% in the case of constant returns to scale; in variable returns to scale, technical efficiency is 94.5%. The efficiency of hotels in the city or near city is higher than that of hotels in rural areas. By using the number of full-time employees, the book value of the assets, number of rooms, the price of labor, the price of capital, and room price as input items, and turnover, number of tourists, and accommodation revenue as output items, Barros and Mascarenhas (2005) used cost DEA to explore the economic efficiency of 43 hotels of the Pousadas de Portugal chain group of Portugal in 2001. According to the research findings, the type of hotel has no impact on economic efficiency. Hotels with more rooms have a higher level of efficiency, and hotels farther from tourism routes have lower efficiency.

In recent years, a number of Taiwanese scholars used DEA to study hotel operating efficiency. By using total operating expenses, number of employees, number of rooms, total floor area of food and beverage sectors, number of employees of the accommodation sector, the number of food and beverage sector staff, and catering cost as the input items, and total operating revenue, number of rooms, average room occupancy revenue, the average salary of the food and beverage sector staff, the accommodation sector's total revenue, and food and beverage sector's total operating revenue as output items, Tsaur (2001) used DEA to study the efficiency of 53 ITHs from 1996 to 1998 in Taiwan. According to the research findings, the average operating efficiency is 87.33%, thus, the operation of Taiwan's tourist hotels is efficient. By using the number of full-time employees, the total area of rooms, food and beverage sector's total floor area, and operating expenses as input items, and room revenue, food revenue, and other revenue as output items, Hwang and Chang (2003) used DEA to measure the operating efficiency of 45 hotels in Taiwan in 1998, and applied Malmquist production variables in order to investigate the change in efficiency during the period between 1994-1998. The research findings suggest that the average efficiency in 1998 was 79.16%.

Hu et al. (2009) also used DEA in analysis of cost efficiency, allocative efficiency, and overall technology efficiency, of ITHs (International Tourist Hotels) in Taiwan during the period of 1997-2006, by using the number of rooms, number of employees, and catering floor area as input items, and the average operating cost of a room, the average annual salary of the employees, and food and beverage average operating cost as the input cost, and food and beverage total revenue, accommodation revenue, and other revenue as the output items. According to the research results, the cost efficiency of these hotels is 57% low efficiency. Chain systems, non-urban locations, and occupancy rates have significant impact. In addition to DEA, Hu et al. (2010) used SFA to discuss cost efficiency and causes of inefficiency of 66 hotels in Taiwan during the period of 1997-2006, using average labor costs, the average of other operating expenses, and the average cost of catering floor area as input items, and room revenue, catering revenue, and other operating revenue as the output items. According to the findings of this study, the average cost efficiency of ITHs in Taiwan is 91.15%. Chain systems, number of tour guides, and international transportation can significantly improve the cost efficiency of Taiwan's ITHs.

By using the number of rooms, number of employees, and catering total floor area as input items, the average operating cost of a room, the average annual salary of the employees, food, and average operating cost, as input costs, and catering total revenue, room revenue, and other revenue, as the output costs, Chen et al. (2010) used DEA to explore the cost efficiency of ITHs in Taiwan during the period of 1997-2007, and explored the impact of nationality of tourists on cost efficiency. According to the research findings, between 2002 and 2005, the average cost efficiency of ITHs in Taiwan tended to gradually decline. In terms of tourist nationality, tourists from Taiwan, North America, Japan, and Australia have a positive impact on the cost efficiency of hotels, while overseas Chinese and European tourists have a negative impact on cost efficiency.

3. Research Method

This study applied the two-stage analysis method to analyze the impact of environmental indicators of tourist hotels in Taiwan on operating cost efficiency. In the first stage, the data envelopment analysis (DEA) is applied in the analysis of the operating efficiency of tourist hotels. In the second stage, using the operating efficiency values of hotels obtained in the first stage as the dependent variables, and environmental indicators as the independent variables, this study applied the regression method to analyze the impact of environmental indicators on operating efficiency.

DEA

The concept of DEA was first proposed by Farrell (1957), who introduced the concept of production frontier into the measurement of relative efficiency. The decision making units of the production frontier are regarded as relatively efficient, while decision making units inside the production frontier are regarded as relatively inefficient. Charnes et al. (1978), based on the Farrell theory, used the linear programming method to add situations with multiple inputs and outputs to measure production efficiency in 1978. This was the first DEA model, known as the CCR model. However, the CCR model is under the assumption of constant returns to scale (CRS). Later, Banker et al. (1984) relaxed CRS assumption into variable returns to scale (VRS) by incorporating the convexity constraint into a DEA model.

At present, DEA models can be divided into the input-oriented model that measures input relative efficiency and the output-oriented model that measures output. This study used the input-oriented VRS-DEA model, which assumes that each hotel has K input items and M output items, using q_i to represent the output vector of the ith decision making unit (M x 1); x_i to represent the input vector of the ith decision making unit (K×1); λ to represent the weight vector of each decision making unit (N×1); w_i to represent the price vector of various input items of the ith decision making unit; and x_i^* to represent the vector of various input items under the minimized input cost of No. i decision making unit, the model is as below:

$$\text{Min } w_i' x_i^*$$
$$x^*, \lambda$$
$$\text{s.t.} \quad -q_i + Q\lambda \geq 0$$
$$x_i^* - X\lambda \geq 0$$

$$\sum \lambda = 1$$
$$\lambda \geq 0 \tag{1}$$

3.1 Truncated Regression

This study adopted the two-stage analysis method. In the second stage, truncated regression is applied. Most previous studies using the two-stage method used the Tobit regression model, as proposed by Tobin (1958). However, DEA estimates the relative efficiency value of the decision making unit according to the characteristics of non-parameters. Tobit regression assumes that efficiency errors are of normal distribution, which will result in the relative bias of Tobit regression results. Simar and Wilson confirmed that, when the efficiency value is an explained variable, truncated regression can produce estimated results closer to the actual parameter values, as compared with the Tobit regression method (Simar and Wilson 2007).

3.2 Date Sources

This study collects various operation-related data of ITHs in Taiwan from 1998 to 2009, taken from the Annual Tourist Hotels Operational Analysis Report, published by the Tourism Bureau, and selected appropriate revenues and input items as the variables for the calculation of efficiency value. In addition, to avoid the problem of different potential measurement benchmarks, as caused by the different monetary supplies of various years, all nominal variables in this study were converted into actual variables by GDP deflator, with 1998 as the benchmark year to avoid distortion by the price level changes. Moreover, the number of official ITHs may be different, due to bankruptcy or ITH standards formulated by the Tourism Bureau, thus, to collect as much data as possible, this study used unbalanced vertical and horizontal data to estimate the relative efficiency of ITHs. The number of observations from 2001 to 2009 is as shown below: 1998 (53 hotels), 1999-2001 (55 hotels), 2002 (56 hotels), 2003-2004 (58 hotels), 2005-2006 (57 hotels), 2007 (58 hotels), 2008 (59 hotels), and 2009 (56 hotels).

The hotel industry's output is mainly measured by revenue. The major sources of revenue for ITHs in Taiwan can be roughly divided into three types of accommodation and beverage revenues, which account for more than 80% of the total. Other revenues, including the revenue of swimming pool, laundry revenue, and rent and services revenues, account for less than 20%. In terms of input, the hotel's input roughly consists of employees, equipment, and capital. The employee salaries represent the human resources invested by a hotel, and the total number of rooms and the catering floor areas can represent the capital input of the hotel (Hwang and Chang 2003). With reference to previous literature, the inputs are divided into the number of rooms, number of employees, and catering floor area, which is multiplied by the average operating cost of a room, the average annual salary of the employees, and food and average operating cost, respectively. Hotel operational revenue, inputs and input costs used by (Hu et al. 2009) are as shown below:

1. Output variables

(1) Food and beverage total revenue

(2) Accommodation revenue

(3) Other revenue

2. Input variables

(1) The number of rooms

(2) Number of employees

(3) Catering floor area

3. Input cost

(1) The average operating cost of a room, meaning the total expenses of a room, as the division of total revenue by the number of rooms;

(2) The average annual salary of the employees;

(3) Food and average operating costs; the division of the relevant expenses of the catering sector by the catering floor area.

Environmental indicators are used as the environmental variables affecting hotel operating efficiency. Regarding environmental indicators, the Council of Sustainable Development, the Executive Yuan, held four rounds of "Ministerial Coordination Conference on Sustainable Development Indicators" since February 2009, inviting representatives from relevant ministries and the private sector to determine the sustainability and accessibility of new indictors. In the No. 29 working meeting of the Council of Sustainable Development, held on December 31st,

2009, the second version of sustainable development indicators of Taiwan, covering 12 themes, 41 subthemes, and 94 indicators (number of indicators increased by 52 as compared to the original version), was approved. The indicators in the environmental perspective can be roughly categorized into four major types, including air and water quality, and waste and environmental management.

Regarding the dependence of the tourism industry on environmental quality, as tourists prefer natural environments, and are concerned with air quality, they account for the greatest percentage. Moreover, it has been mentioned in previous studies that the hotel industry should consider recycling resources, rather than a reduction in the use of resources, for environmental management (Hunter and Green 1995). The purpose of this study is to explore the impact of environmental quality on hotel operating efficiency. Therefore, regarding environmental indicators, the air and waste categories of indicators in the environmental dimension of the sustainable development indicator system, as developed by the Executive Yuan, were used as measurement indicators. The data of various monitoring stations were obtained from Taiwan Environment Data Warehouse (2012), which is maintained by the Environmental Protection Administration in Taiwan.   The variables used in this study are as shown below:

1.   Air

(1) SO_2 density

(2) NO_2 density

(3) CO density

(4) O_3 maximum 8hr average density

2.   Waste

(1) Waste recycling rate

(2) Average daily waste generated per person

In order to avoid the impact of excessive environmental indicators on hotel operating efficiency, which can cause bias, this study adds the variables that commonly affect operating efficiency as the control variables to avoid an overreaction of environmental variables. The control variables used in this study are as shown below:

1.   Chain System (Chain = 1 and non-chain =0)

2.   Rural level (Rural county =2, Metropolitan County = 1, and Metropolitan City = 0)

4. Results

The statistical results of the business operations of Taiwan's ITHs are elaborated upon, as shown in Table 1. The revenue of Taiwan's ITHs mainly comes from catering and accommodation of tourists. The average catering revenue is about 2.74 million NTD, which is slightly more than the accommodation revenue of 2.38 million NTD, possibly because hotel restaurants have become one of the major choices of social gathering for an increasing number of people dining out. Besides accommodation, catering has become the focus of operation for tourist hotels in Taiwan. However, for different business operational patterns, chain or non-chain business models can result in great differences in scale; thus, the standard deviation is considerably large.

4.1 Hotel Efficiency Analysis

In this study, hotel operating efficiency is measured by input-oriented cost DEA. By using the DEAP 2.1 software, as proposed by Professor Coelli (1996), the cost efficiency of various hotels is computed during the period of 1998-2009. The average cost efficiency of ITHs in Taiwan are 0.719 (2001), 0.711 (2002), 0.702 (2003), 0.662 (2004), 0.658 (2005), 0.665 (2006), 0.649 (2007), 0.666 (2008), and 0.667 (2009).   The results showed that the average cost efficiency of ITHs in Taiwan during the period of 1998-2009 was about 67.8%; in other words, the average input cost of the hotel can be further reduced by 32.2%, while maintaining the state of outputs.

Table 1. Average values and standard deviations of input, output, and input prices

Variable	Mean	S.D.
Outputs		
Catering revenue (NTD in 1998 prices)	275,345	274,072
Accommodation revenue (NTD in 1998 prices)	238,404	217,399
Other revenue (NTD in 1998 prices)	98,269	141,214
Inputs		
Actual number of occupied rooms (room)	72,996	43,638
Number of employees (person)	335	229
Catering total floor area (Ping; 1 Ping = 3.305785 square meters)	1,191	1,364
Input Prices		
Average room rate (NTD in 1998 prices)	3,060	1,336
Average salary per employee (NTD in 1998 prices)	526,014	156,231
Catering cost per Ping (NTD in 1998 prices)	106,027	98,404

The trends of cost efficiency during the period of 1999~2007 were roughly downward, as can be seen in the figure. Namely, during these night years, the cost efficiency of Taiwan ITHs continuously declined, and began improving in 2007. The possible cause of such a phenomenon is the unforgettable major earthquake, 921, in 1999. Many hotels in mountainous regions were seriously damaged. This earthquake scared the Taiwanese, depriving them of their interest in tourism, seriously damaging the tourism industry. By 2008, international hotels' operating efficiency improved a little bit. It is believed that the hotel industry actively improved operating performance as the industry awoke from the financial crisis in 2008. Meanwhile, the government approved opening mainland Chinese tourists for sightseeing in Taiwan, thus, facilitating a steady stream of tourists into the tourism industry of Taiwan, which might improve cost operating efficiency of ITHs in Taiwan.

4.2 Truncated Regression Analysis

Following the recommendation of Simar and Wilson (2007), this study used the truncated regression model to replace the traditional Tobit regression, in order to test the impact of environmental indicators on operating efficiency in the second stage. Six environmental indicators are used as the environmental variables, including SO_2 density (SO2), NO_2 density (NO2), CO density (CO), O_3 maximum 8hr average density (O38HR), recycling rate (RE), and average daily waste generated per person (GB), while hotel cost efficiency (CE) is used as the dependent variable, coupled with two control variables, including Rural Level (RURAL) and Chain System (CHAIN).

The truncated regression results on cost efficiency are as shown in Table 2. The average daily waste generated per person, O_3 maximum 8hr average density, recycling rate, and SO_2 density of the six environmental indicators have significant impact on operating efficiency. The average daily waste generated per person has a positive impact on operating efficiency. It can be concluded that higher levels of economic activities can contribute more to the hotel operating efficiency. Indicators of O_3 maximum 8hr average density, recycling rate, and SO_2 density have negative impact on operating efficiency. As O_3 and SO_2 are gases with pungent odors and a toxicity level of O_3, it can lead to headaches, eye burning, and irritation of the respiratory tract, which have a bad influence on tourist accommodation and catering. Therefore, the hotel industry should select locations away from industrial areas prone to SO_2 and O_3. Waste recycling can increase the cost of doing business; therefore, it has a negative impact on cost efficiency.

This paper finds that as the rural level increases, the cost efficiency of an ITH decreases, implying that the degree of urbanization in the local area helps improve the cost efficiency of an ITH, which is consistent with the findings of Hu et al. (2010). The chain store system significantly helps improve the cost efficiency of ITHs, which is consistent with findings of Hu et al. (2009, 2010).

Table 2. Truncated regression results of environmental variables on cost efficiency

Variables	Estimation coefficient	P-value
Rural level (Rural)	-0.07097	<0.0001***
CO density (CO)	-0.06578	0.5165
Average daily waste generated per person (GB)	0.15715	0.0032***
Chain system (CHAIN)	0.09411	<0.0001***
NO2 density (NO2)	-0.00011	0.9659
O3 maximum 8hr average density (O38HR)	-0.00233	0.0145**
Recycling rate (RE)	-0.00145	0.0446*
SO2 density (SO2)	-0.02362	<0.0001***

Note: *** indicates significance at 0.01; ** indicates significance at 0.05; and * indicates significance at 0.10.

5. Conclusions

Taiwan's tourism industry has been rapidly developing in recent years due to the vigorous promotion of the government, and the opening of the tourism market to tourists from mainland China, which brings huge business opportunities to Taiwan's tourism industry. According to the statistics of the Tourism Bureau, about 80% of Taiwanese travelers choose Taiwan as the tourism destination, the average travel time is approximately 1.49 days, and the average annual number of trips is up to about 6 times. It can thus be concluded that Taiwanese people's demand for tours in Taiwan is considerable. In a tour, the choice of accommodation is very important. The catering revenue of Taiwan's ITHs has also begun to surpass the accommodation revenue; therefore, the hotel industry accounts for considerable volume in the overall tourism industry.

The average cost efficiency of ITHs in Taiwan is about 67.8% during 1998-2009. Taiwan's hotel industry has made progress, as compared with hotel cost efficiency at 57% in the period of 1997~2006, as suggested in a study by Hu et al. (2009); however, there still remains space for improvement of cost efficiency. According to the five least efficient hotels, technology efficiency is the major cause for low cost efficiency of hotels with poor performance. Low technology efficiency means some input resources have no corresponding outputs, in other words, there are unnecessary resource inputs. It is recommended that these hotels should re-examine all input resources and observe their industry peers to identify areas of management and resource wastes or seek business consultants to determine which parts of the operating costs can be improved.

This study also finds that the air quality and waste disposal of tourist areas have a definite impact on the operation of local hotels. According to statistical results, tourists are more sensitive to pungent odors in the air, and pungent gases, such as SO_2 or O_3 in the air, can be easily perceived by tourists. Thus, when hotels must spend extra cost to eliminate odors, local hotels' cost efficiency will become lower.

In addition to causing human discomfort, SO_2 will form strong corrosive acids, such as sulfuric and sulfuric acid, when dissolved in water, which may rust metal, fragment plastic, corrupt leather, damage cement walls, thus, shortening the service life of the above items. The above may also be a cause for lower cost efficiency of hotels. The main sources of SO_2 include factories and diesel engine vehicles, including railway diesel-electric locomotives, road buses, city buses, and freight trucks. In the future, operators in the hotel industry should be as far removed as possible from areas that may produce high density of SO_2, and should avoid the use of diesel vehicles or use high grade diesel oil, with 0.5% sulfur content, instead of using the ordinary diesel fuel with 1.2% of ordinary diesel fuel.

Besides pungent odor, O_3 can cause unhealthy conditions if inhaled. These features can be easily perceived and consequently lead to negative emotions. O_3 is formed by NOx and VOCs after a series of photochemical reactions. The sources of NOx and VOCs include industrial emissions, vehicle emissions, and other sources. Areas of high density O_3 include the industrial areas, metropolitan area, traffic arteries, and peri-urban areas. Therefore, hotels should be built far away from areas of high density of O_3. It is recommended that operators in the hotel industry should refer to the air quality data published by the air quality observation stations of the Environmental Protection Administration as a factor of consideration.

Moreover, this study found that the average waste generated per person in an area is positively correlated to the hotel's operating efficiency. The two possible factors for relatively high average daily waste generated per person of an area may include: first, the economic activities of the residents in this area are booming, and the volume of waste generated per person will be relatively more. This suggests the area is high in consuming power and suitable for hotel business. Second, the large number of tourists generates huge volumes of waste and the local residents produce relatively less waste. It results in the inevitable bias in governmental statistics. However, this also suggests that the tourism resources of this area are favored by tourists in large numbers, thus, this area is suitable for development of the tourism industry. No matter the reason, higher levels of the average daily waste generated per person suggest the area is suitable for businesses of the hotel industry. Therefore, operators of the hotel industry are expected to refer to the waste processing data provided by the Environmental Protection Administration as a criterion to determine the appropriate location for a hotel.

This study also finds that the waste recycling rate of the area surrounding a hotel location has a negative impact on hotel cost efficiency. Resource recycling is a long term trend of social development; however, it is actually a cost to the operators of the hotel industry. Therefore, in the implementation of a resource recycling policy, the government should pay special attention to the impact on tourist hotels and shorten the magnitude of the impact, as possible. In the future, resource recycling items and methods should be taken as research subjects to identify which resource recycling items or methods will cause greater burden on the operators of tourist hotels, which would assist the government and operators of tourist hotels to determine the possible optimal resource recycling programs.

Acknowledgements

The second author thanks Taiwan's Minitsry of Science and Technology for financial support (NSC101-2410-H-009-044).

References

Anderson, R. I., Fok, R., & Scott, J. (2000). Hotel industry efficiency: an advanced linear programming examination. *American Business Review, 18*(1), 40-48.

Banker, R. D., Charnes, R. F., Cooper, W. W. (1984). Some models for estimating technical and scale inefficiencies in data envelopment analysis. *Management Science, 30*(9), 1078-1092.

Barros, C. P. (2005). Measuring efficiency in the hotel sector. *Annals of Tourism Research, 32*(2), 456-477. http://dx.doi.org/ 10.1016/j.annals.2004.07.011

Barros, C. P., & Mascarenhas, M. J. (2005). Technical and allocative efficiency in a chain of small hotels. *International Journal of Hospitality Management, 24*(3), 415-436. http://dx.doi.org/ 10.1016/j.ijhm.2004.08.007

Blanke, J., & Chiesa, T. (2007). The travel and tourism competitiveness report 2007: Furthering the process of economic development. Geneva, Switzerland: World Economic Forum.

Bohdanowicz, P. (2003). A study of environmental impacts, environmental awareness and pro-ecological initiatives in the hotel industry. Licentiate Thesis. Department of Energy Technology, Royal Institute of Technology, Stockholm, Sweden.

Bohdanowicz, P. (2006). Environmental awareness and initiatives in the Swedish and Polish hotel industries - survey results. *International Journal of Hospitality Management, 25*(4), 662-682. http://dx.doi.org/ 10.1016/j.ijhm.2005.06.006

Charnes, A. (1994). *Data Envelopment Analysis: Theory, Methodology, and Application.* United States: Springer.

Charnes, A., Cooper, W. W., & Rhodes, E. (1978). Measuring the efficiency of decision making units. *European Journal of Operational Research, 2*(6), 429-444.

Chen, C. T., Hu, J. L, & Liao, J. J. (2010). Tourists' nationalities and the cost efficiency of international tourist hotels in Taiwan. *Africa Journal of Business Management, 4*(16), 3440-3446.

Coelli, T. J. (1996). A guide to DEAP version 2.1: A data envelopemnt analysis (computer) program. CEPA Working Papers, No. 8/96, Armidale: University of New England.

Fairweather, J. R, Maslin, C., & Simmons, D. G. (2005). Environmental values and response to ecolabels among international visitors to New Zealand. *Journal of Sustainable Tourism, 13*(1), 82-98. http://dx.doi.org/ 10.1080/17501220508668474

Farrell, M. J. (1957). The measurement of productive efficiency. *Journal of the Royal Statistical Society, Series A, 120*(3), 253-290.

Font, X., & Tribe, J. (2001). Promoting green tourism: the future of environmental awards. *International Journal of Tourism Research, 3*(1), 9-21. http://dx.doi.org/10.1002/1522-1970(200101/02)

3:1<9::AID-JTR244>3.0.CO;2-Q

Grove, S. J., Fisk, R. P., Pickett, G. M., & Kangun, N. (1996). Going green in the service sector: Social responsibility issues, implications and implementation. *European Journal of Marketing, 30*(5), 56-66.

Hu, J. L., Chiu, C. N., Shieh, H. S., & Huang, C. H. (2010). A stochastic cost efficiency analysis of international tourist hotels in Taiwan. *International Journal of Hospitality Management, 29*(1), 99-107. http://dx.doi.org/ 10.1016/j.ijhm.2009.06.005

Hu, J. L., Shieh, H. S., Huang, C. H., & Chiu, C. N. (2009). Cost efficiency of international tourist hotels in Taiwan: A data envelopment analysis application. *Asia Pacific Journal of Tourism Research, 14*(4), 371-384. http://dx.doi.org/ 10.1080/10941660903310060

Hunter, C., & Green, H. (1995). *Tourism and the Environment: A Austainable Relationship?* London: Routledge.

Huybers, T., & Bennett, J. (2003). Environmental management and the competitiveness of nature-based tourism destinations. *Environmental and Resource Economics, 24*(3), 213-233. http://dx.doi.org/ 10.1023/A:1022942001100

Hwang, S. N., & Chang, T. Y. (2003). Using data envelopment analysis to measure hotel managerial efficiency change in Taiwan. *Tourism Management, 24*(4), 357-369. http://dx.doi.org/ 10.1016/S0261-5177(02)00112-7

Kirk, D. (1995). Environmental management in hotels. *International Journal of Contemporary Hospitality Management, 7*(6), 3-8.

Knowles, T., Macmillan, S., Palmer, J., Grabowski, P., & Hashimoto, A. (1999). The development of environmental initiatives in tourism: responses from the London hotel sector. *International Journal of Tourism Research, 1*(4), 255-265.

10

Process Evaluation of Carbon Dioxide Capture for Coal-Fired Power Plants

Satoshi Kodama[1], Kazuya Goto[2] & Hidetoshi Sekiguchi[1]

[1] Department of Chemical Engineering, Graduate School of Science and Engineering, Tokyo Institute of Technology, Tokyo, Japan

[2] Chemical Research Group, Research Institute of Innovative Technology for the Earth, Kyoto, Japan

Correspondence: Satoshi Kodama, Department of Chemical Engineering, Graduate School of Science and Engineering, Tokyo Institute of Technology, 2-12-1-S4-1 Ookayama, Meguro-ku Tokyo, Japan. E-mail: skodama@chemeng.titech.ac.jp

Abstract

Carbon capture is a promising technology for carbon dioxide (CO_2) removal from large stationary CO_2 sources. The effects of carbon dioxide capture process on output efficiency of fossil power plants were investigated. Supercritical pulverized coal and integrated coal gasification combined cycle (IGCC) were assumed as model coal-fired power plants for this investigation. Heat-driven and pressure-driven CO_2 capture processes such as chemical absorption and physical adsorption were assumed for CO_2 capture process. In this study, these technologies were evaluated and compared under the unified basis and conditions by using the commercial process simulator. For IGCC plant, the efficiency penalty by installing water-gas shift reaction was also investigated. Gross and net power generation, efficiency and the efficiency penalty by CO_2 capture process were calculated. Heat duty for CO_2 capture process and CO_2 compression conditions were varied, and those effects on the efficiency penalty were obtained. The results provide a guideline for development of CO_2 capture process of power plants.

Keywords: carbon dioxide capture, power plant, process simulation

1. Introduction

In working group I contribution to the fifth assessment report of the Intergovernmental Panel on Climate Change (IPCC), it is reported that "Warming of the climate system is unequivocal, and since the 1950s, many of the observed changes are unprecedented over decades to millennia" (IPCC, 2013). Several scenarios which reduce the carbon dioxide (CO_2) emission and stabilize the global climate change are proposed, such as the 450 scenario and the 550 scenario. The 450 scenario can control the average temperature rise within 2 °C by stabilizing the CO_2 concentration in the atmosphere for 450 ppm. In the scenario, 14 GT/y of CO_2 emission must be reduced in 2030. To reduce the CO_2 emission, it is suggested to use renewable energy, biofuels, nuclear power and carbon capture and storage (CCS), as well as energy saving. CCS is expected to remove 2 GT/yr of CO_2 emission (IEA, 2009). CO_2 is generated by various human activities, such as electricity and heat producing, manufacturing, transport etc. The CO_2 from electricity and heat generation and manufacturing are typically emitted from large exhaust stacks, and they can be described as large stationary sources. The large stationary sources represent potential opportunities for the addition of CO_2 capture plants. The properties of each CO_2-containing gas is different, while the CO_2 partial pressure is important for CO_2 capture. Coal for power generation is primarily burnt in pulverized-coal (PC) boilers producing an atmospheric pressure flue gas stream with a CO_2 partial pressure of up to 0.014 MPa. The newer and potentially more efficient integrated coal gasification combined cycle (IGCC) technology has been developed, and CO_2 partial pressure of CO_2 capture target gas is up to 0.014 MPa (post combustion) or 1.4 MPa (pre combustion) (IPCC, 2005).

There are several CO_2 captuure technologies, which use sorbent, solvent or membranes etc. The technologies are also classified as heat and/or pressure driven process. For example, absorption by chemical solvents and temperature swing adsorption are a heat driven process. On the other hand, the physical absorption of physical solvents, pressure swing adsorption and membrane separation are a pressure driven process. Generally, heat-driven CO_2 capture is used for low CO_2 partial pressure on target gas, while pressure-driven process is used for higher CO_2 partial pressure. The combination of heat and pressure-driven CO_2 capture process such as MDEA process is also evaluated. From the point of view of development of CO_2 capture process, it is important to estimate the efficiency penalty of power plants. There are many studies which analyze the effect of operating conditions of CO_2 capture process on power plant efficiency (Abu Zhara, 2011; Cifre, 2009; Goto, 2013; Strube, 2011). However, the relationship between heat and energy duty of CO_2 capture process and compressors on the efficiency of the power plants is not cleared yet.

In this study, PC and IGCC power plants were modelled by using a process simulator. The effect of properties of CO_2 capture process on power plant efficiency was investigated under the unified basis and conditions, such as coal property, efficiency of compressors or pumps etc. For IGCC plant, the efficiency penalty by installing water-gas shift reaction was also investigated. Gross and net power generation, efficiency and the efficiency penalty by CO_2 capture process were calculated. Heat duty for CO_2 capture process and CO_2 compression conditions were varied, and those effects on the efficiency penalty were obtained.

2. Development of Process Simulation Models

A commercial process simulator Aspen Plus 7.3 was used for the process modelling of power plants. The design basis of PC and IGCC power plant was referred from literature (NETL, 2007). The design coal was bituminous (Illinois No. 6) as shown in the reference. The high heat value (HHV) and low heat value (LHV) of the coal is 27,113 KJ/kg and 26,151 KJ/kg, respectively.

2.1 PC Power Plant

For the development of PC power plant with CO_2 removal, "Case12 PC with supercritical case" in the NETL report was referred (NETL, 2007). Process flow diagram of the PC power plant is shown in Figure 1. Pulverized coal was supplied to the burner. In this burner, steam was generated and was supplied to steam turbines. The steam turbines consist of high-pressure (HP), intermediate-pressure (IP) and low-pressure (LP) turbines as shown in Figure 2. The steam temperature generated by the heater was 593 °C and the pressure of superheated steam was 24.1 MPa. The flue gas from the burner was treated by selective catalytic reduction (SCR), bag filter and flue gas desulfurization (FGD). Finally, CO_2 was removed from the gas and treated flue gas was released from the stack. The separated CO_2 was compressed and liquefied by compressors to sequestration-ready pressure, 15.2 MPa. The details of CO_2 compression process and steam turbines are described in 2.3 and 2.4.

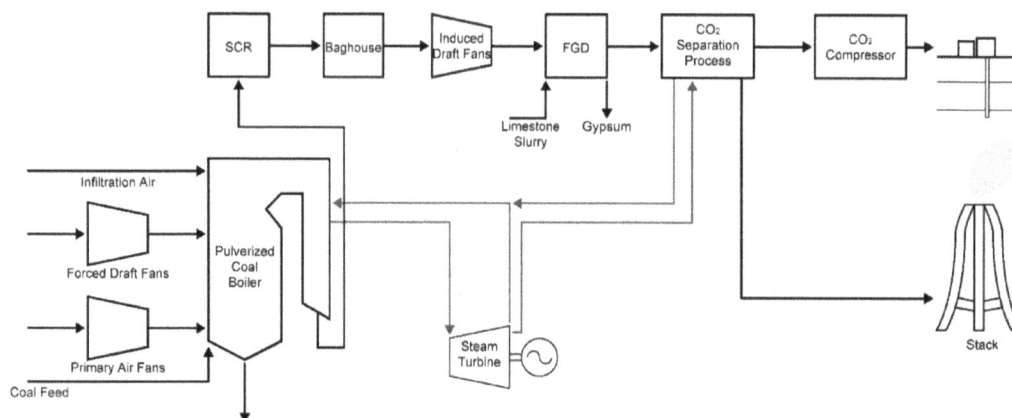

Figure 1. Process flow diagram of PC power plant with CO_2 post-combustion capture (Case12 in the literature (NETL, 2007))

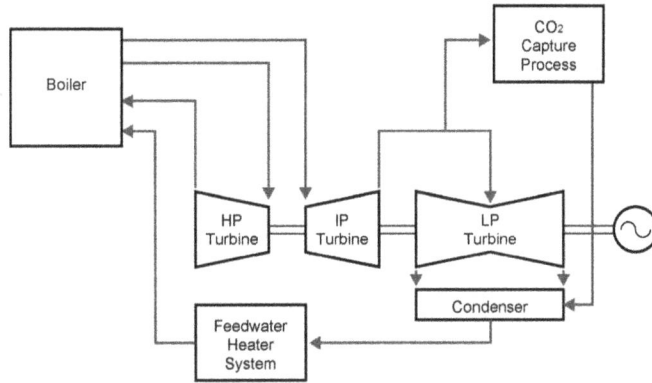

Figure 2. Steam extraction from steam turbines to CO_2 capture process

2.2 IGCC Power Plant

2.2.1 IGCC Without CO_2 Capture

Process flow diagram of the IGCC power plant is shown in Figure 3. Coal was supplied to gasifier and transferred to CO and H_2. The gas was treated by de-SOx reactor and burned in gas turbine engine. The gas turbine engine drove electric generator. The flue gas from the gas turbine was sent to heat recovery steam generator (HRSG) and generate steam which drove steam turbines. The steam temperature from the heater was 538 °C and the pressure of superheated steam was 12.4 MPa. The steam turbines drove electric generators. The fluegas was then sent to stack and emitted to the air.

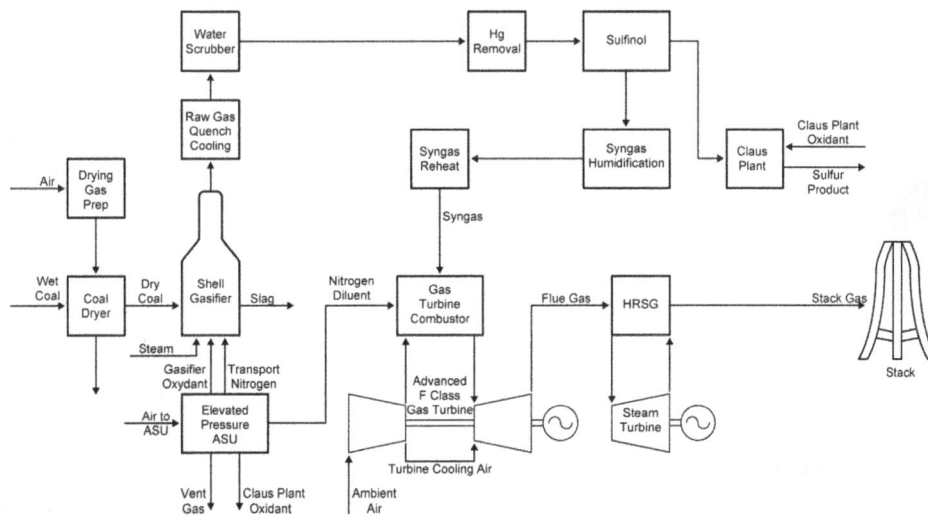

Figure 3. Process flow diagram of IGCC power plant without CO_2 capture (Case5 in the literature (NETL, 2007))

2.2.2 IGCC With CO_2 Capture

Process flow diagram of the IGCC power plant with CO_2 capture is shown in Figure 4. For CO_2 capture case, gasified coal (CO and H_2) was treated by water-gas shift reactor, in which CO reacted with H_2O and yielded CO_2 and H_2. Then CO_2 was removed in CO_2 capture process (shown as Selexol Unit in Figure 4) and liquefied by compressors. The treated gas was burnt in gas turbine engine, yielding hot flue gas. The flue gas was sent to HRSG. The steam temperature from the heater was 566 °C and the pressure of superheated steam was 12.5 MPa.

Generally CO_2 capture from IGCC power plant is performed from the water-shifted gas, and it is called pre-combustion capture. On the other hand, IGCC power plant without CO_2 capture process does not contain water-shift reactor. In this study, post-combustion capture from such process was also estimated. CO_2 partial pressure before CO_2 capture is 1.2 MPa for pre-combustion process and 0.0076 MPa for post-combustion process.

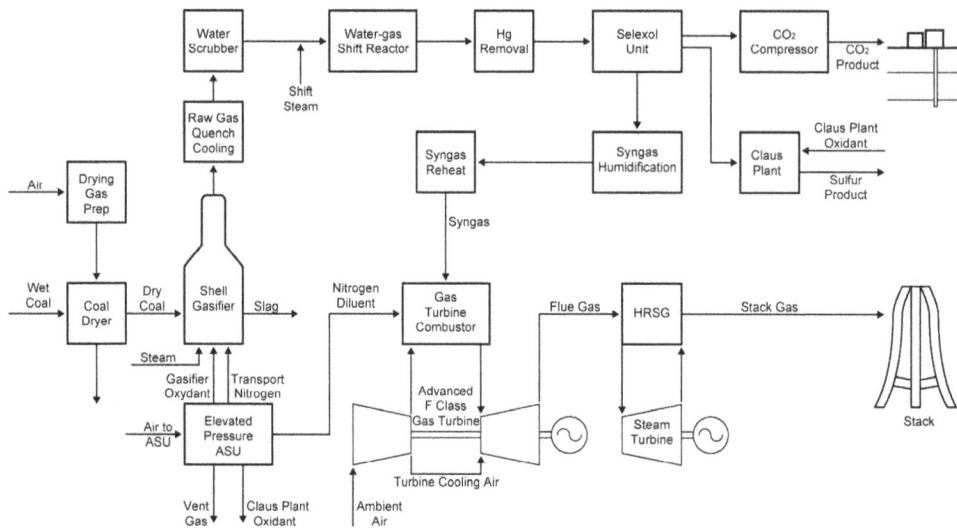

Figure 4. Process flow diagram of IGCC power plant with CO_2 capture (Case6 in the literature (NETL, 2007))

2.3 CO$_2$ Capture Process

2.3.1 Heat-Driven CO$_2$ Capture Process

Chemical absorption solvent or temperature swing adsorption (TSA) process was estimated for the CO_2 capture process and was driven by applying heat to the sorbents. In this study, the detailed reaction in CO_2 capture process such as the reaction of CO_2 with chemical solvents or adsorbents was not considered. It was estimated that CO_2 was separated by applying some amount of heat. The provided heat is consumed for reactions, temperature increase, steam generation, and so on. In this study, such a breakdown was not considered and a total heat requirement was considered for sensitivity analysis. The heat required for CO_2 capture was supplied by bypassing steam from the inlet of LP turbine. The steam was estimated to be cooled to 110 °C in the CO_2 capture process. 90% (for PC) or 95% (for IGCC) of CO_2 from the flue gas was captured. CO_2 was flashed in 0.16 MPa from CO_2 capture process and sent to CO_2 compressor.

2.3.2 Pressure-Driven CO$_2$ Capture Process

Physical absorption solvent, pressure swing adsorption (PSA) and membrane separation process are CO_2 capture process driven by pressure difference. In this process, detailed separation mechanism in CO_2 capture process such as CO_2 absorption, adsorption or membrane transparent was not considered and the outlet pressure of CO_2 capture process was considered as well as that of heat-driven CO_2 capture system. The pressure difference was generated by a pump or compressor driven by electricity generated at the power plant, and was combined with CO_2 compression process.

2.3.3 Heat and Pressure-Driven CO$_2$ Capture Process

In heat-driven CO_2 capture process, CO_2 was released under higher partial pressure than that of CO_2 source. It is expected that combination of CO_2 capture process operated by heat and pressure may reduce the efficiency penalty. Parametric study of heat required to separate CO_2 and inlet pressure on CO_2 compressor on efficiency penalty of the power plant was carried out from the results obtained in 2.3.1 and 2.3.2.

2.4 CO$_2$ Compression

CO_2 separated in the CO_2 capture process was pressurized to 15.2 MPa, so that it could be transferred to the sequestration site by pipeline. CO_2 compressors were connected in series as shown in Figure 5. The compression conditions are listed in Table 1, which were also referred from a literature (NETL, 2007). The compressed CO_2 was cooled to 52 °C (125 °F) between each compressor except the first and second stage (32 °C or 90 °F). The power consumption of the first stage was smaller than the other compressors because it worked as a liquid pump. For the conditions that the inlet pressure was lower than 0.16 MPa, additional compressors were added so that the pressure difference of inlet and outlet would be less than 2.2 times.

Table 1. Outlet pressure for each stage of CO_2 compression process

Stage	Outlet pressure (MPa)
1	15.3
2	8.27
3	3.76
4	1.71
5	0.78
6	0.36

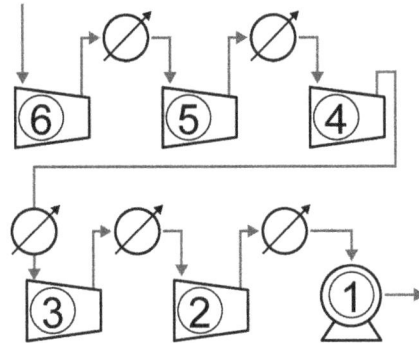

Figure 5. Configuration of CO_2 compression process

The relation between CO_2 inlet pressure P_{CO2} [MPa] and compression energy per CO_2 mass w_{CO2} [kW/t-CO_2] is shown in Figure 6. It shows good relation to semi-log plot and the fitting result is shown in Equation (1).

$$w_{CO2} = -26.37 \ln(P_{CO2}) + 58.16 \tag{1}$$

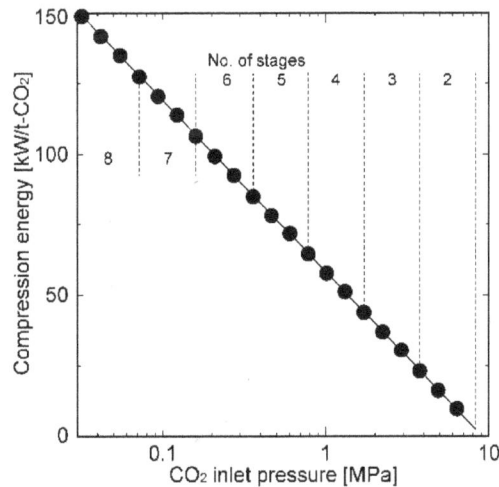

Figure 6. Relation between CO_2 inlet pressure (P_{CO2}) and compression energy (w_{CO2}), outlet pressure = 15.2 [MPa]

2.5 Variations of Process Model Calculation

Power generation and efficiency of the power plants were calculated and verified with reference data. Generated power $\dot{W}$ [kW] and efficiency η (net, gross) [%] are calculated by following equations:

$$\dot{W}_{net} = \dot{W}_{gross} - \dot{W}_{aux} \tag{2}$$

$$\eta_{gross} = (\dot{W}_{gross} \times 3.6) / \dot{Q}_{coal} \times 100 \qquad\qquad (3)$$

$$\eta_{net} = (\dot{W}_{net} \times 3.6) / \dot{Q}_{coal} \times 100 \qquad\qquad (4)$$

Where $\dot{Q}_{coal}$ is thermal input to the plant by coal combustion [GJ/h], 3.6 is a conversion factor from kW to GJ/h. CO_2 recovery ratio was 90% for PC case and 95% for IGCC cases, therefore

$$\dot{m}_{CO2\text{-}sep} = 0.90\ \dot{m}_{CO2\text{-}gen}\ (\text{for PC}),\ \dot{m}_{CO2\text{-}sep} = 0.95\ \dot{m}_{CO2\text{-}gen}\ (\text{for IGCC}) \qquad (5)$$

The data are shown in Tables 2–4. It was confirmed that the calculated data show good agreement with those reference data (NETL, 2007).

Table 2. Reference conditions and calculation results for PC with post-combustion CO_2 capture case

	Reference (Case 12)	This study	w/o CCS
Coal feed [t/h] ($\dot{m}_{coal}$)	266.090	266.090	266.090
Thermal input, LHV [GJ/h] ($\dot{Q}_{coal}$)	6,958,520	6,958,520	6,958,520
Gross power [kW] ($\dot{W}_{gross}$)	663,445	673,181	797,924
Aux power [kW] ($\dot{W}_{aux}$)	117,450	114,212	63,860
Net power [kW] ($\dot{W}_{net}$)	545,995	558,969	734,064
Gross efficiency [%] (η_{gross})	34.3	34.8	41.3
Net efficiency [%] (η_{net})	28.2	28.9	38.0
CO_2 generated [t/h] ($\dot{m}_{CO2\text{-}gen}$)	631.1	622.7	622.7
CO_2 separated [t/h] ($\dot{m}_{CO2\text{-}sep}$)	568.1	560.4	–
Heat duty for CO_2 capture [GJ/h] ($\dot{Q}_{cap}$)	2,067	2,067	–
Heat duty for CO_2 capture per CO_2 mass [GJ/t] (q_{cap})	3.64	3.64	–

Table 3. Reference conditions and calculation results for IGCC case without CO_2 capture

	Reference (Case 5)	This study
Coal feed [t/h] ($\dot{m}_{coal}$)	205.305	205.305
Thermal input, LHV [GJ/h] ($\dot{Q}_{coal}$)	5,368,931	5,368,930
Gross power [kW] ($\dot{W}_{gross}$)	748,020	730,640
Aux load [kW] ($\dot{W}_{aux}$)	112,170	102,012
Net power [kW] ($\dot{W}_{net}$)	635,850	628,628
Gross efficiency [%] (η_{gross})	50.2	49.0
Net efficiency [%] (η_{net})	42.6	42.2
CO_2 generated [t/h] ($\dot{m}_{CO2\text{-}gen}$)	455.2	459.1
CO_2 captured [t/h] ($\dot{m}_{CO2}$)	0	0
Heat duty for CO_2 capture [GJ/h] ($\dot{Q}_{cap}$)	0	0
Heat duty for CO_2 capture per CO_2 mass [GJ/t] (q_{cap})	0	0

Table 4. Reference conditions and calculation results for IGCC case with pre-combustion CO_2 capture

	Reference (Case 6)	This study	w/o CCS
Coal feed [t/h] ($\dot{m}_{coal}$)	214.629	214.606	214,606
Thermal input, LHV [GJ/h] ($\dot{Q}_{coal}$)	5,612,763	5,612,170	5,612,170
Gross power [kW] ($\dot{W}_{gross}$)	693,555	675,845	675,845
Aux load [kW] ($\dot{W}_{aux}$)	176,420	166,359	142,071
Net power [kW] ($\dot{W}_{net}$)	517,135	509,486	533,774
Gross efficiency [%] (η_{gross})	44.5	43.4	43.4
Net efficiency [%] (η_{net})	33.2	32.7	34.2
CO_2 generated [t/h] ($\dot{m}_{CO2-gen}$)	477.0	479.2	479.2
CO_2 captured [t/h] ($\dot{m}_{CO2}$)	453.3	455.3	0
Heat duty for CO_2 capture [GJ/h] ($\dot{Q}_{cap}$)	0	0	0
Heat duty for CO_2 capture per CO_2 mass [GJ/t] (q_{cap})	0	0	0

3. Results

3.1 CO_2 Capture Energy (Heat)

3.1.1 PC Power Plant

The effect of steam extraction on efficiency of PC power plant was estimated. The steam conditions which entered to LP turbine was 414.9 °C, 0.949 MPa, 1,980,000 kg/h and its enthalpy $\dot{H}_{LP-in}$ was −25,003 GJ/h. The enthalpy will be −30,631 GJ/h when it was supplied to CO_2 capture process and cooled to 110 °C ($\dot{H}_{110}$). Thus the maximum CO_2 capture energy to the CO_2 capture process ($\dot{Q}_{cap-max}$) was calculated as;

$$\dot{Q}_{cap-max} = \dot{H}_{LP-in} - \dot{H}_{110} = 5,628 \text{ GJ/h} \tag{6}$$

CO_2 recovered amount ($\dot{m}_{CO2}$) of this condition was 568.6t/h, from Table 2. Therefore, maximum CO_2 capture energy per CO_2 weight $q_{cap-max}$ [GJ/t-CO_2] is;

$$q_{cap-max} = \dot{Q}_{cap-max} / \dot{m}_{CO2-sep} = 9.90 \text{ GJ/t-}CO_2 \tag{7}$$

Power generation by an LP turbine ($\dot{W}_{gen}$) was 339,160 kW, and it was 17.5% of total thermal input to this process ($\dot{Q}_{coal}$ = 6,958,520 [GJ/h] =1,932,922 [kW], LHV).

Efficiency penalty $\Delta\eta$ [%-point] was defined as the difference between the net efficiency with ($\eta_{net-w/CO2cap}$) and without CO_2 capture process ($\eta_{net-w/o\,CO2cap}$). In this condition, $\Delta\eta$ was 17.5% when heat duty for CO_2 capture was 9.90 GJ/t-CO_2. The calculations are summarized in Table 5. From the calculation results obtained above, the relation between CO_2 capture energy per CO_2 mass (q_{cap}) and efficiency penalty ($\Delta\eta$ [%]) was calculated as following equation;

$$\Delta\eta = (\Delta\eta_{max} / q_{cap-max}) \cdot q_{cap} = 1.75\,q_{cap} \tag{8}$$

Table 5. Steam and CO_2 capture conditions for PC power plant

Stream	LP inlet
Enthalpy [GJ/h] ($\dot{H}_{LP-in}$)	−25,003
Enthalpy at 110 °C [GJ/h] ($\dot{H}_{110}$)	−30,631
Maximum heat duty for CO_2 capture [GJ/h] ($\dot{Q}_{cap-max}$)	5,628
Maximum heat duty for CO_2 capture per CO_2 mass [GJ/t-CO_2] ($q_{cap-max}$)	10.0
Power generation by steam turbine [kW] ($\dot{W}_{gen}$)	339,160
Maximum efficiency penalty [%] ($\Delta\eta_{max}$)	17.5%
Constant of proportionality	1.75

3.1.2 IGCC Power Plant (Post-Combustion)

The effect of steam extraction on efficiency for post-combustion CO_2 capture for IGCC power plant (i.e., without water-gas shift reactor) was also examined, and the results are shown in Table 6. As shown in the table, the maximum CO_2 capture energy was 1,833 GJ/h under this condition. The CO_2 generation was 453.3 t/h from Table 3, therefore maximum CO_2 capture energy per CO_2 mass was 4.12 GJ/t-CO_2. If the heat duty exeed the value, steam should be supplied from the inlet of IP turbine. When steam is supplied from the inlet of IP turbine, maximum CO_2 capture energy per CO_2 mass is 4.95 GJ/t-CO_2.

Table 6. Steam and CO_2 capture conditions for IGCC power plant with post-combustion CO_2 capture process

Stream	IP inlet	LP inlet
Enthalpy [GJ/h] ($\dot{H}$)	−7,327	−7,649
Enthalpy at 110 °C [GJ/h] ($\dot{H}_{110}$)	−9,481	−9,482
Maximum heat duty for CO_2 capture [GJ/h] ($\dot{Q}_{cap\text{-}max}$)	2,154	1,833
Maximum heat duty for CO_2 capture per CO_2 mass ($q_{cap\text{-}max}$)	4.94	4.20
Power generation by steam turbine [kW] ($\dot{W}_{gen}$)	202,025	112,884
Efficiency penalty [%] ($\Delta\eta$)	13.5	7.57
Constant of proportionality	2.74	1.80

In this condition, the relation between CO_2 capture energy per CO_2 mass (q_{cap} [GJ/t-CO_2]) and efficiency penalty ($\Delta\eta$ [%]) is given by following equations.

$$\Delta\eta = 1.80\, q_{cap} \qquad (0 \leq q_{cap} \leq 4.20) \tag{9}$$

$$\Delta\eta = 2.74\, q_{cap} \qquad (4.20 < q_{cap} \leq 4.94) \tag{10}$$

3.1.3 IGCC Power Plant (Pre-Combustion)

CO_2 capture from pre-combustion CO_2 capture from IGCC (i.e., with water-gas shift reactor) was considered. Calculation results are shown in Table 7. The power generated in steam turbines was smaller than that of the post-combustion case because water-gas shift reactor consumes energy.

Table 7. Steam and CO_2 capture conditions for IGCC power plant with pre-combustion CO_2 capture process

Stream	IP inlet	LP inlet
Enthalpy [GJ/h] ($\dot{H}$)	−5,880	−6,128
Enthalpy at 110 °C [GJ/h] ($\dot{H}_{110}$)	−7,287	−7,287
Maximum heat duty for CO_2 capture [GJ/h] ($\dot{Q}_{cap\text{-}max}$)	1,406	1,159
Maximum heat duty for CO_2 capture per CO_2 mass ($q_{cap\text{-}max}$)	2.93	2.41
Power generation by steam turbine [kW] ($\dot{W}_{gen}$)	156,528	87,852
Efficiency penalty [%] ($\Delta\eta$)	10.0	5.64
Constant of proportionality	3.42	2.33

In this condition, the relation between CO_2 capture energy per CO_2 mass (q_{cap} [GJ/t-CO_2]) and efficiency penalty ($\Delta\eta$ [%-point]) is given by following equations.

$$\Delta\eta = 2.33\, q_{cap} \; (0 \leq q_{cap} \leq 2.41) \tag{11}$$

$$\Delta\eta = 3.42\, q_{cap} \; (2.41 < q_{cap} \leq 2.93) \tag{12}$$

3.2 CO₂ Capture Energy (Pressure)

As discussed in 2.4, there is a linear semi-log relation between CO_2 compression energy and CO_2 inlet pressure to the compression process. The CO_2 compression energy w_{CO2} can be converted to efficiency penalty $\Delta \eta$ for PC and IGCC (post and pre-combustion CO_2 capture) processes by following equation.

$$\Delta \eta = w_{CO2} \times (\dot{m}_{CO2} / \dot{Q}_{coal}) \times 100 \qquad (13)$$

The y-axis intercept and constant of proportionality of Equation (1) are converted as shown in Table 8.

Table 8. Slopes and y-intercepts of the relation between CO_2 inlet pressure P_{CO2} and compression energy w_{CO2} for PC and IGCCpower plants

	Slope	y-intercept
PC with post-combustion CO_2 capture	−0.765	1.69
IGCC with post-combustion CO_2 capture	−0.771	1.70
IGCC with pre-combustion CO_2 capture	−0.770	1.70

4. Discussion

4.1 CO₂ Capture Energy (Combination of Heat and Pressure)

The results obtained in 3.1 and 3.2 were combined and the efficiency penalty ($\Delta \eta$) was obtained as a function of CO_2 capture energy (q_{cap}) and CO_2 inlet pressure for compressors (P_{CO2}). For PC with post-combustion CO_2 capture, the relation was obtained from Equation (8) and Table 8, and is expressed as Equation (14)

$$\Delta \eta = 1.75 \, q_{cap} - 0.765 \ln(P_{CO2}) + 1.69 \qquad (14)$$

The relation is plotted in three-dimensional graph shown in Figure 7.

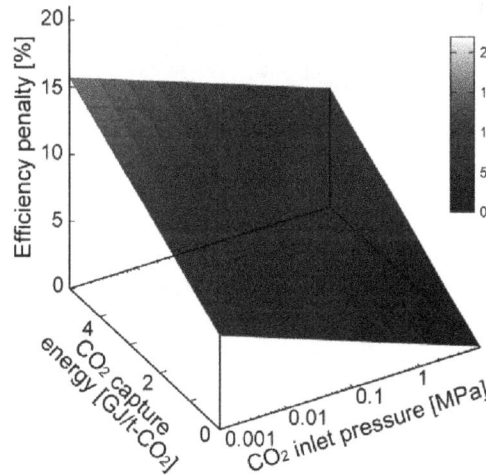

Figure 7. Relation between CO_2 capture energy (q_{cap}), CO_2 inlet pressure for compressors (P_{CO2}) and efficiency penalty ($\Delta \eta$) for PC with post-combustion CO_2 capture

For IGCC with post-combustion CO_2 capture case, the efficiency penalty ($\Delta \eta$) was obtained from Equation (9), (10) and Table 8, and is expressed as Equations (15), (16) and Figure 8.

$$\Delta \eta = 1.80 \, q_{cap} - 0.771 \ln(P_{CO2}) + 1.70 \quad (0 \leq q_{cap} \leq 4.21) \qquad (15)$$

$$\Delta \eta = 2.74 \, q_{cap} - 0.771 \ln(P_{CO2}) + 1.70 \quad (4.21 < q_{cap} \leq 4.95) \qquad (16)$$

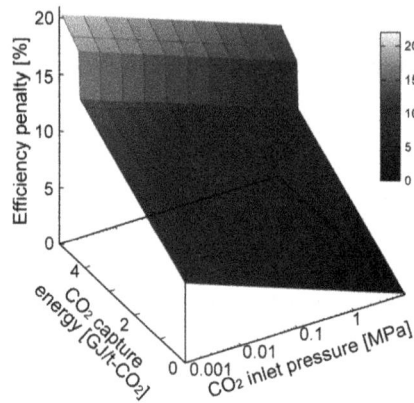

Figure 8. Relation between CO_2 capture energy (q_{cap}), CO_2 inlet pressure for compressors (P_{CO2}) and efficiency penalty ($\Delta\eta$) for IGCC with post-combustion CO_2 capture

For IGCC with pre-combustion CO_2 capture case, the efficiency penalty ($\Delta\eta$) was obtained from Equations (11), (12) and Table 8, and is expressed as Equations (17), (18) and Figure 9.

$$\Delta\eta = 2.33\ q_{cap} - 0.770\ \ln(P_{CO2}) + 1.70 \quad (0 \leq q_{cap} \leq 2.41) \tag{17}$$

$$\Delta\eta = 3.42\ q_{cap} - 0.770\ \ln(P_{CO2}) + 1.70 \quad (2.41 < q_{cap} \leq 2.93) \tag{18}$$

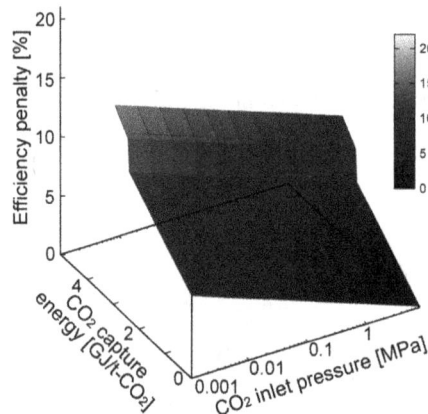

Figure 9. Relation between CO_2 capture energy (q_{cap}), CO_2 inlet pressure for compressors (P_{CO2}) and efficiency penalty ($\Delta\eta$) for IGCC with pre-combustion CO_2 capture

4.2 Comparison With Other Studies

The comparison of efficiency penalty reported in other studies with calculation by this study is summarized in Table 9. The calculated results show a good agreement with the reference data.

Table 9. CO_2 capture conditions and efficiency penalty of literatures and calculation results

Reference	CO_2 capture Technology	CO_2 capture energy [GJ/t-CO_2]	Inlet pressure [MPa]	Efficiency penalty [%-points]	This study [%-point]	Note[*]
NETL, 2007	selexol	0	1.2	1.6	1.6	(b)
NETL, 2007	Econamine	3.6	0.16	9.1	9.4	(a)
Dave, 2011	MEA	4.0	0.16[**]	11.1	10.1	(a)
Stöver, 2011	MEA	3.6	0.16[**]	8.0-9.8	9.4	(a)
Stöver, 2011	H3	2.8	0.16[**]	9.9-11.2	8.0	(a)

[*]: (a) PC with post combustion CO_2 capture, (b) IGCC with pre combustion CO_2 capture.

[**]: Estimated value from the stripper temperature.

4.3 The Effect of Water-Gas Shift Reaction on Efficiency in IGCC

The difference in efficiency of post- and pre-combustion CO_2 capture was considered. The difference in those processes was the existence of water-gas shift reaction. There was 5.6%-points of difference in the gross efficiencies of without CO_2 capture (Table 3) and pre-combustion CO_2 capture (Table 4) case. This was attributed to the loss of water-gas shift reaction. In case 6 or pre-combustion case, the compression energy of CO_2 was 24,288 kW, which corresponds to 0.4%-points of efficiency penalty. Therefore the total efficiency penalty by CO_2 capture and compression will be 6.0%-points. Therefore, Equation (15) will be;

$$q_{cap} \leq 0.428 \ln(P_{CO2}) + 2.39 \quad (0 \leq q_{cap} \leq 4.21 \text{ or } 0.00378 \leq P_{CO2} \leq 70.2) \tag{19}$$

By solving the equation, the condition for the case 5 which would be lower than the efficiency penalty in Case 6 was obtained as Figure 10. Therefore, it can be concluded that the efficiency of IGCC with post-combustion CO_2 capture will be better than that of IGCC pre-combustion process if the CO_2 capture process which satisfies the painted region in Figure 10 of CO_2 capture energy and CO_2 inlet pressures to CO_2 compressor.

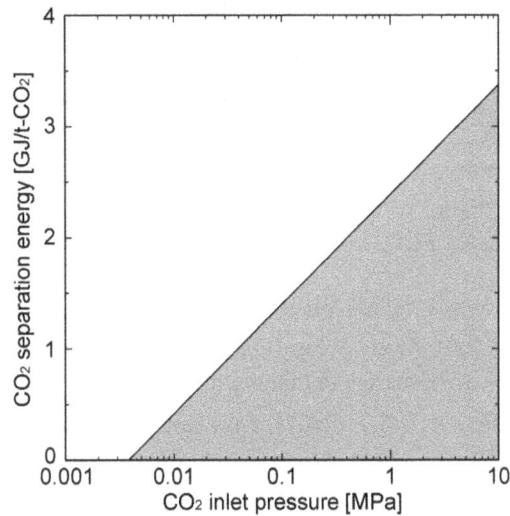

Figure 10. Relation between CO_2 capture energy (q_{cap}), CO_2 inlet pressure for compressors (P_{CO2}) and efficiency penalty ($\Delta\eta$) for IGCC with pre-combustion CO_2 capture

5. Conclusion

The effects of carbon dioxide capture process on output efficiency of fossil power plants were investigated aiming to obtain a performance guideline of CO_2 capture process for coal-fired power plants. The power plant models were developed by using a commercial process simulator Aspen Plus 7.3 and plant data reported from NETL. It was found that the efficiency penalty $\Delta\eta$ [%-point] can be estimated from the CO_2 capture energy (q_{cap} [GJ/t-CO_2]) and CO_2 inlet pressure to the CO_2 compressors (P_{CO2} [MPa]) by the following equations;

for PC power plant,

$\Delta\eta = 1.75\, q_{cap} - 0.765 \ln(P_{CO2}) + 1.69$

for IGCC power plant with pre-combustion CO_2 capture,

$\Delta\eta = 1.80\, q_{cap} - 0.771 \ln(P_{CO2}) + 1.70 \quad (0 \leq q_{cap} \leq 4.21)$ or

$\Delta\eta = 2.74\, q_{cap} - 0.771 \ln(P_{CO2}) + 1.70 \quad (4.21 < q_{cap} \leq 4.95)$

for IGCC power plant with post-combustion CO_2 capture,

$\Delta\eta = 2.33\, q_{cap} - 0.770 \ln(P_{CO2}) + 1.70 \quad (0 \leq q_{cap} \leq 2.41)$ or

$\Delta\eta = 3.42\, q_{cap} - 0.770 \ln(P_{CO2}) + 1.70 \quad (2.41 < q_{cap} \leq 2.93)$.

The calculated results were compared with the reference data, and they showed a good agreement.

For IGCC power plants, the effect of the installation of water-gas shift reactor on the efficiency penalty was investigated. The net efficiency was reduced 5.4 points by installing water-gas shift reactor. It was calculated that the efficiency of IGCC with post-combustion CO_2 capture will be better than that of IGCC pre-combustion process if the CO_2 capture process which satisfies the following condition;

$q_{cap} \leq 0.428 \ln(P_{CO2}) + 2.39$ ($0 \leq q_{cap} \leq 4.21$ or $0.00378 \leq P_{CO2} \leq 70.2$)

References

Abu Zahra, R. M. M., Fernandez, S. E., & Goetheer, L. V. E. (2011). Guidelines for Process Development and Future Cost Reduction of CO_2 Post-Combustion Capture. *Energy Procedia, 4,* 1051-1057. http://dx.doi.org/10.1016/j.egypro.2011.01.154

Cifre, G. P., Brechtel, K., Hoch, S., García, H., Asprion, N., Hasse, H., & Scheffknecht, G. (2009). Integration of a chemical process model in a power plant modelling tool for the simulation of an amine based CO_2 scrubber. *Fuel, 88,* 2481-2488. http://dx.doi.org/10.1016/j.fuel.2009.01.031

Dave, N., Do, T., Palfreyman, D., & Feron, P. M. H. (2011). Impact of liquid absorption process development on the costs of post-combustion capture in Australian coal-fired power stations. *Chemical Engineering Research and Design, 89,* 1625-1638. http://dx.doi.org/10.1016/j.cherd.2010.09.010

Goto, K., Yogo, K., & Higashii, T. (2013). A review of efficiency penalty in a coal-fired power plant with post-combustion CO_2 capture. *Applied Energy, 111,* 710-720. http://dx.doi.org/10.1016/j.apenergy.2013.05.020

Intergovernmental Panel on Climate Change (IPCC). (2005). *IPCC Special Report on Carbon Dioxide Capture and Storage.* New York, NY: Cambridge University Press. Retrieved from http://www.ipcc.ch/pdf/special-reports/srccs/srccs_wholereport.pdf

Intergovernmental Panel on Climate Change (IPCC). (2013). *Climate Change 2013 The Physical Science Basis Working Group I Contribution to the Fifth Assessment Report of the Intergovernmental Panel on Climate Change Summary for Policymakers.* Retrieved from http://www.climatechange2013.org/images/uploads/WGI_AR5_SPM_brochure.pdf

International Energy Agency (IEA). (2009). *World Energy Outlook. Paris, France. International Energy Agency (IEA).* Retrieved from http://www.worldenergyoutlook.org/media/weowebsite/2009/WEO2009.pdf

National Energy Technology Laboratory (NETL). (2007). *Cost and Performance Baseline for Fossil Energy Plants DOE/NETL-2007/1281.* Retrieved from http://www.netl.doe.gov/energy-analyses/pubs/BitBase_FinRep_2007.pdf

Stöver B., Bergins, C., & Klebes, J. (2011). Optimized Post Combustion Carbon capturing on Coal fired Power Plants. *Energy Procedia, 4,* 1637-1643. http://dx.doi.org/10.1016/j.egypro.2011.02.035

Strube, R., Pellegrini, G., & Manfrida, G. (2011). The environmental impact of post-combustion CO_2 capture with MEA, with aqueous ammonia, and with an aqueous ammonia-ethanol mixture for a coal-fired power plant. *Energy, 36,* 3763-3770. http://dx.doi.org/10.1016/j.energy.2010.12.060

11

Quantitative Impacts of Solar PV on Television Viewing and Radio Listening in Off-grid Rural Ghana

George Yaw Obeng[1] & Ebenezer Nyarko Kumi[2]

[1] Technology Consultancy Centre, College of Engineering, Kwame Nkrumah University of Science and Technology, Kumasi, Ghana

[2] The Energy Center, College of Engineering, Kwame Nkrumah University of Science and Technology, Kumasi, Ghana

Correspondence: George Yaw Obeng, Technology Consultancy Centre, College of Engineering, Kwame Nkrumah University of Science and Technology, Kumasi, Ghana. E-mail: geo_yaw@yahoo.com

Abstract

The use of solar photovoltaic (PV) for powering electronic devices such as radio and television can contribute to increase access to information and entertainment in off-grid rural communities. However, there is a lack of quantitative data on impact of solar PV electrification on television viewing and radio listening. This paper relied on primary data from cross-sectional surveys of solar-electrified and non-electrified households in rural Ghana using questionnaires which were developed into a database. The study results showed that solar-electrified households could view television for 2.5 hours/day, while in non-electrified households it was 1.5 hours/day. The avoided cost of television viewing using solar PV instead of car battery was US$ 1-3/month. The study found a linear relationship between incomes above US$ 1.08/day and television ownership. Further, the results showed that on average radio listening in solar-electrified households was 5 hours/day, while in non-electrified households it was 6.3 hours/day. The avoided cost of radio listening using solar PV instead of drycell batteries was US$ 1.08/month. We conclude that the difference in the results suggests an overall impact of solar PV on television viewing and radio listening. Once quantitative data are made available, the decision to use solar PV for off-grid electrification will be apparent.

Keywords: off-grid electrification, quantitative data, fee-for-service, avoided cost, Ghana

1. Introduction

It is globally accepted that electrification does not only stimulate economic growth at a broader level but also can enhance quality of life at the household level. It is generally agreed that the immediate benefit of solar phovoltaic (PV) electrification comes through improved lighting. All the same, in off-grid rural communities, the use of solar PV for powering electronic devices such as radio and television can significantly contribute to increase access to information and entertainment at the household level and this is reported in several studies (Wamukonya & Davies, 2001; Obeng & Evers, 2009; Bahauddin & Salah, 2010). But, there appears to be a lack of data on the quantitative impacts of solar PV on hours of usage and costs of television viewing and radio listening, particularly in off-grid rural communities, where access to electricity is relatively low.

While the electricity access of Ghana was about 72% in 2011, rural access was about 35% (Ministry of Energy and Power, 2013). For communities without access to the national grid, off-grid electrification using solar home systems, biogas, wind etc are being explored as alternative technologies (Ministry of Enery, 2010; Ghana Publishing Company, 2011). Since the technology radar of Ghana demonstrates the inclusion of solar PV, there is the need to quantify the potential contribution that can be derived from its application. The main objective of this paper therefore was to analyse the quantitative impacts of solar PV electrification on television viewing and radio listening focusing on hours of usage, costs of usage, income and ownership, avoided costs and any other relevant factors.

2. Methods

Cross-sectional survey was the main data collection method. The survey was conducted in sixteen rural communities located in seven districts of six regions in Ghana: Northern, Upper East, Upper West, Brong Ahafo, Volta and Greater Accra regions. The study locations were off-grid rural communities with relatively high poverty incidence (Ghana Statistical Service, 2002; World Bank, 2003). They include: Kpentang, Kpenbung, Kambatiak, Bamong, Kintango, Chintilung, Tojing, Gbetmanpaak, Jimbali, Najong No. 1 and Pagnatik in Bunkpurungu Yunyoo district (Northern region); Kpalbe in East Gonja district (Northern region); Tengzuk in Talensi-Nabdam district (Upper East region); Wechiau in Wa-West district (Upper West region); Nkoranza in Nkoranza district (Brong Ahafo region), Kpassa in Nkwanta district (Volta region); and Apollonia in Tema district (Greater Accra region). Pre-testing of the questionnaires was carried out in the Nkoranza district of Brong-Ahafo region. Figure 1 is the map of Ghana showing the study regions and electricity access in 2010. The entire land surface of Ghana receives solar radiation ranging between 4.4 kWh/m²-day and 5.6 kWh/m²-day and sunshine duration of about 1800-3000 hours per year (Forson et al., 2004; Energy Commission, 2009; Kemausuor et al., 2011).

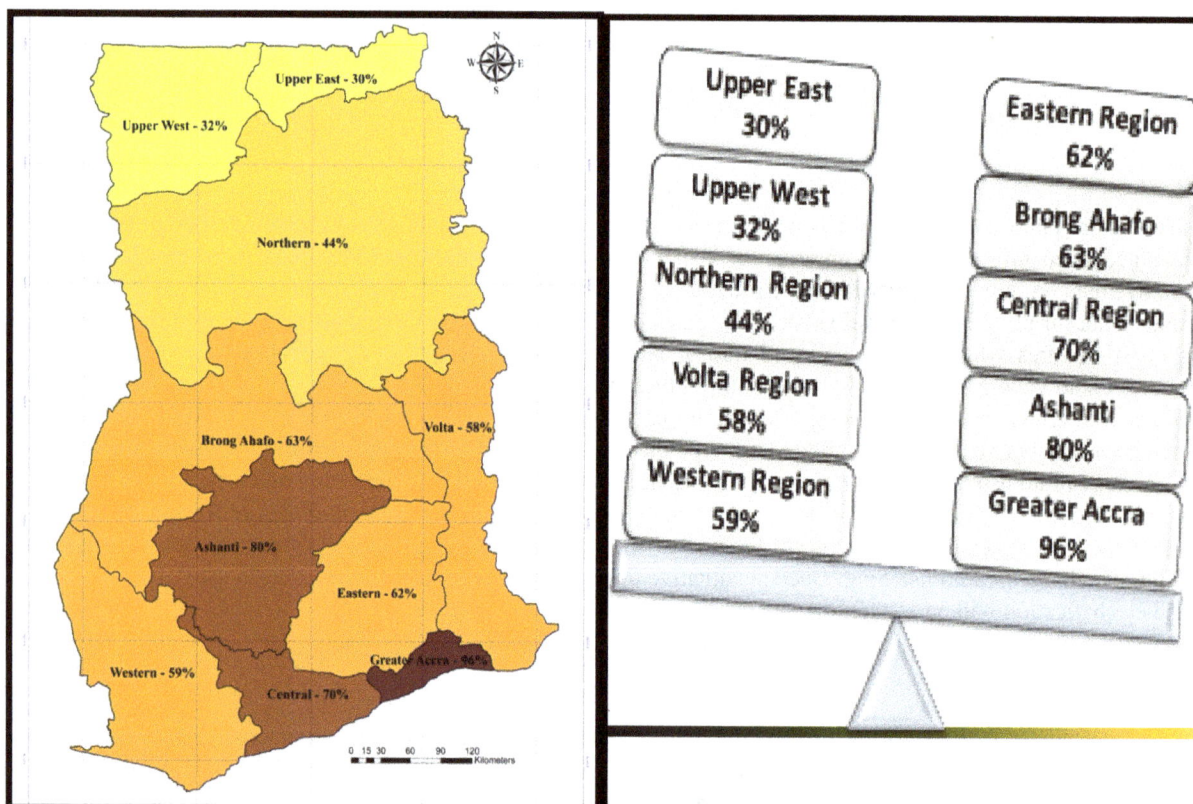

Figure 1. Map of Ghana showing the study regions and electricity access rate in 2010

A total of 209 household heads were randomly selected for the study: 96 solar-electrified households and 113 non-electrified households. In each of the regions research assistants from the Kwame Nkrumah University of Science and Technology, Kumasi who speak the local language were engaged in the administration of the questionnaires. A questionnaire contained 192 variables including indicators on demographic and socio-economic impacts, technology functionality and environmental impacts. A list of project beneficiaries (solar-electrified households) and incoming clients (non-electrified households) were used to select the households in a systematic sampling.

From the lists of the beneficiaries of solar PV electrification projects in rural Ghana, solar home systems (50Wp and 100Wp of both polycrystalline silicon and amorphous silicon module types) that have been operational for over three years and had not been earmarked by the Ministry of Energy for relocation were selected. This criterion was based on the assumption that over a three-year period, PV systems and components (car battery, regulator and fluorescent lamp) would have gone through a cycle of operation and maintenance and end-users would have learned some impact lessons. With an estimated average daily sunshine hours of 6 hours, each of the 50Wp and

100Wp panels could generate 300Wh and 600Wh of energy daily respectively (Energy Commission, 2009). Incoming clients (non-electrified households) were used as the comparison or control group because their lists were available for systematic sampling and they also appeared to be similar to the beneficiary group than random selected non-beneficiaries. The purpose of the questionnaire was to gather ex-post information that indicates change in the responses between households with and without solar PV.

2.1 Underlying Assumption and Data Analysis

The underlying assumptions that governed the interpretation of the study results were as follows:

- In the absence of the solar PV, solar-electrified households would have depended on car batteries and drycell batteries like the non-electrified households.

- The analysis relied on fee-for-service of US$ 1.63/month for a 50Wp solar home system and US$ 2.72/month for 100Wp system in the surveyed communities.

- A zero transportation cost was assumed in the analysis of non-electrified households because car batteries were sometimes carried on bicycles or transported free-of-charge to nearby towns for recharging.

- Access to solar PV might have contributed to the difference (or change) in response between the two groups (solar-electrified and non-electrified).

Using central tendency measures, dispersion and cross-tabulation the variables of interest were analysed in the relationship between solar-electrified and non-electrified. The difference in the responses between the two groups provided the basis for explaining how solar PV electrification has been responsible for the observed differences. Significant differences were analysed using statistics with p-values ($p<0.05$) considered statistically significant.

3. Results

3.1 Electrification Status and Television Viewing

Television provides access to information and entertainment to improve quality of life. The results in Table 1 revealed that, in the solar-electrified households, about 22% used solar PV to power their television, while in the non-electrified households (Note 1), about 5% used car battery to watch television and only 1% used a diesel generator. An overwhelming majority of 94% of the non-electrified households did not have access to television. A two-sided chi-square asymptotic significance (Sig. = 0.000), indicated significant difference in access to television between the two groups.

Table 1. Access to television by household

Access to TV	No Access	Access by Solar PV	Access by car battery	Access by generator**	Total
Solar-electrified household	73 (76%)	21 (22%)	2 (2%)	-	96 (100%)
Non-electrified household	106 (94%)	-	6 (5%)	1 (1%)	113 (100%)
Total	179 (86%)	21 (10%)	8 (3.5%)	1 (0.5%)	209 (100%)
	Pearson Chi-Square = 28.892		df = 3	Asymp. Sig. = .000	

Note: **Using "with" and "without" comparative method, the 1% user of diesel were placed under non solar-electrified (non-electrified households)

The study results presented in Table 2 revealed that on average, solar-electrified households could view television for 2.5 hours/day (median), while non-electrified households who used car batteries could view television for an average of 1.5 hour/day (median).

Table 2. Number of hours of watching television per day

	Mean	Median	Std. Deviation	Min.	Max.	n
Solar-electrified						
Farming	2.05	2.00	0.63	1	3.5	69
Teaching	2.20	3.00	0.45	1.5	3.0	12
Public Service	1.50	1.50	0.75	0.5	2.5	10
Social Worker	2.10	2.00	0.65	1.5	3.0	5
(Average)	2.25	2.50	-	-	-	
TOTAL						96
Non-electrified						
Farming	1.20	1.40	0.38	0.5	1.5	78
Artisan	1.30	1.30	0.54	0.5	2.0	10
Teaching	1.50	1.50	0.53	1	2	10
Public Service	1.00	1.00	0	1	1	10
(Average)	1.00	1.50	-	-	-	
TOTAL	1.20	1.34				108

3.2 Ownership of Television Because of Energy Services and Income

The study further examined whether the surveyed households owned their television because of the availability of energy services - solar PV, car battery, or generator. The results in Table 3 showed that about 12% of the solar-electrified households owned television because of the availability of solar PV. Households that acquired television due to the availability of solar PV were headed by people (4.2%) whose incomes fell within US$ 1.08–2.17/day, and household heads (7.3%) whose income are likely to exceed US$ 2.17/day. Non-electrified households that acquired television as a result of the availability of car battery were headed by people (1 percent) whose income fell between US$ 1.08–2.17/day. None of the households headed by people with income below US$ 1.08/day owned a television. The low significance values of 0.001 ($p<0.05$) for both Goodman & Kruskal Tau and uncertainty coefficients, indicate an association between income levels above US$ 1.08/day and television ownership.

3.3 Avoided Cost of Using Solar PV for Television Instead of Car Battery

The study analysed the avoided cost of using solar PV for television instead of car battery. Adapted from Maine State Planning Office, the avoided cost represents the amount of money that is not spent (therefore 'saved') by a household to recharge a car battery when solar PV option exists. According to Schmidt (2014) when an action prevents a future cost, the result is called an avoided cost if it is reasonably certain that the cost would have appeared without the action

For this purpose two indicators were analysed: (1) number of hours of television viewing per day; and (2) monthly costs of viewing television with solar PV and with car battery. Table 5 showed the average hours of television viewing per day and the cost avoided by the use of solar PV instead of a car battery. The results indicated that on average, solar-electrified households gained about 1.0 hour/day (30 hours/month) of television viewing than non-electrified households, who use car batteries. To determine the monthly cost of viewing television, the data presented in Table 4 were used to estimate the daily energy consumption and hence the cost of viewing television with a 50Wp solar PV.

Table 3. Assets owned because of energy services and income

Expenditure (Note 2) /Income Per day (USD)	Assets owned because of energy services	Solar-electrified Household		Non-electrified Household		Total	
Up to US$1.08	Radio	2	2.0%	5	4.4%	7	3.2%
US$1.08 – 2.17		4	4.2%	5	4.4%	9	4.3%
>US$ 2.17		7	7.3%	6	5.4%	13	6.2%
Up to US$1.08	Television	0	0	0	0	0	0
US$1.08 – 2.17		7	7.3%	1	0.9%	8	3.8%
>US$ 2.17		4	4.2%	0	0	4	1.9%
Up to US$1.08	Tape player	4	4.2%	3	2.7%	7	3.2%
US$1.08 – 2.17		11	11.4%	7	6.1%	18	8.6%
>US$ 2.17		12	12.5%	0	0	12	5.7%
Up to US$1.08	Table lamp	1	1.0%	0	0	1	0.5%
US$1.08 – 2.17		0	0	0	0	0	0
>US$ 2.17		0	0	0	0	0	0
Up to US$1.08	Fan	0	0	0	0	0	0
US$1.08 – 2.17		0	0	0	0	0	0
>US$ 2.17		1	1.0%	0	0	1	0.5%
Up to US$1.08	None	10	10.4%	32	28.4%	42	20.1%
US$1.08 – 2.17		17	17.8%	33	29.2%	50	24.0%
>US$ 2.17		13	13.5%	18	15.9%	31	14.8%
Up to US$1.08	Other	0	0	0	0	0	0
US$1.08 – 2.17		1	1.0%	0	0	1	0.5%
>US$ 2.17		2	2.0%	3	2.7%	5	2.4%
Total		96	100%	113	100%	209	100%

US$ 1.08 – 2.17 per day
Goodman & Kruskal Tau value = 0.441 Sig. = .002; Uncertainty coeff = 0.266 Sig.=0.001
US$ 2.17 per day
Goodman & Kruskal Tau value = 0.249 Sig. = 0.008; Uncertainty coefficient = 0.253 Sig.=0.001

Table 4. Daily energy consumption of appliances (50Wp solar PV household)

Load	Daily Use		Wattage		Total Energy Consumption (watt-hrs)
Radio	5 hours	x	20 watts	=	100
Television (black & white)	2.25 hours	x	30 watts	=	67.5
Lighting	6 hours	x	2 x 7 watts	=	84
Total Daily Energy Consumption					251.5 watt-hrs

Using a monthly fee of US$ 1.63 being the fee-for-service for a 50Wp solar home system in the surveyed communities, the cost of television viewing was calculated as: [(67.5 watt-hrs/day) ÷ (251.5 watt-hrs/day)] x US$1.63 = US$ 0.44. On average the non-electrified households recharged their car batteries every week mainly for television viewing and they paid a fee of about US$ 0.54-1.00 per week. A zero transportation cost is also assumed because car batteries were sometimes carried on bicycles or transported free-of-charge to nearby towns for recharging. In Table 4 the avoided cost of television viewing by a household using solar PV instead of car battery is demonstrated.

Table 5. Avoided cost of T.V Viewing by solar PV instead of car battery

	Solar-electrified Households (with solar PV)	Non-electrified Households (with car battery)
No. of hours of TV viewing per day (black/white)	2.0 - 2.5	1.0 - 1.5
Average hours of TV viewing per month	2.25 x 30 = 67.5	1.25 x 30 = 37.5
Cost of battery charging for TV viewing per week	-	US$ 0.5-0.6
Monthly costs of viewing TV (with solar /with car battery)	US$ 1.00	US$ 2.17
Cost of TV viewing per hour	US$ 0.015	US$ 0.058
Avoided cost per hour	US$ 0.043	
Avoided cost per day (@ 2.25 hours/day)	US$0.10-0.117	
Avoided cost per month	US$ 3-3.51	

3.4 Electrification Status and Energy Device for Radio

In off-grid rural communities radios are normally powered by dry-cell batteries, car batteries, solar PV, or generators. In Table 6 the proportions of energy services used for powering radios in the surveyed communities are shown. While the non-electrified households relied mainly on dry-cell batteries (81%) and car batteries (4%); solar-electrified households used mainly solar PV systems (69%) and dry-cell batteries (24%). Only 11 % of the respondents did not use a radio. In all about 89% used radio.

Table 6. Type of energy services for powering radio in households

Type	Solar-electrified	Non-electrified	Total Respondents
No radio	7 (7%)	16 (14%)	23 (11%)
Solar PV for radio	66 (69%)	-	66 (31%)
Drycell battery for radio	23 (24%)	92 (81%)	115 (55%)
Generator for radio	-	1 (1%)	1 (1%)
Car battery for radio	-	4 (4%)	4 (2%)
Total	96 (100%)	113 (100%)	209 (100%)

Note: Some of the percentages do not add up to 100 because they were rounded up.

3.5 Number of Hours of Radio Listening

In this analysis the number of hours of radio listening per day is used as a proxy for measuring the impact on households' radio information acquisition. The assumption is that consumers benefit if they obtain more listening time at a lower cost per hour. The results in Table 7 indicated that on average radio listening in solar-electrified households was 5 hours/day (mean), while in non-electrified households it was 6.31 hours/day (mean). A significant value of 0.001 ($p<0.05$) indicates a significant difference in radio usage per day between households with and without solar PV.

Furthermore, by introducing the type of occupation as a layer (intermediary) variable, the results showed that on average, farmers in non-electrified households listened about 2 hours/day more than farmers in solar-electrified households. Teachers in non-electrified households listened about 2 hours/day more than teachers in solar-electrified households. In non-electrified households, artisan(s) listened to radio for about 12 hours/day. There was no significant difference in the radio usage of charcoal burners in the two groups of households. Data for public workers were inconsistent. A significant value of 0.000 ($p<0.05$) for the linearity and association measures, indicate a likely linear association between occupation and radio usage per day.

Table 7. Hours of radio listening per day by occupation

	Mean	Median	Std. Deviation	Min.	Max.	n
Solar-electrified						
Farming	5.00	5.00	1.67	3	9	69
Teaching	5.00	4.00	3.95	1	11	12
Public Service	5.00	5.00	0.00	5	5	10
Social Worker	5.00	5.00	0.00	5	5	5
(Average)	5.00	5.00	-	3.5	7.5	
TOTAL						96
Non-electrified						
Farming	7.90	7.00	3.23	5	19	78
Artisan	12.20	12.00	1.75	10	15	10
Teaching	7.00	7.00	0.00	7	7	10
Public Service	9.50	9.00	1.51	8	12	10
Charcoal Burner	6.20	5.00	2.77	3	11	5
(Average)	6.31	7.00	-	6.6	12.8	
TOTAL						113

3.6 Hours and Costs of Radio Listening and Television Viewing

In off-grid rural communities, dry-cell batteries are among the major sources of energy for powering radio for entertainment and information. Though radio provides valuable information on health, education, politics etc to improve the quality of life of families and individuals, its recurrent cost can be relatively high if one depends solely on dry-cell batteries.

The results in Table 8 indicated a difference in average expenditure on dry-cell of about US$ 1.0 per month being the amount avoided by using household solar PV instead of depending solely on dry-cell batteries for radio listening. Knowing the average hours of radio listening per day, the number of listening hours per month was calculated. Dividing the total cost of dry-cell used per month in both households by radio listening hours per month, the cost per listening hour was relatively less in the solar-electrified households than non-electrified households (Table 9). A significance value of 0.000 (p<0.05), indicate a statistically significant difference between the households with and those without solar PV. Table 10 and Table 11 provide summary data on hours and costs of television viewing and radio listening.

Table 8. Total cost of dry-cell batteries used per month by households

	Mean	Median	Std. Deviation	Minimum	Maximum	N
Solar-electrified	US$ 2.70	US$ 1.77	2.24	0.00	8.48	84
Non-electrified	US$ 3.78	US$ 3.26	2.54	0.33	10.09	98
Total	US$ 3.28	US$ 2.61	2.46	0.00	10.09	182
F-value = 9.296 df = 1 Sig. = 0.003						

Table 9. Price and quantity of radio listening in surveyed households

	Hrs/day	Hrs/month	Cost/month	Cost/hr	Assumption (average)
Solar-electrified	5.00	150	(US$ 2.70)	(US$ 0.018)	Cost per listening hour using solar PV
Non-electrified	6.31	189	(US$ 3.78)	(US$ 0.020)	Cost per listening hour using drycell radio

Table 10. A matrix of hours and costs of television viewing

	Solar-electrified household	Non-electrified household
Hours (mean) of television viewing per day	2.25	1.25
Hours (median) of television viewing per day	2.50	1.50
Cost of television viewing per hour	US$ 0.015	US$ 0.058
Avoided cost per month of television viewing using solar PV instead of car battery	US$ 3.00	

Table 11. A matrix of hours and costs of radio listening

	Solar-electrified household	Non-electrified household
Hours (mean) of radio listening per day	5	6.31
Hours (median) of radio listening per day	5	7
Cost of radio listening per hour	US$ 0.018	US$ 0.020
Cost of radio listening per month	US$ 2.70	US$ 3.78
Avoided cost per month of radio listening using solar PV instead of drycell battery	US$ 1.08	

In Figure 2 the costs of television viewing with and without solar PV are analysed. The results showed relatively low cost of viewing televison with solar PV compared to that of using car battery or a few cases using a generator. Again in Figure 3 the costs of radio listening using solar PV electricity were relatively low compared to that of using drycell batteries. Whereas solar PV systems have components to regulate usage and hence household users have limitations with respect to hours of usage. In the case of non-electrified households users usually contnue using their radio until their batteries gradually get discharged, hence they could have longer hours of usage that resulted in higher cost.

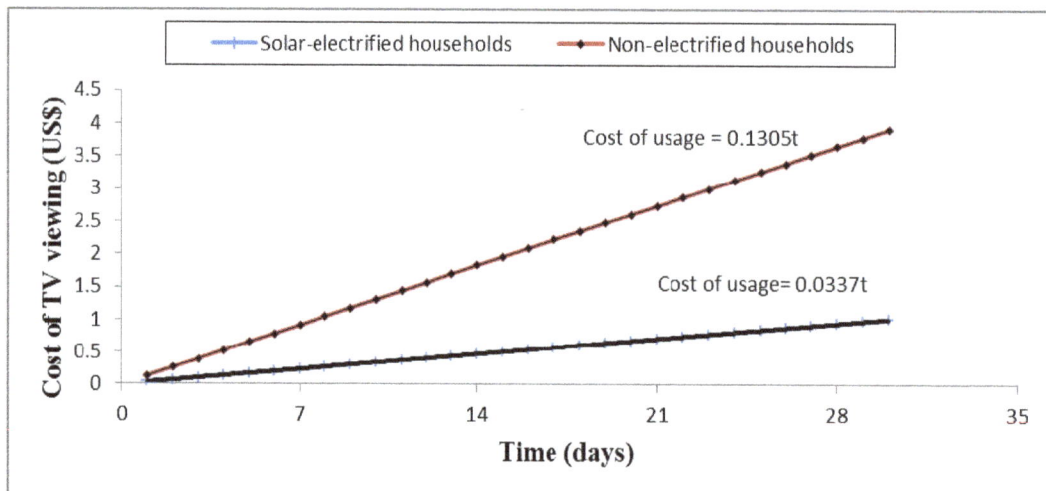

Figure 2. Costs of television viewing in solar-electrified and non-electrified households

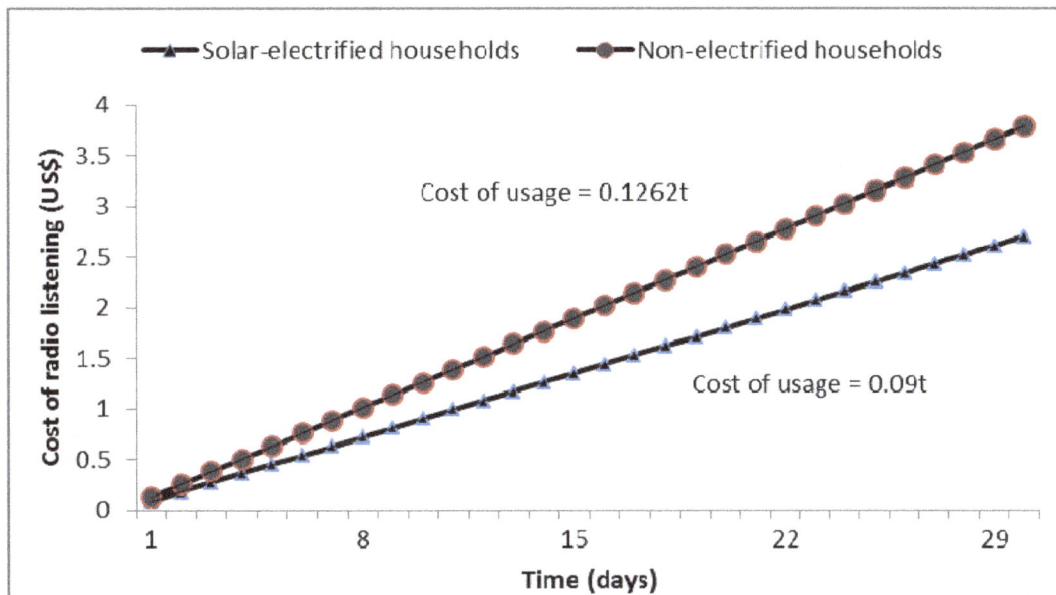

Figure 3. Costs of Radio Listening in Solar-electrified and Non-electrified Households

4. Discussion

With regard to television viewing, the results revealed that the proportion of solar-electrified households who used television was about four times that of the non-electrified households. The results further revealed that the availability of solar PV in the households was an important factor that influenced ownership of television. This finding confirms a study on the socio-economic impact of rural electrification in Namibia, which found that both grid and solar PV electrification enabled more households to own television and this was an important benefit (Wamukonya & Davies, 2001). Assets ownership is considered as a measure of wellbeing and living standards of a household or an individual (Ghana Statistical Service, 2005 a,b,c,d).

Household income emerged as a factor, which influenced television ownership. The results revealed that in both solar-electrified and non-electrified households, where televisions were used the incomes of the household heads exceeded US$ 1.08/day. None of the households headed by people with income below US$ 1.08/day owned a television. The study found a linear relationship between income levels above US$ 1.08/day and television ownership. The association between television ownership and higher income is reported by other studies (Madon & France, 2003). This result suggests that though the availability of solar PV enables households to have access to electricity for television viewing, the income factor is likely to eliminate the poor from the benefit of access to solar PV electricity for television information and entertainment. Households headed by the poor whose income fell below the US$ 1.08 (Note 3) /day poverty line (Reddy & Minnoiu, 2006; Zeller, 2004), are less likely to own a television and hence may receive relatively less information and entertainment through television viewing.

In terms of monetary benefits, the study found that in the surveyed communities solar-electrified households could avoid a monthly expenditure of about US$ 3.00 per household, being the estimated monthly cost of charging a wet battery for television viewing. According to Plastow and Goldstone (2001) rural households in developing countries typically spend between US$3 and US$20 per month on kerosene, candles, or other energy products. It is reported that in Sri Lanka and Indonesia, recurrent costs on kerosene, candles and batteries could reach $10-$30 per month (Cabraal et al., 1996).

Additionally, the solar-electrified households gained more television viewing hours (1 hour/day) or 30 hours/month than non-electrified households, who depended mostly on car batteries. These findings suggest that in off-grid rural communities: first, households with solar PV are likely to gain more hours of television viewing time; and second, spend relatively less money on the energy source for television viewing than households without, and hence achieve greater benefits and impact. Solar-powered television benefits rural households by providing access to health, education, business and environmental information to improve their standard of living. Rural solar PV electrification programmes must therefore be linked to programmes aimed at providing knowledge and information on agriculture, health, education etc. for quality of life improvements. Public programmes that help to expand access to and proper utilization of basic amenities assist in reducing poverty (Ghana Statistical Service, 2001).

Dry-cell batteries and solar PV were the two main energy carriers used for powering radio. However, the estimated consumption of dry-cell batteries differed between the solar-electrified and non-electrified households. The results revealed that the non-electrified households consumed more radio listening hours (mean of 6.3 hours) than solar-electrified households (mean of 5 hours) and hence paid more price per listening hour of radio. Average radio listening of 6.8 hours on weekdays and 6.4 hours on weekends were reported in a study in rural Kenya (Gathigi, 2009). High dependence on non-rechargeable dry-cell batteries is associated with high recurring cost. The impact of solar PV electrification on radio information is much felt if it improves listening hours at a lower cost compared to the use of dry-cell batteries. For this reason the number of hours of listening to radio was used as a proxy indicator for measuring impact on radio information acquisition.

Knowing the total hours of radio listening per month and the monthly cost of dry-cell batteries, the costs per hour of households with and without solar PV were compared. The results indicated a lower price per hour of radio listening in solar-electrified households than non-electrified households. In dollar equivalent, the costs per listening hours of radio obtained in both solar-electrified and non-electrified households are fairly consistent with the results of (Barnes, 2002); though there were little variations. The results suggest that rural households using solar PV electricity to power their radio are likely to benefit from lower cost per listening hour. This is a significant benefit, which can contribute to household savings to improve quality of life.

5. Conclusion

The study had shown that solar-electrified households achieved greater benefits by gaining more hours of television viewing and at the same time spent relatively less money compared to the non-electrified households. The results further demonstrated that the costs per hour of radio listening and television viewing were significantly lower in the solar-electrified households than non-electrified households who used dry-cells for radio listening and car batteries for television viewing. It was found that the availability of solar PV in households greatly influenced television ownership. Furthermore, the study revealed a linear relationship between household incomes above the international poverty line and television ownership. The difference in the responses provided by households with and without solar PV suggests an overall impact of solar PV on television viewing and radio listening. We conclude that once quantitative data are made available, the decision on the use of solar PV as a useful alternative for off-grid electrification will be apparent.

References

Bahauddin, K., & Salah, U. T. (2010). Impact of Utilization of Solar PV Technology among Marginalized Poor People in Rural Area of Bangladesh. *Proceedings of International Conference on Environmental Aspects of Bangladesh (ICEAB'10)*, September, Japan.

Barnes, D. F. (2002). Rural Electrification and Development in the Philippines: Measuring the Social and Economic Benefits. *UNDP/ESMAP Report*. Washington: World Bank.

Cabraal, A., Cosgrove-Davies, M., & Schaeffer, L. (1996). Best Practices for Photovoltaic Household Electrification Programs. Lessons from Experiences in Selected Countries. *World Bank Technical Paper 324*. Asia Technical Department Series. Washington: World Bank. http://dx.doi.org/10.1596/0-8213-3728-9

Energy Commission. (2009). Renewable Energy Policy Framework for Climate Change Mitigation in Ghana: Review of Existing Renewable Energy Resource Data, Energy Policies, Strategies, Plans and Projects, Accra: Energy Commission.

Forson, F. K., Agbeko, K. E., Edwin, I. A., Sunnu, A., Brew-Hammond, A., & Akuffo, F. O. (2004). Solar Energy Resource Assessment for Ghana. *Journal of the Ghana Institution of Engineers, 2*(1), 31-34.

Gathigi, G. W. (2009). *Radio Listening Habits among Rural Audiences: An Ethnographic Study of Kieni West Division in Central Kenya*. A Doctoral Dissertation Presented to the Faculty of the Scripps College Of Communication of Ohio University, USA.

Ghana Publishing Company. (2011). Renewable Energy Act, 2011. Act 832, GPCL/A663/350/12?2011. Assembly Press, Accra, Ghana.

Ghana Statistical Service. (2001). *Core Welfare Indicators Questionnaire (CWIQ) Survey (1997)*. CIWQ Regional Profiles and CD-Rom, Accra: Statistical Service.

Ghana Statistical Service. (2002). Population and Housing Census. Summary Report of Final Results Statistical Service, Accra: Statistical Service.

Ghana Statistical Service. (2005a). 2000 Population and Housing Census, Upper West Region. Analysis of District data and Implications for Planning, Accra: UNFPA funded Project GHA/01/P070.

Ghana Statistical Service. (2005b). 2000 Population and Housing Census, Volta Region. Analysis of District data and Implications for Planning, Accra: UNFPA funded Project GHA/01/P070.

Ghana Statistical Service. (2005c). 2000 Population and Housing Census, Northern Region. Analysis of District data and Implications for Planning, Accra: UNFPA funded Project GHA/01/P070.

Ghana Statistical Service. (2005d). 2000 Population and Housing Census, Upper East Region. Analysis of District data and Implications for Planning, Accra: UNFPA funded Project GHA/01/P070.

Ghana Statistical Service. (2005e). Core Welfare Indicator Questionnaire (CWIQ II) Survey: Statistical Abstract, Accra: Ghana Statistical Service.

Ghana Statistical Service. (2005f). Population Data Analysis Report: Policy Implications of Population Trends (Vol. 2), Accra: Ghana Statistical Service.

Ghana Statistical Service. (2007). Pattern and Trends of Poverty in Ghana 1991-2006, Accra: Ghana Statistical Service

Kemausuor, F., Obeng, G.Y., Brew-Hammond, A., & Duker, A. (2011). A Review of Trends, Policies and Plans for Increasing Energy Access in Ghana. *Renewable and Sustainable Energy Reviews, 15*(2011), 5143-5154. http://dx.doi.org/10.1016/j.rser.2011.07.041

Madon, G., & France, M. (2003). *Energy, Poverty, and Gender. Impacts of Rural Electrification on Poverty and Gender in Indonesia.* Volume 1: Facts, Analysis and Recommendations, Washington D.C: World Bank.

Maine State Planning Office. (n.d.). Retrieved from http://www.state.me.us/spo/recycle/docs/avoidedcosts.PDF

Ministry of Energy. (2010). National Electrification Scheme (NES) Master Plan Review (2011-2020), Ministry of Energy, Ghana.

Ministry of Energy and Power. (2013). Achievement in the Power Sector, Ministry of Energy and Power, Ghana. Retrieved from http://www.energymin.gov.gh/?page_id=183

Obeng, G. Y., & Ever, H. D. (2009). *Solar Photovoltaic Rural Electrification and Energy Poverty: A Review and Conceptual Framework With Reference to Ghana.* Center for Development Research, University of Bonn. Working Paper Series No. 36: 1-20. ISSN 1864-6638. (Munich Personal RePEc Archive (MPRA) Paper No. 17136). Retrieved from http://mpra.ub.uni-muenchen.de/17136/

Plastow, J., & Goldsmith, A. (2001). *Investigating in Power and People. A Global action Plan.* Renewable Energy World, November-December 2001. James and James Publication (pp. 47-59).

Reddy, S. G., & Minnoiu, C. (2006). *Chinese Poverty: Assessing the Impact of Alternative Assumptions.* Working Paper No. 25, July. Brazil: UNDP International Poverty Centre.

Schmidt, M. (2014). Avoided Cost, Cost Savings, and Opportunity Cost Explained Definitions, Meaning, and Example Calculations, Business Encyclopedia, ISBN 978-1-929500-10-9. Revised on 2014-04-17. Retrieved April 2014, from http://www.business-case-analysis.com/avoided-cost.html

Wamukonya, N., & Davies, M. (2001). Socio-economic Impacts of Rural Electrification in Namibia: Comparisons between Grid, Solar and Non-electrified Households. *Energy for Sustainable Development, V*(3), 5-13. http://dx.doi.org/10.1016/S0973-0826(08)60272-0

Zeller, M. (2004). *Review of Poverty Assessment Tools.* Accelerated Micro-enterprise Advancement Project (AMAP), Report submitted to IRIS and USAID.

Notes

Note 1. Non-electrified households were those "without solar PV" and therefore the very few users of diesel generators without solar PV were included in this category for the "with" and "without" comparative analysis.

Note 2. Expenditure was used as a proxy for income. The computed values were: US$ 1-1.08/day; US$ 1.08-2.17/day; and above US$ 2.17/day. See Table 3.

Note 3. The US$ 1 per day poverty line is actually US$ 1.08 per day (Reddy and Minnoiu, 2006; Zeller, 2004).

12

Economic Analysis of a Fuelwood Consuming Activity: Empirical Evidence for Traditional Red Sorghum Beer Producers in Ouagadougou, Burkina Faso

Boukary Ouédraogo[1] & Patrick Point[2]

[1] Université Ouaga2, Burkina Faso

[2] Université Montesquieu Bordeaux IV, France

Correspondance: Boukary Ouédraogo, Université Ouaga, 04 BP 8938 Ouagadougou 04, Burkina Faso. E-mail: boukary_ouedraogo2003@yahoo.fr

Abstract

This paper aims to analyze the determinants of input demand for local red sorghum beer production in Ouagadougou, which is an activity conducted mainly by women from these areas. The descriptive analysis in the paper emphasizes on the extent of this activity in the national economy, and namely in firewood consumption. The producer theory approach through an econometrical method using a cost translog function also helps to derive conditional input demand namely for firewood. The estimated coefficients of firewood conditional demand of local beer producers, "dolotières", have shown weak price elasticity hampering the fuelwood price policy. However, the simulation regarding taxation of fuelwood price effect on quantity demanded consequently states an overall reduction in demand for fuelwood in Ouagadougou. This induces a decrease in the local beer producer's profit for a given level of production. In addition, the paper deals with short term analyses.

Keywords: Translog Cost Function, return to scale, dolo supply, elasticities, red sorghum beer, producer's surplus, firewood demand

1. Introduction

Most studies on the demand for wood energy in developing countries are referred to as household demand. Several authors have studied the determinants of household energy choices and the transition between wood energy and other conventional sources of energy: Hosier (1988), Ferari (1990), Hosier and Dowd (1997), Ouedraogo (2002; 2006), and Fawelinmi Oyerinde (2002), Campbell et al (2003). Some authors were also interested in describing wood energy sector and quantifying the volume and value of timber produced on certain markets (Chavin, 1981; Sow, 1990; Banks et al., 1996; Alam et al., 1998).

These studies have either emphasized the scale of the problems of urban and rural households' wood energy supply, or analyzed the relationship between the demand for wood energy and income levels of households. They have also described the difficulties to substitute other energy sources to firewood and have contributed to the evaluation of the determinants of household energy choices.

This attention to consumption of wood for domestic use obscures the usage of wood in crafts. However, it is clear that the majority of artisans in Sub Saharan Africa uses wood energy as input in their production, so that today, due to growth of certain craft activities (various foundries, forges, jewelry, traditional brewery, restaurant etc.), this demand especially in urban areas, far exceeds domestic demand. In Ouagadougou, the aggregate demand of wood energy in 2000 is shared by 59% as intermediate consumption for artisans and 41% as household consumption. The five major categories of artisans whose demand for fuelwood is very marked in this city are the bronze foundries, aluminum smelters, restaurant owners, barbecue and local red sorghum beer producers namely "dolotières".

In the foregoing contribution (of artisan usage of firewood), it is the "dolotières" who catch our attention. They produce "dolo", a traditional beer made from red sorghum. This activity is conducted by more than 1,000 of women in Ouagadougou. *Dolo* preparation requires two days of cooking and uses an average of 458.17 kg of firewood, that is, about a half ton of wood. In 2000, wood consumption for the production of *dolo* amounted to

48,269 tons, representing 28% of craftsmen's total consumption and 16.5% of wood total consumption. Note that this activity generates significant revenue for very vulnerable social strata. Since the devaluation of the CFA franc in January 1994 in West African Economic and Monetary Union (WAEMU) countries, higher prices for industrial beer resulted in increased dolo demand, which induces an increase in the production of this traditional beer and an increase in the number of production units.

As it is difficult to conduct effective policies for firewood substitution by other sources (of energy) for households mainly due to the heterogeneity in this population, public interventions on a production sector such as dolo production which is relatively homogeneous could be easier and more effective to design and implement.

The objective of this work is to bring together elements of an economic analysis in order to provide better insight into the possibilities to limit firewood consumption in a sector that heavily relies on it.

After delivering pieces of information on the processing and marketing of *dolo*, we will econometrically estimate the function of total production cost, and we analyze the properties of production technology. We will then go through an aggregate approach of supply function for the city of Ouagadougou. The results obtained will allow us to outline effective policies to reduce wood consumption in this activity.

2. Notes on *Dolo* Production and Data Set

We consider the description given by Marie-Michele Ouedraogo (1974, p. 284). "Red Sorghum purchased on the market or in the countryside, is put to germinate, [and] then dried up before being crushed. Thus, it is quenched in large jars half stuck in the ground. The mixture is stirred and bailed out, and transferred into cooking jars put on wood fireplaces. After a first cooking of about three hours, a new transfer occurs in the jars stuck in the ground, where the mixture cools for hours. A second cooking of seven or eight hours completes the preparation of *dolo*. We then have a sweet non-alcoholic juice called "Biss-Kom" or water of germinated red sorghum. From that moment, the *dolo* is drinkable, but it is never marketed at this stage. The liquid is again cooled in jars stuck in the ground and then fermented for a long time with a handful amount of sourdough incorporated by the "dolotière". In Ouagadougou, plants of all kinds can be introduced into the *dolo* at that moment to activate fermentation and spice *dolo* taste. *Dolo* must be kept the night of the second day of preparation and the morning of the third day it is ready [for consumption] …" So is summarized the process of making this traditional beer that lasts about ten hours of preparation.

Women who produce red sorghum beer are called "dolotières". *Dolo* has a very important cultural value and occurs largely in the tradition of the Burkinabe people (customary rites and festivals). It is also a highly priced drink in both urban and rural areas. The CFA franc's devaluation in January 1994 and the country's fiscal policy regarding taxation of alcoholic beverage greatly contributed to raising the price of industrial beer. As a consequence, demand for traditional beer has increased, resulting in new entry of "dolotières" in the *dolo* industry and in increasing the number of producers. "Dolotières" are present in all districts and neighborhoods of Ouagadougou, though it should be noted that their density is higher in the suburbs (newly parceled and non-parceled areas).

The data used in this study were collected in 2000 during extensive surveys on *"craftsmen woodfuel demand in Ouagadougou"*. These surveys were funded by the Coordination Unit of the Regional Program of Traditional Energy Sector (UC-RPTES). A specific questionnaire was developed for "dolotières". Of 1013 dolotières located in Ouagadougou, 154 were interviewed during the survey, that is, 15% of these producers. Table 1 provides details of key economic data listed.

Table 1. Key economic variables collected on "dolotières"

Variables	Units
Production of « Dolo »	Liter « Dolo » or value in CFA Franc
Price of « Dolo »	CFA F/Liter
Cost of red sorghum (ferment)	CFA F/Kilogram
Price of red sorghum	CFA F/Kilogram
Firewood cost	CFA F/Kilogram
Firewood price	CFA F/Kilogram
Water cost	CFA F/Liter
Water price	CFA F/Kilogram
Wage / production	CFA F/Employee/production (cooking)
Firewood quantity	Kilogram

| Water quantity | Liter |
| Red sorghum ferment quantity | Kilogram |

Source: RPTES (2000).

It should be noted that this information relates mainly variable costs. This leads us to confine ourselves to a short term approach, without direct consideration of fixed costs. These data will be used to develop a total restricted variable cost function. We will use a flexible functional form: the translog.

3. A Translog Model of Total *Dolo* Production Costs

The non-homothetic translog function is a second order Taylor series of the logarithm of the cost with respect to the logarithm of the explanatory variables. It gives an approximation of an unknown function *a priori* and imposes relatively few restrictions on production technology. We can add that cost shared equations derived are linear in parameters. Unlike homothetic functions, conditional demands for input are function of production level. Given its properties, it is the most common functional form used to specify cost functions (Note 1) nowadays.

3.1 Model Functional Specification

The non-homothetic Translog Cost Function is written as follows:

$$\ln C = \ln \alpha_0 + \sum_{i=1}^{n} \alpha_i \ln P_i + \frac{1}{2} \cdot \sum_{i=1}^{n} \sum_{j=1}^{n} \gamma_{ij} \ln P_i \ln P_j + \alpha_y \ln Y + \frac{1}{2} \gamma_{yy} (\ln Y)^2 + \sum_{i=1}^{n} \gamma_{iy} \ln P_i \ln Y \qquad (1)$$

where $\gamma_{ij} = \gamma_{ji}$,

i, j = firewood (B), germinated red sorghum (M), labor (L) and water (E); C is the total cost of production, Pi the price of the ith input, Y the quantity of *dolo* produced.

Factor prices and production levels are exogenous variables. α and γ are parameters to be estimated (Note 2).

In order to adequately describe producer's behavior, cost function must be, among other things, homogeneous of degree 1 with respect to prices of inputs used, for a given level of production. This implies the following restrictions:

$$\sum_{i=1}^{n} \alpha_i = 1, \quad \sum_{i=1}^{n} \gamma_{ij} = \sum_{i=1}^{n} \gamma_{ji} = \sum_{i=1}^{n} \gamma_{iY} = 0 \qquad (2)$$

Other restrictions inherent to Translog Cost Function could be added:

- For this cost function to be homothetic, it is necessary and sufficient that the parameters $\gamma_{iY} = 0 \; \forall \; i = 1,....,n.$ Homogeneity of constant degree with production function takes place if in addition to the foregoing restrictions on homotheticity, we set: $\gamma_{YY} = 0$; in which case the degree of homogeneity is equal to $1/\alpha_Y$.

- The constant returns to scale of dual production function will appear if, in addition to the restrictions of homotheticity and homogeneity, $\alpha_Y = 1.$ Finally, translog function reduces to Cobb-Douglas function with constant returns to scale if, besides all the above-mentioned limitations, we require that each of $\gamma_{ij} = 0, i, j = 1,....,n.$

3.2 Estimation Procedure

The Translog Cost Function can be directly estimated from equation (1), but for more efficient estimators, it can be achieved by estimating the optimum demand equations of factors resulting from cost minimization, transformed here into cost shares of factors equations (E. R. Berndt, 1991). Thus, by deriving the equation (1) and using Shephard's lemma, we obtain the following cost shares of factors equations:

$$\frac{\partial \ln C}{\partial \ln P_i} = \frac{P_i}{C}\frac{\partial C}{\partial P_i} = \frac{P_i X_i}{C} = \alpha_i + \sum_{j=1}^{n} \gamma_{ij} \ln P_j + \gamma_{iY} \ln Y \tag{3}$$

where $\sum_{i=1}^{1} P_i X_i = C$, total cost of factors.

The cost shares of factors S_i, are defined as follows:

$$S_i \equiv P_i X_i / C, \text{ et} \sum_{i=1}^{n} S_i = 1.$$

For this study, "dolotières" use four factors of production. So, factor demand equations or cost shares of factor equations will be as follows in the system of equations (4):

$$S_i = \alpha_i + \sum_{j=1}^{n} \gamma_{ij} \ln P_j + \gamma_{iY} \ln Y \quad , \quad i = 1, 2, .., 4 \tag{4}$$

Note that without symmetry restrictions there are 24 parameters to be estimated, six in each of four cost shares equations. If the constraints of symmetry between the four equations are imposed $(\gamma_{12} = \gamma_{21}, \gamma_{13} = \gamma_{31}, \gamma_{14} = \gamma_{41}, \gamma_{23} = \gamma_{32}, \gamma_{24} = \gamma_{42}$, and $\gamma_{34} = \gamma_{43})$, the number of parameters to be estimated is reduced to 18. As previously posed in equation (2), economic theory requires that the Translog Cost Function be homogeneous of degree 1 with respect to factor prices: the inclusion of this constraint results in the following restrictions in the group of equations (5):

$$\sum_{i=1}^{4} \alpha_i = 1 \quad , \quad \sum_{i=1}^{4} \gamma_{iY} = 0, \quad \sum_{j=1}^{4} \gamma_{ij} = 0 \quad , \quad i = \{1, 2, 3, 4\} \tag{5}$$

These six restrictions will further reduce the number of parameters to estimate from 18 to 12. The system of equations of factors' demand has the special property that for each observation, the sum of the dependent variables (the share of cost) of all equations is always equal to one. Indeed, if there are n units of factor cost (or factor demand equations), only n-1 are linearly independent. This property of factor demand equations has several econometric implications on which we will focus our attention.

First, since the sum of cost shares is always equal to unity and only n-1 of factor demand equations are linearly independent, then system (5) relations imply that the sum of Ordinary Least Squares (OLS) residuals u_i across all factor demand equations is always equal to 0 for each observation:

$\sum_{i=1}^{4} u_i = 0$, and errors matrix A* is singular and non-diagonal.

Second, because of the unit amount of cost shares of factors for each observation, when symmetry restrictions are not imposed, estimation equation by equation using least squares (OLS) method gives parameters that always obey property of homogeneity in the system of equations (5). Berndt and Wood (1975.1979) have empirically shown that the parameters estimated by OLS, when the constraint of symmetry among the demand for factors equations is not imposed, still obey the restrictions of the system (5). However, the cross product of OLS residues matrix is not singular. Thus, OLS method produces good estimators.

Third, the fact that the covariance and errors cross-product matrices must both be singular, the estimation by the method of maximum likelihood, which minimizes the cross product of errors, is not feasible. Also, the procedure most commonly used to solve matrix singularity problem is to eliminate arbitrary one of the n demand equations and proceed to the estimation of the n-1 remaining equations by the ML method. In this case, we will eliminate the fourth equation and directly estimate, equation by equation, the 12 parameters of the three equations shown in system (6):

$$S_1 = \alpha_1 + \gamma_{11} \ln(P_1 / P_4) + \gamma_{12} \ln(P_2 / P_4) + \gamma_{13} \ln(P_3 / P_4) + \gamma_{1Y} \ln Y$$
$$S_2 = \alpha_2 + \gamma_{21} \ln(P_1 / P_4) + \gamma_{22} \ln(P_2 / P_4) + \gamma_{23} \ln(P_3 / P_4) + \gamma_{2Y} \ln Y \tag{6}$$
$$S_3 = \alpha_3 + \gamma_{31} \ln(P_1 / P_4) + \gamma_{32} \ln(P_2 / P_4) + \gamma_{33} \ln(P_3 / P_4) + \gamma_{3Y} \ln Y$$

Eliminated equation parameters are deduced indirectly from restrictions tied to the property of homogeneity in system (5), which allows them to be written in terms of estimated parameters of the first three equations. These parameters are expressed as follows:

$$\alpha_4 = 1 \quad -(\alpha_1 + \alpha_2 + \alpha_3)$$
$$\gamma_{14} = \quad -(\gamma_{11} + \gamma_{12} + \gamma_{13})$$
$$\gamma_{24} = \quad -(\gamma_{12} + \gamma_{22} + \gamma_{23})$$
$$\gamma_{34} = \quad -(\gamma_{13} + \gamma_{23} + \gamma_{33})$$
$$\gamma_{4Y} = \quad -(\gamma_{1Y} + \gamma_{2Y} + \gamma_{3Y})$$
$$\gamma_{44} = \quad -(\gamma_{14} + \gamma_{24} + \gamma_{34}) \Rightarrow \gamma_{44} = \gamma_{11} + \gamma_{22} + \gamma_{33} + 2(\gamma_{12} + \gamma_{13} + \gamma_{23})$$

(7)

Note that the parameters indirectly estimated are linear combinations of parameters estimated directly from the three equations selected. Likewise, variances of these parameters are linear combinations of those of the first equations, which facilitate their computation.

4. Results and Economic Analysis of Dolo Production

The sample used includes 149 "dolotières" of the 154 respondents. For the estimation, several methods are used (OLS, 2SLS, 3SLS and ML). We will favor the method of Maximum Likelihood, since it not only gives invariant parameters, but also provides statistics that directly help test our hypotheses.

4.1 Variable Total Cost Function of Folo Production

We have simultaneously estimated the demand functions for input and the cost function under the constraints of symmetry, of homothety and of homogeneity.

Table 2. Estimation results by the method IZEF / LM (SUR) of cost shares equations and of the Translog Cost Function. 149 sample observations. T-statistics are in parentheses

Translog Cost Function			
Variables	Parameters	Variables	Parameters
$\alpha(0)$	10,942 (616,9)***	$\gamma(LY)$	-0,027 (-2,5)**
$\alpha(B)$	0,312 (45,0)***	$\alpha(B)$	0,312 (45,0)***
$\alpha(M)$	0,583 (89,6)***	$\gamma(BB)$	0,134 (5,0)***
$\alpha(E)$	0,053 (23,5)***	$\gamma(BM)$	-0,128 (-5,5)***
$\alpha(L)$	0,052 (0,1)	$\gamma(BE)$	0,0003 (0,0)
$\gamma(BB)$	0,134 (5,0)***	$\gamma(BL)$	-0,006 (-0,2)
$\gamma(MM)$	0,204 (8,5)***	$\gamma(BY)$	-0,039 (-5,2)***
$\gamma(EE)$	0,035 (12,4)***	$\alpha(M)$	0,583 (89,6)***
$\gamma(LL)$	0,053 (1,1)	$\gamma(MM)$	0,204 (8,5)***
$\gamma(BM)$	-0,128 (-5,5)***	$\gamma(ME)$	-0,032 (-5,6)***
$\gamma(BE)$	0,0003 (0,0)	$\gamma(ML)$	-0,044 (-1,3)
$\gamma(BL)$	-0,006 (-0,2)	$\gamma(MY)$	0,072 (10,2)***
$\gamma(ME)$	-0,032 (-5,6)***	$\alpha(E)$	0,053 (23,5)***
$\gamma(ML)$	-0,044 (-1,3)	$\gamma(EE)$	0,035 (12,4)***
$\gamma(EL)$	-0,003 (-0,3)	$\gamma(EL)$	-0,003 (-0,3)
$\alpha(Y)$	0,801 (39,7)***	$\gamma(EY)$	-0,006 (-2,7)***
$\gamma(YY)$	-0,064 (-2,4)**	$\alpha(L)$	0,052 (0,1)
$\gamma(BY)$	-0,039 (-5,2)***	$\gamma(LL)$	0,053 (1,1)
$\gamma(MY)$	0,072 (10,2)***	$\gamma(LY)$	-0,027 (-2,5)***
$\gamma(EY)$	-0,006 (-2,7)**		

Statistics	Functions				
	Firewood	Red Sorghum	Water	Labor	Log (Variable Cost)
R^2	0,20	0,48	0,52	0,40	0,95
Ajusted-R^2	0,18	0,46	0,51	0,38	0,94
Durbin-Watson	1,87	1,86	2,06	1,58	1,80
Mean dependent var.	0,33	0,57	0,05	0,06	10,61

B = Firewood, M =Germinated Red Sorghum (ferment), L = Labor, E = Water.

*** implies that the coefficient is statistically significant at 1%,

** implies that the coefficient is statistically significant at 5%

* implies that the coefficient is statistically significant at 10%.

IZEF/LM = Iterative Seemingly Unrelated Regression

Parameters estimated by IZEF/LM method are consistent with symmetry constraint, and are linearly homogeny to prices.

Graph n°1 shows adjustment of cost function to data.

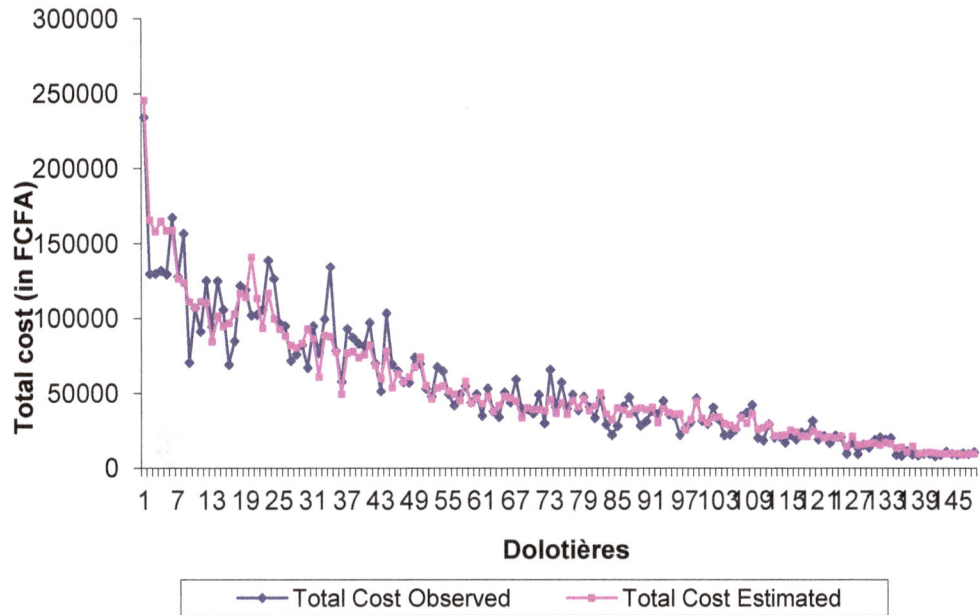

Figure 1. Total cost of dolo production

If one focuses on the relationship between quantity produced and total production costs, by setting the variables to their average price, we obtain the line represented by the graph n° 2

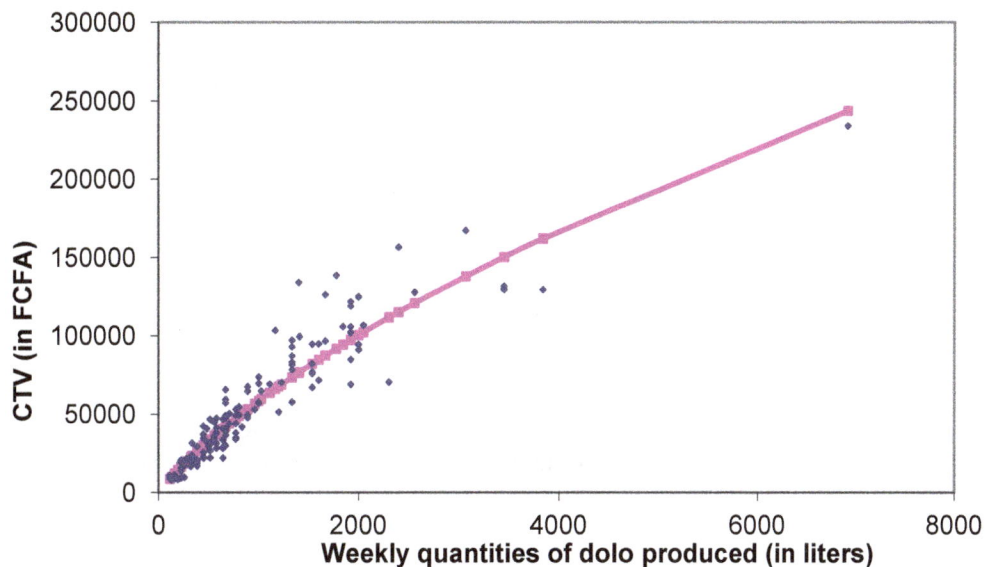

Figure 2. Links between quantities produced of dolo and total cost (setting prices to the sample average)

4.2 Economic Analysis of Dolo Production Technology

4.2.1 Cost Shares

Estimation procedure of total cost bases upon the computation of cost shares. These are obtained par the formula (4 bis):

$$S_i = \alpha_i + \sum_{j=1}^{n} \gamma_{ij} \ln P_j + \gamma_{iY} \ln Y, \quad i, j = B, M, L, E. \quad (4\,bis)$$

The leading share is from millet, and then wood with respectively on average, 57% and 32%. The other two factors weigh very little since water is 5% and wages 1% of the variable total cost.

Since our focus is on wood consumption, it is interesting to examine the relationship between the share of costs devoted to wood and the level of *dolo* production. Table 2 shows that the coefficient γ_{BY} is negative. This effect is clearly illustrated on graph 3 which connects the cost shares of wood to production level.

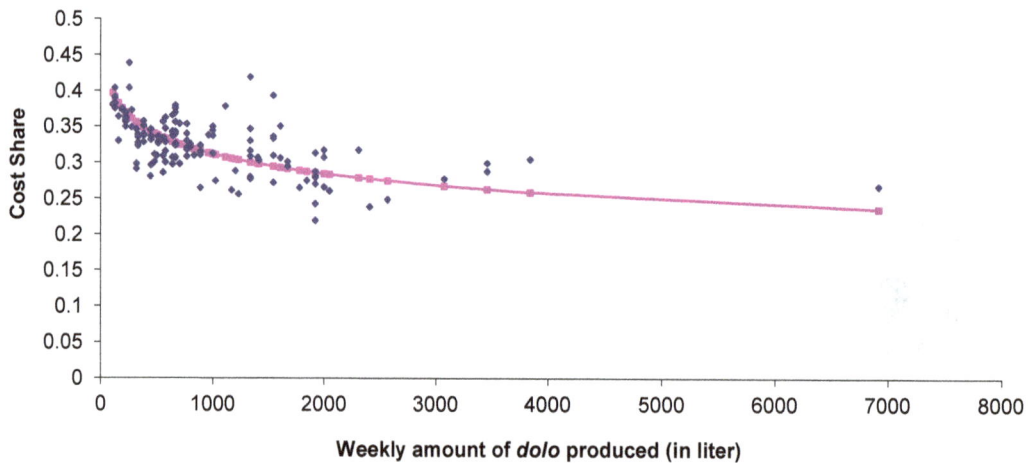

Figure 3. Trend of wood cost share with respect to dolo production

Production is characterized by a decrease of costshare devoted to the factor of production, wood, when the quantities produced increase. This assumes the existence of production technology of increasing returns to scale as indicating an improvement in inputs (firewood) productivity when stepping dolo production.

4.2.2 Return to Scale

The whole process seems to present returns to scale. As Giora Hanoch (1975) showed it, returns to scale (presented here by μ) are computed as the inverse of elasticities of cost function with regard to production (Note 3) (εCY):

$$\mu = 1/\varepsilon CY, \text{ where } \quad \varepsilon CY = \partial \ln C / \partial \ln Y = \alpha Y + \gamma BY \ln PB + \gamma MY \ln PM + \gamma LY \ln PL + \gamma EY \ln PE + \gamma YY \ln Y \quad (8)$$

εCY is computed using average prices of factors or by approximating this quantity relative to the average marginal cost and the mean of average costs observed.

Calculation gives the value: $\varepsilon cy = 0{,}801$ and a return to scale: $\mu = 1/\varepsilon cy = 1{,}25$. So, technology shows an increasing return to scale.

4.2.3 Marginal Cost of Production

We scrutinize production marginal cost taking into account input prices set to the sample average. To the extent that variable total cost has been estimated by normalizing the variables, terms linked to prices cancel out, and average variable total cost (CTVM) becomes:

$$CTVM = \frac{e^{\alpha_0}}{\overline{Y}^{\alpha_Y}} Y^{\alpha_Y} e^{\gamma_{YY} \left[\ln(Y) - \ln(\overline{Y}) \right]^2} \quad (9)$$

Thus, marginal cost gives:

$$CmVM = \frac{\partial CTVM}{\partial Y} = -\frac{e^{\alpha_0}}{\overline{Y}^{\alpha_Y}}\Big[2\gamma_{YY}\ln(\overline{Y}) - \alpha_Y - 2\gamma_{YY}\ln(Y)\Big]Y^{(\alpha_Y - 1)}e^{\gamma_{YY}\big[\ln(\overline{Y}) - \ln(Y)\big]^2} \qquad (10)$$

The following graph 4 shows marginal and average costs curves.

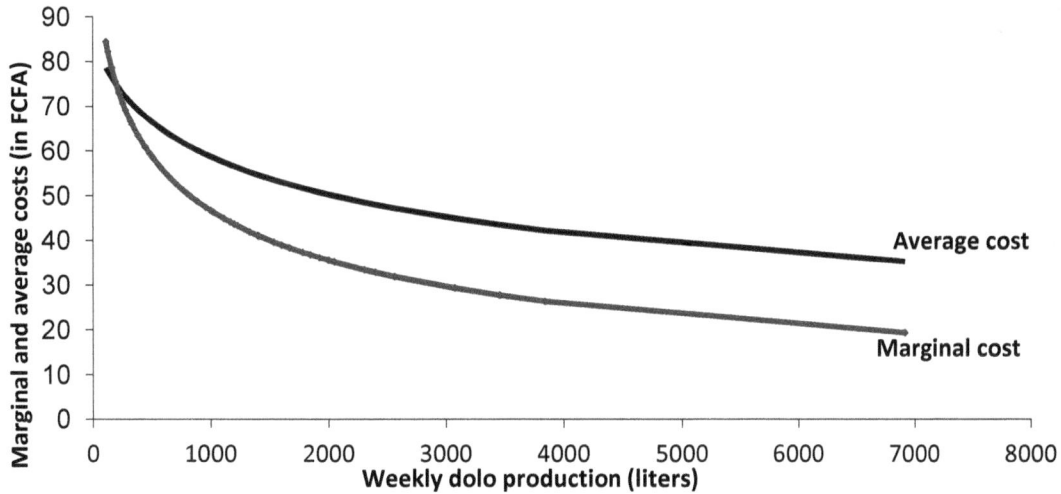

Figure 4. Marginal and mean costs of dolo production (setting inputs prices to sample average)

Marginal cost is a decreasing function of the production. This does confirm the existence of the previously identified increasing return to scale.

We have computed marginal costs relating to each "dolotière" k considering input prices that each of them faces. The formula used is then:

$$CmVM_k = \frac{\partial CTVM_k}{\partial Y_k} = -\frac{e^{b_k}}{\overline{Y}^{c_k}}\Big[2\gamma_{YY}\ln(\overline{Y}) - c_k - 2\gamma_{YY}\ln(Y_k)\Big]Y_k^{(c_k - 1)}e^{\gamma_{YY}\big[\ln(\overline{Y}) - \ln(Y_k)\big]^2}, \text{ avec}$$

$$b_k = \alpha_y + \sum_i \gamma_{iY}\ln\left(\frac{P_{ik}}{\overline{P}_i}\right) \text{ et } c_k = \ln\alpha_0 + \sum_{i=1}^{n}\alpha_i \ln\left(\frac{P_{ik}}{\overline{P}_i}\right) \qquad (11)$$

Graph 5 shows marginal costs distribution according to production level.

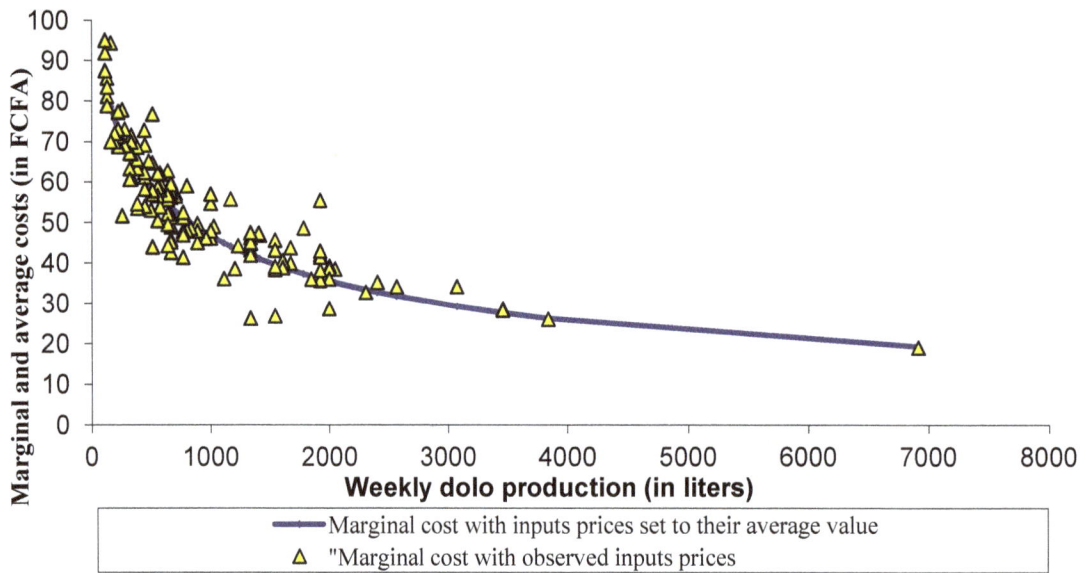

Figure 5. Estimated and trend of marginal costs of dolo production (setting prices to sample average)

4.2.4 Technical Substitution Elasticities

Allen's partial substitution elasticities allow to measure technical substitution between factors of production. These crossed between one another indicate which ones are substitutable or complementary. A positive (negative) value of σ_{ij} implies that factors i and j are substitutable (complementary); if $\sigma_{ij} = 0$, then the two factors are independent (Field and Berndt, 1981, p. 55). These elasticities are computed through the following formula:

$$\sigma_{ij} = \frac{\gamma_{ij} + S_i S_j}{S_i S_j}, \qquad i,j = 1,....,n, \quad mais \quad i \neq j$$

$$\sigma_{ii} = \frac{\gamma_{ii} + S_i^2 - S_i}{S_i^2}, \quad i = 1,....,n \quad avec \quad i = j$$

(12)

γ_{ij}, γ_{ii}, S_i, S_j are respectively coefficients of factors prices, and cost shares of different factors in the production.

Table 3. Allen's partial technical substitution elasticities

	Firewood	Red Sorghum	Water	Labor
Fuelwood	**-0,8021**			
Red Sorghum	0,1004	**-0,1277**		
Labor	0,0023	-0,0004	**-4,8037**	
Water	0,0021	-0,0012	0,0000	**-0,1506**

Germinated millet, water and labor appear weakly substitutable to firewood. We observe a complementarity between germinated millet and water, and between germinated millet and labor.

4.2.5 Input Price-Elasticities

Direct and cross price-elasticities of input conditional demand are given by the following formula:

$$\varepsilon_{ij} = \frac{\gamma_{ij} + S_i S_j}{S_i}, \quad i = 1,....,n \ , \quad avec \quad i \neq j$$

$$\varepsilon_{ii} = \frac{\gamma_{ii} + S_i^2 - S_i}{S_i}, \quad i = 1,....,n \ , \quad avec \quad i = j$$

(13)

A given factor demand sensitivity to changes occurring in its own price is measured thanks to direct price-elasticities. Existing economic relation between factors is quantified by cross price-elasticities.

Table 4. Input demands price-elasticities

	Firewood	Red Sorghum	Water	Labor
Firewood	-0.26241			
Red Sorghum	0.17672	-0.07257		
Labor	0.04845	-0.00864	-0.22905	
Water	0.03723	-0.02054	-0.00047	-0.00857

From all factors, firewood is the most responsive to its own price. A 1% increase of its price induces a 0.26% decrease in its demand. We find that geminated millet of which the cost share is prevailing is found to have a relatively low own price elasticity (0.07%). Cross-price elasticities with firewood are positive, the highest being the one between firewood and germinated millet. A price increase of these factors leads to an increased demand for firewood. As for sprouted millet, its technical substitutability with firewood, observed above, is legitimated. For the other two factors, we keep commenting the figures due to their very low coefficients.

4.3 Results for Dolo Production Industry in Ouagadougou

4.3.1 Supply Function of Dolo

Total production per week of the surveyed sample is 142,021 liters for the 149 "dolotières". The sampling procedure allows us to assume these samples as representative. We can then estimate the total production of dolo for the city of Ouagadougou provided by the 1013 "dolotières" to: 142021 × (1013/149) = 965 550 liters. To establish the supply curve, we aggregate the productions according to the level of marginal cost by assigning each sample production the multiplicative coefficient of that one (1013/149). Graph No. 6 restores the supply curve.

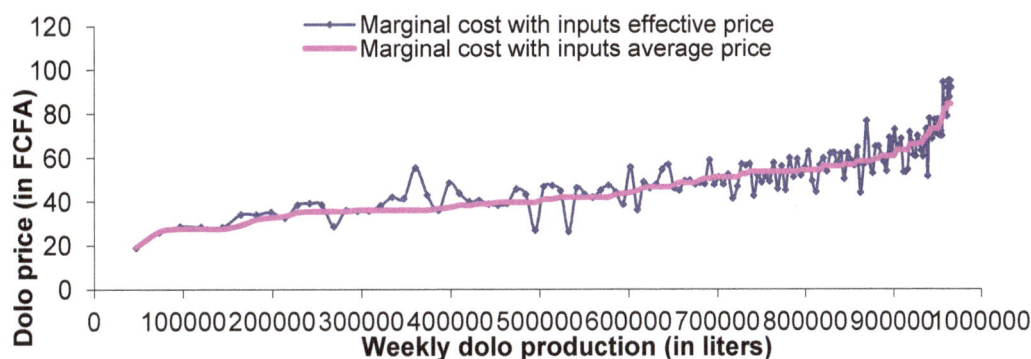

Figure 6. Dolo weekly supply curve

A review of producers' staff according to their production capacity shows that a relatively small number of "dolotières" provides a very significant share of production. Graph No. 7 describes this situation.

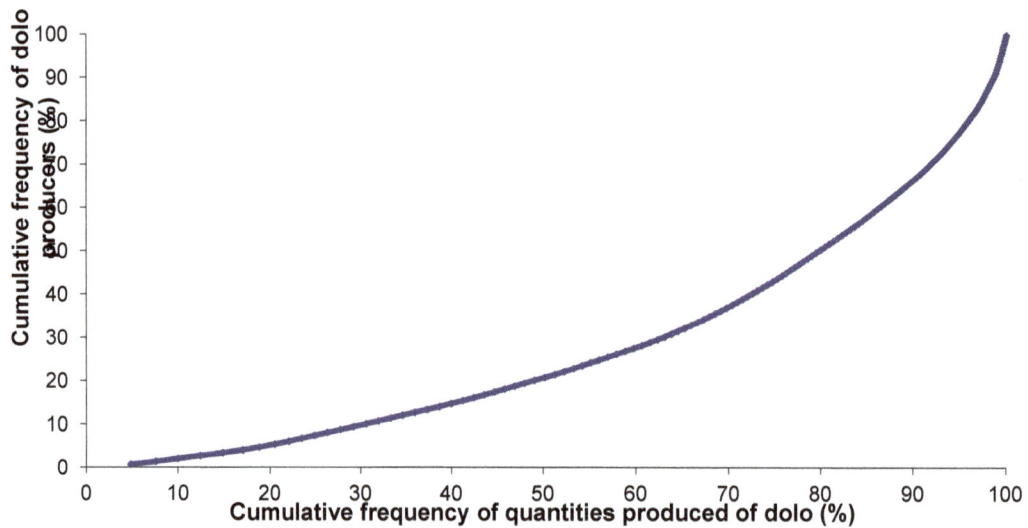

Figure 7. Concentration of dolo production's activity

We remark that 50% of production is provided by 20% of "dolotières", and that 50% of "dolotières" supply 80% of the market. This relative concentration is to be taken into account in any policy designed to promote a reduced use of firewood.

4.3.2 Dolo Producers' Total Surplus

The price per liter of *dolo* is uniform over the city of Ouagadougou. It is 100 CFA Franc (Note 4). Examination of the supply curve and the level of price 100 CFA F show the producer surplus. To the extent that we do not have individualized information on fixed costs, we will not calculate the profit, but producer surplus which is the difference between their turnover and their variable total cost.

Here, the total surplus per week establishes at 42.09 million of CFA Franc. That is, 2,188.72 million of CFA F per year. It is found that this activity is highly profitable.

Graph 8 shows the evolution of total producer's surplus as a function of total production.

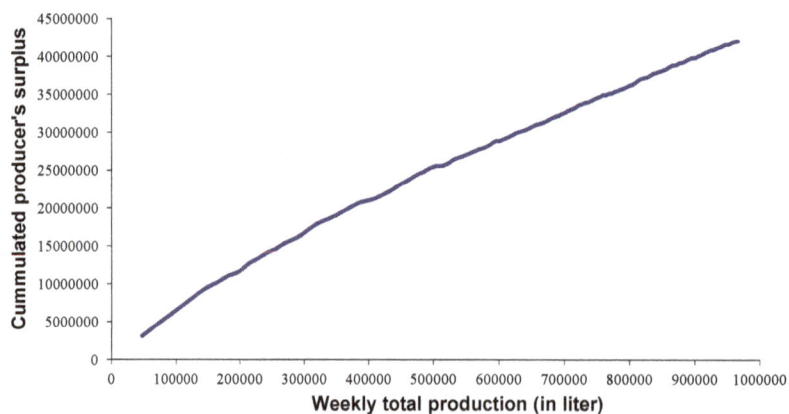

Figure 8. Trend of cumulated producers's surplus as a function of dolo total production

Production concentration, as above observed, add to the existence of economy returns of scale contribute to magnify concentration in producers' surplus mobilization.

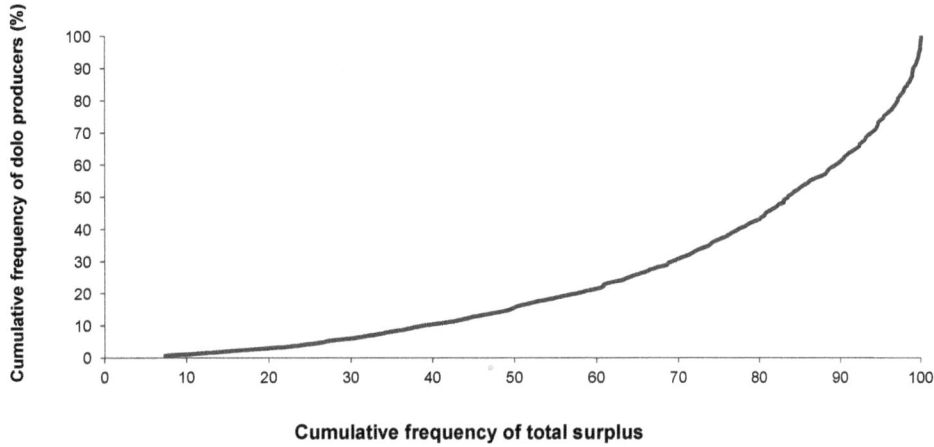

Figure 9. Concentration of total surplus between dolo producers

Graph N° 9 reveals that almost 60% of the total surplus is mobilized by 20% of producers.

4.3.3 Firewood Demand for Dolo Production

Shepard's lemma allows getting directly conditional demand functions from cost function. As for firewood, we get:

$$Q_B(P_B, P_M, P_E, P_L, Y) = \frac{\partial CTV(P_B, P_M, P_E, P_L, Y)}{\partial P_B} \tag{14}$$

Fixing values, excepted wood price to sample average, induces variable total cost to become:

$$\ln(CTV) = \ln(\alpha_0) + \alpha_B \ln(\frac{P_B}{\bar{P}_B}) + 1/2\alpha_{BB}\left[\ln(\frac{P_B}{\bar{P}_B})\right]^2 \tag{15}$$

Or:

$$CTV = e^{\alpha_0}\left(\frac{P_B}{\bar{P}_B}\right)^{\alpha_B} e^{\left[1/2\alpha_{BB}\left(\ln(\frac{P_B}{\bar{P}_B})\right)^2\right]} \tag{16}$$

$$\frac{\partial CTV}{\partial P_B} = \left(\frac{e^{\alpha_0}}{\bar{P}_B^{\alpha_B}}\right)\left(\alpha_B - \alpha_{BB}\ln(\frac{P_B}{\bar{P}_B})\right)P_B^{(\alpha_B-1)}e^{1/2\alpha_{BB}\left(\ln\frac{P_B}{\bar{P}_B}\right)^2} \tag{17}$$

Relying on firewood prices paid by "dolotières" in the sample, we can plot demand function derived from millet beer production wood as in graph no. 10.

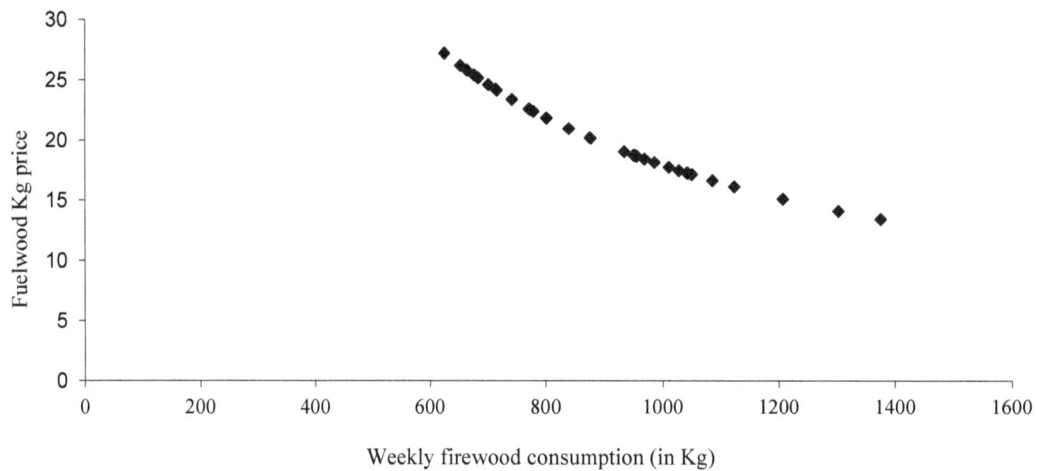

Figure 10. Firewood demand for dolo production

Examination of direct-price elasticity for firewood ε_{BB}= -0,262 confirms that firewood demand is not rigid, even though this value is quite low.

4.4 Simulations for a Lower Usage of Firewood in Dolo Production

We will develop only analyses with regard to actions regarding variable factors. This should be supplemented by a combined action on the fixed factors.

4.4.1 Effects of an Environmental Levy on Firewood Demand

To restore a proper pricing of the resource and to get virtuous results in saving resource, in alternating source of energy and in incentivizing an increase of supply, it is often advised to build on taxation through the re-adjustment of existing taxes or charges, or by creating new taxes.

It is known that the price of the standing resource is on average very low and often below the cost of rebuilding the resource. A study of firewood price structure sold at retail shows that the share of taxes related to the resource itself (environmental taxes and charges related to the regulation of uses) represents 20.4% of the price. The effectiveness of this tax is problematic, particularly with regard to environmental taxes. Arguably, 1,167,610 cubic meters get beyond environmental taxes. That is, a loss of 1.051 billion CFA francs. This represents a percentage of 85% of tax avoidance. Since the industry supplying "dolotières" uses large motorized transport (trucks with a capacity of 12 to 35 cubic meter of wood) sourcing generally in arranged and controlled timber yards, it is assumed that the degree of avoidance is lower than the observed one throughout firewood sector in Ouagadougou, and this rate of avoidance is at an average level of about 50% for the sector affecting "dolotières".

Hypothesis:

Tax avoidance = 50%

Eco tax = 20.4%.

If there is more evasion, the average price is × 2.

p the price of firewood

H the percentage increase

$H \times 0.204 \times 2p = 1.25\, p$, equally $h = 1.25 / 0.204 \times 2 = 3$

A multiplier coefficient of 3, associated with a general implementation led to an increase in retail prices of wood by 25%.

We will show what effects of such an increase are on firewood consumption for *dolo* production, on *dolo* production itself and on producers.

Conditional demand for timber factor will be modified by the reassessed price of wood. Starting from demand function derived from equation 17, we can draw the new profile.

The movement of wood demand curve resulting from changing of wood supply conditions (charging the price at 25%) is given by graph 11.

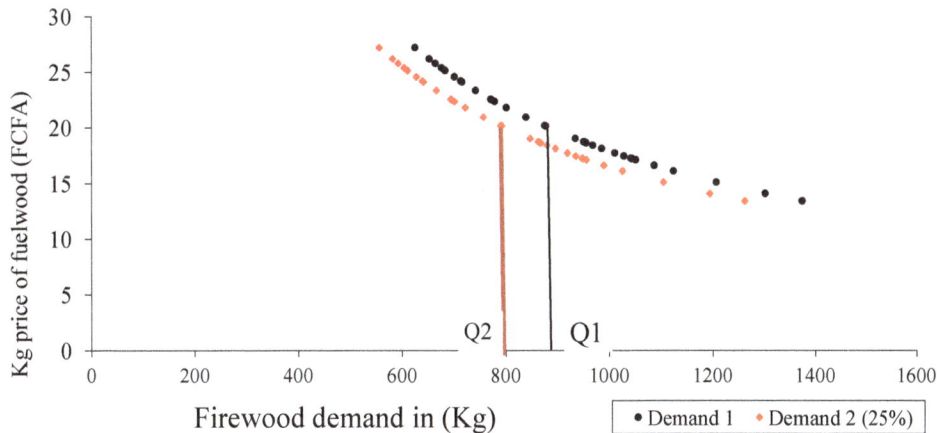

Figure 11. Effects of firewood demand of an environmental levy of 25% on firewood prices

This is achieved by assuming that the prices of all other factors are fixed at their average. We clearly observe a downward shift of the derived demand for wood, meaning that for the same price level the quantity demanded of firewood is always less on disturbed demand than on the initial one. For the same price level, the quantity required from the disturbed demand curve is always less than that of the original demand curve (Q2 <Q1).

We will more accurately calculate the change in quantity consumed, considering the factor price for each producer. The introduction of a 25% tax on timber prices will generate on average and by "dolotière" a weekly decline in quantity demanded of firewood of 85 kg, representing *dolotière*'s weekly saving of wood generated. On a global scale of all "dolotières" of this town, an annual saving of firewood resulting from this tax will be around a volume of 4,915 tons of firewood.

4.4.2 Effects of an Environmental Levy on Dolo Production

Chart No. 12 shows the movement direction of the aggregate supply curve by dolotières following a rise in firewood price 25%.

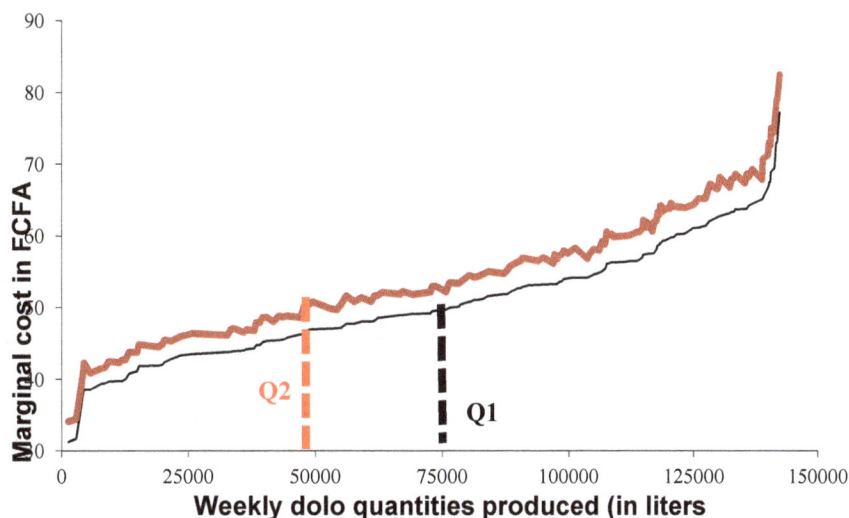

Figure 12. Effects of an environmental levy on dolo total supply in Ouagadougou

Rising timber prices will lead to a change in *dolo* supply conditions. The rising of firewood price will cause an upward movement of the supply curve to reflect naturally the increase in marginal cost due to this price increase. This upward shift of the aggregate supply curve represents a drop in *dolotière*'s profit, since for a given level of output, marginal cost is higher for supply curve 2 (*Offer* 2). This decrease in the *dolotière*'s profit reflects a loss of *dolotière*'s production surplus due to firewood high price.

The annual loss of production surplus resulting from this 25% price increase for all *dolotières* is given by the following formula:

$\Delta SP = \Sigma$ Initial production surplus - Σ Production surplus with rising prices

$\Delta SP = (6,191,047$ CFA F - $5570\ 158.5$ CFA F) x $(1013/149)$ x 52

$\quad = 620\ 888.5$ CFA F x $(1013/149)$ x $52 = 219\ 502\ 836$ CFA Franc.

An annual loss of production surplus of about CFA F 219 502 836 will be due to this price increase, for an annual production level of 50,208,713 liters of *dolo* produced by the whole industry in this city. This decline of surplus will result in a medium and long term decrease in *dolo* production and therefore, a lower demand for firewood. For the same level of marginal cost, production level of the sector will always fall behind the initial one after the price increase (Q2<Q1). This increase could also result for *dolotières* in coping behaviors through saving attitudes and efficiency in energy usage, or through increases in *dolo* cost to offset the loss of surplus. In the latter case, it is the consumer who will pay the firewood higher price.

5. Conclusion

This work estimated the cost function and derived the conditional inputs' demands of the local red sorghum beer production in Ouagadougou. The calculation of inputs' technical substitution elasticities helped to highlight the technical characteristics of this activity. These elasticities show a low substitutability of other inputs to firewood. Thus, they highlight complementary links between water and firewood on the one hand, so for labor and firewood on the other.

The analysis of the relationship between the dolo production share costs and quantity produced shows a decline in firewood share cost when the quantities produced increase. The results predict increasing returns to scale for this activity. The formula Giora Hanoch (1975) has highlighted a dolo production technology with increasing returns to scale with a coefficient equal to 1.25 performance.

Studying the concentration of dolo production and producers' surplus among *dolotières* reveals a pronounced inequality in the distribution of production and surplus. While 50% of production is provided by 20% of dolotières, 60% of the total surplus amounts to 20% of producers. There exists a few large dolotières who produce only for wholesalers account for the largest share of the surplus. This concentration of dolo production and its surplus among producers is a major asset for any policy to reduce the consumption of firewood.

This paper also stated the economy of wood energy and the loss of dolo producers' surplus induced by an environmental levy on wood energy. This results in an annual saving of 4,915 tons of workable wood energy on this activity with an annual decline of *dolotières* total surplus of about 219,502,836 FCFA.

Craft industry, user of firewood as input, represents 59% of total demand in Ouagadougou and includes a limited number of operators (about 10,000), and as such, should be a prime target for policy regarding demand for firewood management. Given the price inelastic craft demand for firewood observed and the lack of real alternatives, a pure tax policy would not be effective. On the other hand, support for suitable equipment (Note 5) establishment might be effective. One could imagine a system like the one adopted by water agencies in France. It collects charges on craftsmen, and it uses the amount of these fees to subsidize the acquisition of better equipment and / or training craftsmen on techniques or processes of production more efficient in terms of wood consumption. This could be implemented by a specialized agency that would manage a fund from fees and redistribute according to the development of projects that would get real savings on firewood. Such a specialized agency could be the Applied Sciences and Technology Research Institute (Note 6) (IRSAT). The I.R.S.A.T., indeed, is the best suited institute for such an operation since it has available skills at its disposal in this area to propose or evaluate technically and economically feasible projects thereto. It might require some financial contribution to the beneficiaries. "Dolotières" should be a pilot class of artisans for testing such an operation or policy with regard to the importance of their consumption coefficient.

References

Anderson, D. (1996). Energie et environnement: Possibilités techniques et économiques. Revue Finances & Développement: Paris.

Babin, G. F., Willis, C. E., & Allen, P. G. (1982). Estimation of Substitution Possibilities between Water and Other Production Inputs, American Agricultural Economics Association, USA.

Berndt, E. (1991). The practice of econometrics: Classic and contemporary, Addison-Wesley Publishing Company, Reading.

Berndt, E., & Christensen, L. R. (1973). The translog function and the substitution of equipment, structures and labor in U.S. manufacturing, 1929-1968. *Journal of Econometrics, 1,* 81-363. http://dx.doi.org/10.1016/0304-4076(73)90007-9

Bousquet, A., & Ivaldi, M. (1998). An individual choice model of energy mix. Resource and Energy Economics (pp. 263-286). Santa Barabara,

Garcia, S. (2002). Rendements et efficacité dans les industries en réseau: le cas des services d'eau potable délégués. *Economie & Prévision, 154*(3), 123-138.

Grebenstein, C. R., & Field, B. C. (1979). Substitution for Water Inputs in U.S. Manufacturing. *Water Resources Research, 15*(2), 228-232, New York. http://dx.doi.org/10.1029/WR015i002p00228

Greene, H. W. (1993). Econometric Analysis, Macmillan Publishing Company; New York, USA.

Mas-Colell, A., Whinston, M. D., & Green, J. R. (1995). Microeconomic theory, Oxford University Press. New York.

Mcelroy, M. B. (1987). Additive error models for production, cost and derived demand of share system. *Journal of Politic Economics, 95,* 737-757. http://dx.doi.org/10.1086/261483

Naief, A. M., & Nadeem, A. B. (2002). Factor substitution, and economies of scale and utilisation in Kuwait'scrude oil industry. *Energy Economics, 24,* Elsevier, Pages 337-354, North-Holland.

Ouedraogo, B. (2002). Eléments Économiques pour la Gestion de l'Offre et de la Demande du Bois-énergie dans la région de. Thèse de Doctorat Unique, Université de Ouagadougou & Université Montesquieu-Bordeaux IV.

Ouedraogo, M. M. (1974). L'Approvisionnement de Ouagadougou en Produits Vivriers, en Eau et en Bois., Thèse de Doctorat de 3° Cycle, C.V.R.S.: Université Bordeaux III. 353 pages.

Panayotou, T., & Sungswuan, S. (1996). An econometric analysis of the causes of tropical deforestation: The case of Nordeast Thailand. In K. Brown, & D. W. PEARCE (Eds.), *The Causes of Tropical Deforestation* (pp. 192-210). ULC Press. London.

RPTES (mai 2000). Demande Artisanale du bois-énergie comme Facteur de Production: cas des dolotières, restaurateurs, grilleurs de viande, fondeurs de bronze et d'aluminium, Ouagadougou.

Sadoulet, E., & De, J. A., (1995). Quantitative Development Policy Analysis. The Johns Hopkins Press Ltd, London.

Teeples, R., & Glyer, D. (1987). Production Functions for Water Delivery Systems: Analysis and Estimation Using Dual Cost Function and Implicit Piece Specification. *Water Resources Research, 23*(5), 765-773, California. http://dx.doi.org/10.1029/WR023i005p00765

Urga, G., & Walters, C. (2003). Dynamic translog and linear logit models :a factor demand analysis of interfuel substitution in US industrial energy demand. *Energy Economics, 25*(I), 1-21, Elsevier. http://dx.doi.org/10.1016/S0140-9883(02)00022-1

Wooldridge, J. M. (2000). Introductory Econometrics: A Modern Approach, South-Western College Publishing, Michigan State University.

Zellner, A. (1962). An Efficient Method of Estimating Seemingly Unrelated Regression and Tests for aggregation Bias. *Journal of the American Statistical Association, 57,* 348-368. http://dx.doi.org/10.1080/01621459.1962.10480664

Notes

Note 1. Many studies related to water and telecommunication manufacture industries demand for energy in the United States, have used translog as functional form [Hudson and Jorgenson (1974), Berndt and Wood (1975), Christensen and Green (1975), Berndt and Khaled (1979), Berndt and Morrison (1979), C. R. Grebenstein and B. C. Field (1979), Anderson (1980), Berndt et al. (1981), Evans and Heckman (1984), R. Teeples and D. Glyer

(1987), Berndt (1991) and Garcia S. (2002)].

Note 2. Equation (1) is the most general and flexible cost function specification which can be used to test diverse restrictions on production structure such as homotheticity, homogeneity, substitution possibilities, etc.

Note 3. Recall that the term YY =0 and that as such YY.lnY=0 in the elasticity expression.

Note 4. In reality, dolo is consumed per calabash.

Note 5. Nonetheless, one should note that reducing cooking time and / or achieving savings on fuelwood could come from artisans' production processes. Our thought is directed toward the specific case of "dolotières" where research has revealed the possibility of reducing quite considerably the cooking time or preparation time while obtaining the best beer.

Note 6. Applied Sciences and Technology Research Institute (IRSAT) is the former "Burkinabé Institute of Energy" which has already experienced the adaptation of gas fireplaces and studied yields on improved stoves with firewood.

13

Production of Pyrolysis Oil with Low Bromine and Antimony Contents from Plastic Material Containing Brominated Flame Retardants and Antimony Trioxide

Hu Wu[1], Yafei Shen[1], Noboru Harada[1], Qi An[1] & Kunio Yoshikawa[1]

[1] Department of Environmental Science and Technology, Tokyo Institute of Technology, Yokohama, Japan

Correspondence: Hu Wu, Department of Environmental Science and Technology, Tokyo Institute of Technology, Yokohama, Japan. E-mail: wuhu2010wuhu@gmail.com

Abstract

Thermal degradation of high impact polystyrene (HIPS) containing brominated flame retardants and antimony trioxide (Sb_2O_3) was conducted at different temperatures with the presence of various additives (red mud, limestone and natural zeolite) in a fixed-bed reactor. The effect of the pyrolysis temperature on the product yield and the bromine content in the oil product was investigated. It was found that the maximum oil yield (84.38 wt.%) was obtained at the pyrolysis temperature of 500 °C. The pyrolysis temperature had no significant impact on the bromine reduction in the oil products. The bromine in the flame retardant was mainly transferred into the oil products, where the bromine content was in the range of 7.96-8.56 wt.%. With the aim of removing bromine and antimony from the oils, three additives (red mud, limestone and natural zeolite) was used to investigate the influence on the product yield and composition, especially on the bromine and antimony removal ability from the oil products. In this study, it was found that all of the additives could significantly lower the bromine and antimony content in the oils and the red mud was the most effective. The presence of red mud could reduce the bromine and antimony content from 8.21 and 1.84 wt.% when no additive was employed to 0.84 and 0.35 wt.%, respectively. In addition, the distribution and fate of bromine and antimony in the residues were also studied by the SEM-EDX and XRD analysis in detail.

Keywords: E-waste plastic, Br-HIPS, Sb_2O_3-synergist, catalytic pyrolysis, debromination, fuel oil

1. Introduction

Waste electrical and electronic equipment (WEEE) are currently considered to be one of the fastest growing solid waste streams in the world. According to a report of UNEP (2009) about the recycling from WEEE to resources, 40 million tons of WEEE were generated and discharged annually in the world and it was expected an alarming growth per year in the future (United Nations Environment Programme, 2009). It was well known that there are lots of valuable metals and plastics contained in WEEE, which are worthy recyclable feedstock and could be converted into important mineral resources, fuel and chemical feedstock if recycled scientifically. On the other hand, WEEE also contained certain dangerous and hazardous substances, such as toxic metals and brominated flame retardants, which will pose considerable environmental pollution and health risks if treated inadequately (Yang, Sun, Xiang, Hu & Su, 2013; Ongondo, Williams & Cherret, 2011). Therefore, how to scientifically and cost-effectively reuse, recycle and recover WEEE has drawn plenty of attentions through the world.

WEEE plastics, which account for about 30% of the total weight of WEEE, are worthwhile recyclable parts of WEEE (Yang et al, 2013). One of the most popular plastics widely used in electrical and electronic equipment (EEE) is high-impact polystyrene (HIPS), because of its low cost and excellent impact resistance and machinability properties. HIPS is a composite material composed of a polystyrene phase and a dispersed polybutadiene rubber phase (Bhaskar et al, 2003). Polybrominated compounds and antimony trioxide (Sb_2O_3), as synergistic flame retardants, are frequently added to HIPS to reduce its flammability (Jakab, Uddin, Bhaskar & Sakata, 2003). Because of the presence of brominated flame retardants (BFRs), the traditional methods of dealing with WEEE plastics, such as land-filling and incineration, will produce secondary pollution on the ecological environment and endanger human health as well as being a waste of resource (Yang et al, 2013). For instance, the direct incineration of WEEE plastics containing brominated flame retardant will produce some highly toxic brominated dioxins and dibenzofurans (Ni et al, 2012). With a view to the environmental protection and saving non-regeneration resources, feedstock recycling technologies have been proposed as a viable processing route for converting WEEE plastics into useful fuels and chemical feedstock.

The pyrolysis and the catalytic pyrolysis are the most widely used feedstock recycling technologies for the conversion of WEEE plastics into valuable chemicals and fuel oil. Because of its low cost and easy operation, pyrolysis used for recycling WEEE plastics have been intensively investigated under different feedstocks, operation conditions, reactors, and increasing temperature stages (Yang et al, 2013). However, there were lots of organic brominated compounds remaining in the pyrolysis oils, which would reduce the quality and hinder the reuse of them. Compared with the thermal pyrolysis, the catalytic pyrolysis not only reduces the degradation temperature, but also obtains comparatively high-grade oil fuel without bromine. A wide range of catalysts have been tested for upgrading the quality and debromination of the pyrolysis oil derived from WEEE plastics and brominated flame retardants, such as zeolites (Hall & Williams, 2008; Bozi & Blazsó, 2009), FCC catalysts (Hall, Miskolczi, Onwudili & Williams, 2008), metallic oxides (Jung, S. Kim & J. Kim, 2012; Terakado, Ohhashi & Hirasawa, 2011; Terakado, Ohhashi & Hirasawa, 2013), etc. For example, Hall and Williams (2008) investigated the catalytic pyrolysis of brominated flame-retarded HIPS and ABS by using HY Zeolite and HZSM-5 Zeolite, which could effectively remove the organobromines from the pyrolysis products. Terakado and Hirasawa (2011; 2013) used the metal oxides in the pyrolysis of TBBPA and printed circuit boards including brominated flame retardants, respectively. They concluded that the addition of metal oxides could suppress the formation of HBr and brominated organic compounds.

However, from a practical industrial application point of view, the use of expensive catalysts, such as commercial HY and HZSM-5 zeolites, would increase the operation cost, due to a large amounts of catalyst demand and the deactivation of catalysts in a large continuously operating plant. Ali et al. (2002) indicated that the main factor is the catalyst cost for the economic comparison of catalytic cracking and thermal cracking technologies. In addition, Cardona et al. (2000) concluded that a plastic waste pyrolysis process could only be supported if the catalyst cost was practically zero (Lópeza et al, 2011). Therefore, it is necessary to develop the low-cost catalysts or additives for pyrolysis of WEEE plastic and debromination of brominated flame retardant.

Red mud, a solid waste product of the bauxite processing through the Bayer process, is mainly composed of Fe_2O_3, Al_2O_3, SiO_2 and TiO_2. Because of its special physicochemical properties, such as a high content of Fe_2O_3, a high surface area, the sintering resistance and a low cost, red mud have been used as catalysts for the hydrodechlorination and hydrogenation reactions (Sushil & Batra, 2008; Ordóñez, Sastre & Deíz, 2001; Álvarez, Ordóñez, Rosal, Sastre & Díez, 1999). In addition, it was found that, in the presence of Al_2O_3, SiO_2 and TiO_2, the red mud with some acidity could contribute to the catalytic cracking of plastic wastes (Lópeza et al, 2011). Limestone, as a common natural mineral, was also widely employed as a catalyst or an additive in the pyrolysis and gasification of biomass and plastic for the dehalogenation and tar removal (Hinz et al, 1994; Jung et al, 2012). The natural zeolite was also widely studied as catalysts for the catalytic pyrolysis of biomass and plastic to upgrade the liquid product due to its large surface area, acidity and sintering resistance property (Lee, Yoon, Kim & Park, 2002; Lee, Yoon, Kim & Park, 2001).

To our knowledge, there were few reported works on the use of red mud, limestone and natural zeolite for the WEEE plastic pyrolysis (Yanika, Uddinb, Ikeuchib & Sakata, 2001; Vasile et al, 2008). In this study, thermal degradation of HIPS containing BFRs and Sb_2O_3 was carried out at different temperatures with the presence of three additives (red mud, limestone and natural zeolite) in a fixed-bed reactor. The influence of the pyrolysis temperature on the product yield and bromine distribution in the oil products was evaluated systematically. In addition, the effect of three additives (red mud, limestone and natural zeolite) on the yields and compositions of products were investigated in details. Furthermore, the special attention was paid to study the content and distribution of bromine and antimony in the oils and residues, respectively, when the additives were used.

2. Experimental

2.1 Materials

The feedstock sample was high-impact polystyrene containing brominated flame retardant and antimony trioxide as a synergist, which was supplied by the PS Japan Corporation, Japan. The sample will be referred to as Br-HIPS. In order to mix the Br-HIPS sample with an additive uniformly, the pellet-type Br-HIPS sample was ground and sieved to obtain the powder Br-HIPS sample with a diameter smaller than 0.5mm. The proximate analysis and ultimate analysis of the Br-HIPS sample was presented in Table 1.

Table 1. Proximate analysis and ultimate analysis of the Br-HIPS sample

Proximate analysis	wt.%	Ultimate analysis	wt.%
Moisture	0.00	C	78.61
Volatile matter	98.13	H	7.11
Fixed carbon	0.34	O	0.75
Ash	1.53	N	0.10
		Br	9.30
		Sb	3.77
		Ti	0.36

The additives employed in this study were red mud, natural zeolite and limestone, which were obtained from the local company of Indonesia. Before the experiments, three additives were ground, sieved and then dried at 110 °C over the night. The red mud and natural was calcined at 500 °C for 2 hours to remove the organic impurities while the natural limestone was calcined at 900 °C for 4 hours to obtain CaO from the limestone. There were no any other activation operations for calcined red mud, natural zeolite and limestone. The additives will be referred to as RM, NZ and CL, respectively. Their chemical components and BET surface area are shown in Table 2.

Table 2. Chemical components and BET surface area of RM, NZ and CL (wt.%) (dry basis)

Additives	Fe_2O_3	SiO_2	Al_2O_3	CaO	TiO_2	Na_2O	MgO	Others[a]	BET surface area (m^2/g)
Red Mud	51.30	13.00	13.90	3.31	10.40	4.93	0.00	3.16	21.26
Calcined Limestone	0.00	0.97	0.42	97.60	0.00	0.00	0.32	0.69	7.49
Natural Zeolite	3.58	77.30	9.62	5.32	0.64	0.69	0.00	2.85	60.16

[a] By difference.

2.2 Experimental Setup

Fig. 1 is the schematic diagram of the experimental apparatus. The pyrolysis reactor was made of quartz. The inner diameter and inner height were 50 mm and 280 mm, respectively. In the thermal pyrolysis experiments without additives, the powder sample was fed into the reactor and was pyrolyzed over a range of temperature (450-550 °C) under the carrier gas of N_2 with a flow rate of 50 ml/min. When the additives were used, 40 g samples and 8g of each additive (20 wt.%) were mixed well, respectively, and fed into the reactor, which was heated to 500 °C at a heating rate of 50 °C/min. In the all experiments, after the reactor temperature reached the target temperature, it was held at this temperature for 2 hours and was then cooled quickly. The oil product was condensed in an oil collector, which was cooled by dry ice and ethanol mixture solution. In order to capture HBr, the gaseous products were scrubbed with 1 mol/L NaOH solution in the second trap before being collected in a tedlar bag. The every part was connected by silicon tubes.

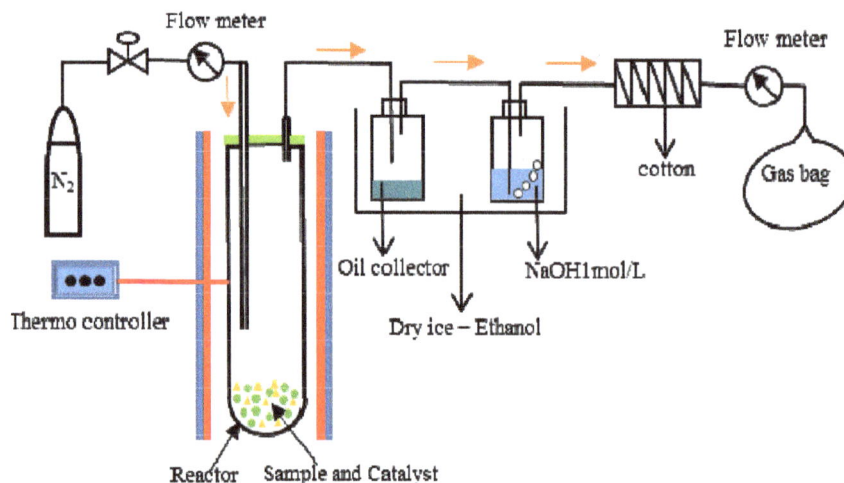

Fig. 1. Schematic diagram of the pyrolysis apparatus

2.3 Analytical Methods

The CHN element analysis was conducted by using a Micro Corder JM 10 Elemental Analyzer. The bromine contents of the sample and the pyrolysis oils were determined by using an air combustor coupled with Dionex ICS-1100 ion chromatography fitted with a Shodex IC S1-904E column according to JIS K 7392. The amount of antimony in the Br-HIPS sample and the oil products were determined by the inductively coupled plasma mass spectrometry (ICP-MS). The solution for the ICP-MS analysis was prepared by digesting 10 mg of sample or product in the mixed concentrated nitric and sulfuric acid in a sonicator for 3 hours in 200 °C. After digestion, the solution was diluted with distilled water to 50 ml for the ICP-MS analysis. The antimony and bromine contents in the pyrolysis residues were measured by a scanning electron microscopy with energy disperse X-ray analysis (SEM-EDX).

The chemical compositions of the catalysts were determined by an energy dispersive X-ray fluorescence spectrometer (XRF) under vacuum mode for precise measurement. A powder X-ray diffraction (XRD) analysis was carried out for the verification of the crystallinity of the fresh additives and used additives. XRD measurements were performed using a Rigaku Ultimal V diffractometer with the CuKα radiation (λ = 1.540) at 40kV and 40 mA. The XRD patterns were accumulated in the range of 5–80° every 0.02° (2θ) with the counting time of 1 s per step. The XRD patterns of three additives were presented in Fig. 2. The surface structure property was analyzed by the SEM. Surface area and textural properties of the used catalysts were determined by N_2 physical adsorption at 77 K, applying the Brunauer–Emmett–Teller (BET) method, using a Micromeritics Tristar 3020 equipment.

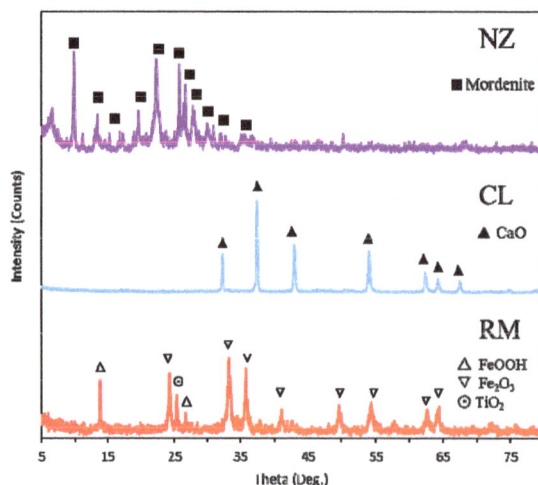

Fig. 2. The XRD patterns of RM, CL and RM

The yield of the oil product was measured by the weight difference of the silicon tube and the oil collector after and before the experiments. The yield of the solid product was determined by the weight difference of the reactor after and before the experiments. The yield of the gas product was calculated by subtracting the weight of the solid and oil products from the total weight of the sample.

The composition of the oil product was analyzed by a gas chromatograph coupled with a mass spectrometer (GC-MS) (Agilent 6890N, GC-MSD 5973N). The column was an HP5 (5% Ph-Me-Siloxane) capillary column, 30 m length with 0.25 mm diameter and 0.25 μm film thickness. Helium was used as the carrier gas. The injector temperature was 250 °C. The temperature program used was the initial temperature of 40 °C for 10 minutes followed by the heating rate of 5 °C /min to 300 °C and held at 300 °C for 20 minutes. The ion source and Quadrupole temperatures were 230 °C and 150 °C, respectively. The organobrominated compound in the oil products were measured by the gas chromatograph fitted with electron capture detectors (GC-ECD). In addition, the liquid pyrolysis products were further characterized using Fourier Transform Infrared Spectroscopy (FT-IR) with a JIR-SPX200 FT-IR spectrometer. The oil samples were mixed with KBr and pelletized. Then, it was scanned from 400 to 4000 cm^{-1} with the resolution of 4 cm^{-1}.

The composition of the produced gases in the experiments was measured by a micro gas chromatograph fitted with a thermal conductivity detector (GC-TCD) (Agilent Micro 3000), and the yield of each component was calculated by the following equation.

$$m_i = M_i \times \frac{C_i}{C_{N_2}} \times \frac{V_{N_2}}{22.4} \tag{1}$$

Where m_i = the yield of each gas product, M_i = the molar mass of each gas product; C_i = the concentration of the gas i in the gas products; C_{N_2} = the concentration of N_2 in the gas products; V_{N_2} = N_2 volume = the N_2 flow rate × the flow time.

3. Results and Discussion

3.1 TGA Results

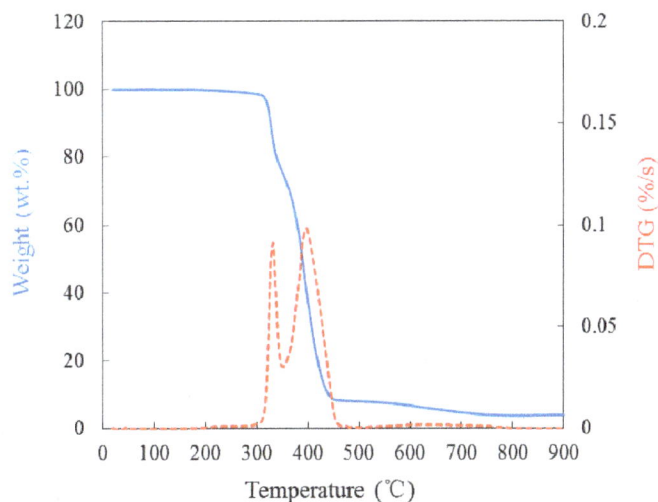

Fig. 3. TGA and DTG curve of Br-HIPS at 5°C/min

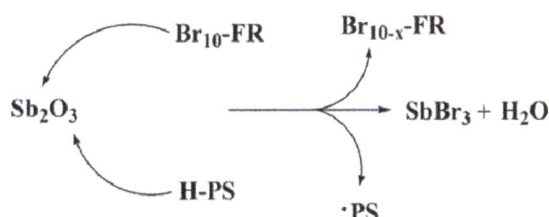

Fig. 4. First-stage degradation mechanism of Br-HIPS proposed by Bhaskar et al. (2003)

In order to investigate the thermal decomposition property of Br-HIPS, the thermogravimetric analysis (TGA) was conducted with a thermogravimetric analyzer (Shimadzu D50). 10 mg of HIPS-Br sample was loaded into the alumina crucible and heated from the ambient temperature to 900 °C with a heating rate of 5 °C/min. The flow rate of the carrier gas (N_2) was 150 mL/min. The TGA and DTG curves of Br-HIPS were shown in Fig. 3. The TGA curves indicated that the decomposition of Br-HIPS mainly occurred between 310 and 450 °C, followed by a stable mass reduction at higher temperatures. The DTG curve obviously illustrated that the pyrolysis of Br-HIPS mainly included two distinct decomposition stages from 310 to 350 °C and from 360 to 450 °C, respectively. This result corresponds to the previous reported works (Bhaskar et al, 2003; Jung et al, 2012) that the first decomposition stage was attributed to the presence of Sb_2O_3 as a synergist, which will reacted with BFRs and partial polystyrene by the dehydrogenation and debromination reactions, as presented in Fig. 4. The decomposition of the majority of Br-HIPS took place in the second stage by the β-scission reaction (Lee et al, 2001). Furthermore, the Br-HIPS pyrolysis residues were about 7 wt.% and 2 wt.% at 600 and 900 °C, respectively, which indicated that the remaining residue could be further decomposed at a higher temperature. These thermogravimetric analysis results could be used for determining the suitable pyrolysis temperatures for the following experiments in the fixed bed reactor.

3.2 Effect of the Pyrolysis Temperature on the Product Yields and the Bromine Content in the Produced Oils

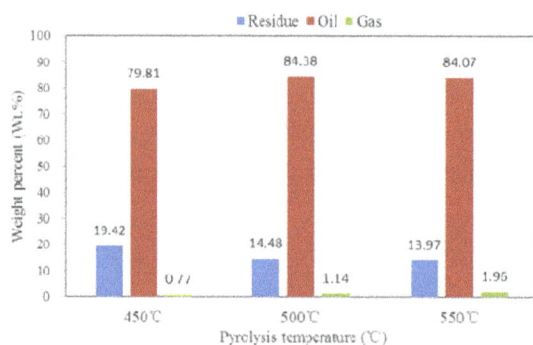

Fig. 5. Effect of the pyrolysis temperature on the product yield

It is well known that the reaction temperature is an essential parameter for both the plastic pyrolysis degradation and the catalyst activation (Park, Kang & Kim, 2008). Therefore, a preliminary study of Br-HIPS pyrolysis at different temperatures was conducted in an attempt to investigate the impact of the pyrolysis temperature on the product yield and the bromine content in the oil products. The effect of the pyrolysis temperature on the product yield is shown in Fig. 5. It indicated that as the increasing of the pyrolysis temperature from 450°C to 500°C, the yield of the oil and gaseous products increased, accompanied by the decrease of the yield of the solid residue, which was attributed to the intense cracking of Br-HIPS under the higher pyrolysis temperature resulting in a higher conversion ratio of samples into the gaseous and liquid products (Aguado, Serrano, Miguel, Castro & Madrid, 2007; Syamsiro et al, 2013). When the pyrolysis temperature increased from 500°C to 550°C, the yield of the gaseous product increased at the expense of the liquid product and the solid residue, which implied that a higher pyrolysis temperature would further convert some liquid products into small-molecular gaseous products. However, the change was slight, which indicated that the pyrolysis temperature of 500°C was high enough for the thermal degradation of Br-HIPS. The maximum oil product yield (84.4 wt. %) was obtained at the pyrolysis temperature of 500°C.

The total bromine content in the pyrolysis oils was measured using a bomb calorimeter equipped with an ion chromatograph (JIS K 7392), and the result is shown in Table 3. This result clearly shows that there was a high bromine content in each oil product, which means large amounts of bromine was transferred to the oil products during the pyrolysis of Br-HIPS. With the increase of the pyrolysis temperature, the bromine content increased slightly. When the pyrolysis temperature was 550°C, 77.4 wt. % of the total bromine in the Br-HIPS sample was transferred into the oil products. Furthermore, it should be mentioned that there are some solid precipitates existed in the oil phase. Because Sb_2O_3 might react with HBr derived from the thermal degradation of BFRs and then generate $SbBr_3$ with a low boiling point, which easily evaporated from the reactor and condensed in the oil section, resulting in the high bromine contents in the oil products, as shown in Fig. 4.

Table 3. Total bromine contents in the oil products derived from the pyrolysis of Br-HIPS at different temperatures.

	450°C	500°C	550°C
Br in oil (wt. %)[a]	7.96	8.21	8.56
Yield of bromine in the oil (%)[b]	68.31	74.49	77.38

[a] Br in oil (wt. %) = mass of bromine in oil (g) /mass of oil (g) ×100;

[b] Yield of Br (%) = mass of Br in oil/ mass of Br in plastic ×100.

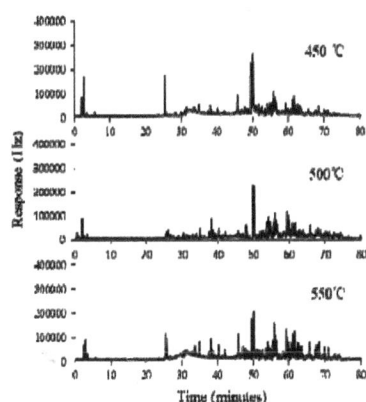

Fig. 6. GC-ECD chromatograms of the oils produced by the pyrolysis of Br-HIPS at different temperatures

The total bromine content in the oil products contained the organobrominated and inorganic brominated compounds, such as antimony bromide. As for the organobromine content in the pyrolysis oils, it was determined by the GC-ECD analysis, which only responds to organo-halogenated compounds, in terms of the organobrominated compounds in this study. In each analysis, the same volume of oils was injected and analyzed at the same GC conditions (Miskolczi, Hall, Angyal, Bartha & Williams, 2008). Therefore, the chromatographs can be generally compared with each other to reveal the variation of the organobrominated compounds by the number, size and intensity of ECD peaks. Fig. 6 shows the results of the GC-ECD analysis of the oil products obtained at different pyrolysis temperatures. It apparently illustrated that a large amount of organobrominated compounds also existed in the pyrolysis oil. As the increase of the pyrolysis temperature, the intensity of the ECD peak increased, especially the peaks in the later retention time.

3.3 Effect of Additives on the Pyrolysis of Br-HIPS

3.3.1 Effect of Additives on the Gas and Oil Products

As stated above, a high total bromine content and lots of organobrominated compounds retained in the oil products, and the effect of the pyrolysis temperature on them was limited. Therefore, the effective catalysts or additives were essential for the debromination to obtain valuable oil products. RM, CL and NZ were industrial solid wastes or natural mineral. Because of their low-cost, high metal oxides content and catalytic cracking properties, they were frequently applied as catalysts or additives for the pyrolysis of biomass and plastics (Yathavan & Agblevor, 2013; Lópeza et al, 2011; Lee, 2001; Zhang, Zhang, Yang & Yan, 2014).

Table 4. Main components of gas and oil products (wt.%/sample mass)

Main products	Thermal	+RM	+CL	+NZ
Main gaeous components				
Hydrogen	0.02	0.00	0.00	0.05
Methane	0.15	0.29	0.16	0.27
Ethene	0.14	0.35	0.15	0.49
Ethane	0.16	0.43	0.18	0.60

Propene	0.21	0.49	0.22	0.64
Propane	0.17	0.65	0.14	0.69
Main oil components				
Single ring compounds				
Toluene	6.96	3.16	3.78	3.39
Ethylbenzene	28.60	11.80	11.03	16.20
Styrene	0.48	25.10	31.17	19.69
Cumene	4.29	2.42	1.96	3.86
α-Methylstyrene	0.92	1.47	3.05	1.37
Two ring compounds				
1,3-Diphenylpropane	18.02	8.98	6.93	11.43
1,3-diphenyl-1-butene	4.44	1.03	1.49	2.29
2-Methylnaphthalene	0.14	0.10	1.32	1.37
1-Methyl-3-phenylindan	0.47	1.86	0.79	2.60
Multi ring compounds				
2-Benzylnaphthalene	0.92	1.19	0.89	1.41
Anthracene	n.d.[a]	1.01	0.87	1.49
9-Ethenylanthracene	0.28	0.41	0.22	0.83
1,3,5-Triphenylbenzene	0.85	0.47	0.60	0.35
Brominated compounds				
(1-Bromoethyl)benzene	2.75	0.53	0.51	0.97
3-Methylbenzyl bromide	0.47	n.d.	n.d.	0.10
9,10-Dibromoanthracene	0.11	0.15	0.15	0.24

[a] Not detected.

In this study, in order to investigate the effect of RM, CL and NZ additives on the degradation of Br-HIPS and the debromination characteristic, the pyrolysis experiments were carried out in presence of each additive at the pyrolysis temperature of 500°C. Because of the reaction of the additives and the evolved HBr, the accurate mass balance and the product yields of Br-HIPS degradation were difficult to be calculated. As for the gaseous products, they were measured by the GC-TCD and the results were listed in Table 4. It was found that when the Br-HIPS sample was pyrolyzed without additives, propene and propane were the main gaseous products. When CL was used in the reaction system, there was no obvious change in the gaseous product. When the Br-HIPS sample was pyrolyzed with RM and NZ, respectively, the yields of gaseous products increased significantly, especially in the case of NZ. It was presumably because of the zeolite property (acidity and large surface area) of RM and NZ, which could exert a catalytic cracking effect during the pyrolysis of Br-HIPS and promoted the gaseous products. In addition, the bromine content of the gas products was almost zero, regardless of the presence or absence of each additive. This result was consistent with the results reported by other researchers (Hall & Williams, 2008).

The oil products evolved from the pyrolysis of Br-HIPS in the absence and presence of additives were analyzed by GC-MS, and the main components of them were also listed in Table 3. It illustrated that toluene, ethylbenzene, styrene and 1,3-diphenylpropane were the main components in oil products, over 50 wt. % of Br-HIPS sample weight. It was in well agreement with the previous findings (Hall & Williams, 2008; Miskolczi et al, 2008; Jung et al, 2012), which was a consequence of the structure property and the thermal degradation property of the HIPS matrix. In addition, in the case of without additives, the ethylbenzene, cumene and 1,3-diphenylpropane were the main components of the oil products while the small amount of styrene existed in the oil. It was attributed to the presence of Sb_2O_3 that reacted with HBr derived from the decomposition of BFRs to produce $SbBr_3$ with an acidity nature, which played a catalytic role to convert the styrene and α-methylstyrene into ethylbenzene and

1,3-diphenylpropane by the intermolecular H transfer and the intermolecular carbanion.

However, when the additives were added, the yield of styrene increased a lot, accompanied by the decrease of ethylbenzene and 1,3-diphenylpropane. On the one hand, it partially originated from the fact that the main components of these three additives were metal oxides, such as Fe_2O_3 and CaO, which would exert basic-catalytic impact on the degradation of HIPS to form carboanions by the H abstraction, resulting in the increase of styrene in the oil products (Lee et al, 2001). On the other hand, the existence of metal oxides in the additives would inhibit the formation of $SbBr_3$ due to the competing reaction with HBr generated by the degradation of BFRs. In addition, owing to the pore property, the additives also could capture the produced volatile $SbBr_3$, subsequently reducing its amount in the oil.

In addition, when the Br-HIPS sample was pyrolyzed in the presence of RM, CL and NZ, respectively, the amount of styrene in the oil products was obviously different. It was probably caused by the differences in the composition and the surface area of additives, as shown in Table 1. In the case of RM, it was mainly attributed to the fact that red mud also contained the SiO_2, Al_2O_3 and TiO_2 components and forms the Si-O-T units (T = Si or Al), which makes RM to have zeolite nature (Brønsted acid) and play catalytic action to increase the amount of ethybenzene by further hydrogenation of styrene (Wang, Ang & Tadé, 2008). As for NZ, a high ratio of Al_2O_3 and SiO_2 in NZ will also improve the acidity of NZ. Therefore, the oil derived from the pyrolysis of Br-HIPS and NZ contained smaller amounts of styrene and larger amounts of ethylbenzene. Besides, the high amounts of 2-methylnaphthalene, anthracene and 9-ethenylanthracene compounds in the oil were detected due to the Diels-Alder reaction, which was proposed by Koo et al (1991). It also illustrated the stronger acidity of NZ than the other two additives.

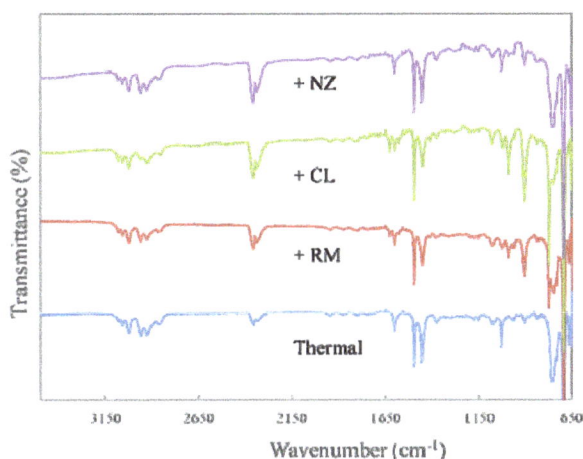

Fig. 7. FT-IR spectra of oil products evolved from the pyrolysis of Br-HIPS without additive, with RM, CL and NZ additives (thin film/KCl)

For the purpose of verifying the molecular structure and the organic functional groups, the oil products were further characterized by FT-IR analysis and the result is presented in Fig. 7. Generally, the band in the 3150-3000 cm^{-1} range is associated with the vibration of C-H bond in the aromatic ring. The strong bands observed in the interval of 900-675 cm^{-1} were assigned of C-H out-of-plane vibration of the mono-substituted aromatic ring, which further confirmed the aromatic characteristics of the oil products. Additionally, the vibration bands in the region of 3000-2900 cm^{-1} correspond to the C-H asymmetric stretching of CH_3, CH_2 and CH groups. The vibration bands at 992 and 906 cm^{-1} are the typical vibration mode of $CH=CH_2$ in the vinyl compounds and the bands at 1450 cm^{-1} are ascribed to the asymmetric bending vibration of ethyl groups. The aliphatic functional groups observed in the FTIR spectra are probably attributed to the alkyl groups attached to the aromatic rings.

Comparing the FTIR spectra of the oil products derived from Br-HIPS pyrolysis in the absence and presence of additives, it was observed that the vibration bands at 992 and 906 cm^{-1}, corresponding to $CH=CH_2$ bonds in the vinyl compounds, in the oil from the Br-HIPS pyrolysis with additives are more intensive than that of oil from the Br-HIPS pyrolysis without additives. Whereas, the bands at 1450 cm^{-1} are ascribed to the asymmetric bending vibration of ethyl groups and become weaker than that when no additive was used. The above variations corroborated the GC-MS analytical results that when additives were used, the amount of ethylbenzene decreased, accompanied by the increase of styrene. In addition, it should be noted that the bond vibration bands at 1030 cm^{-1} were typical of aryl-bromines (Hall & Williams, 2008; Miskolczi et al, 2008) and when the additives were used, the intensity of this band became weakened, which was also well associated with those obtained by the GC-MS analysis.

3.3.2 Effect of Additives on Bromine and Antimony Contents in the Oil Products

The total bromine and antimony contents in the pyrolysis oils were determined using a bomb calorimeter equipped with an ion chromatograph and ICP-MS after digestion of oil samples, respectively. The bromine and antimony contents of each oil product were listed in Table 5. It was apparent that the additives played a positive role in the removal of bromine and antimony contents from the pyrolysis oils. At the same pyrolysis temperature of 500°C, the bromine content of 8.21wt.% in the pyrolysis oil of Br-HIPS degradation without additives was reduced dramatically to 0.84wt.%, 0.91wt.% and 3.47wt.%, when RM, CL and NZ were used as the additive, respectively. RM was more effective than the other two additives in the respect of bromine removal, which was attributed to the fact that, on the one hand, Fe_2O_3 in the red mud would react with HBr derived from the degradation of BFRs and hinder the formation of volatile $SbBr_3$, on the other hand, the zeolite property of RM (Lópeza et al, 2011) could catalytically degrade the organobromine compounds. The effect of NZ on the bromine removal was weakest. It was mainly due to the fact that Al_2O_3 and SiO_2 were the main components, which could not react with bromine radicals derived from the degradation of BFRs. Therefore, it illustrated that the reaction between bromine and metal oxides was the dominant mechanism of the bromine fixation when the additives were used. In addition, the antimony content of the oil product was measured and the result is also shown in Table 5. When additives were employed, the antinomy content in the oils was also reduced a lot. The main reasons were already explained in the 3.3.1 session.

Table 5. Total bromine and antimony contents in the oil products derived from the pyrolysis of Br-HIPS with and without additives at the temperature of 500°C

	Thermal	+RM	+CL	+NZ
Br in oil (wt. %)	8.21	0.84	0.91	3.47
Sb in oil (wt. %)	1.84	0.35	0.52	0.69

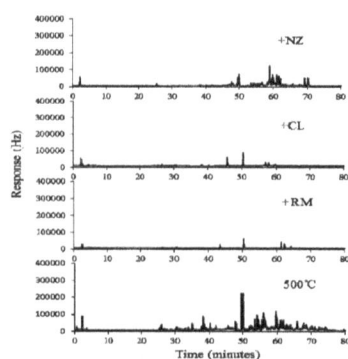

Fig. 8. GC-ECD chromatographs of the oils produced from the pyrolysis of Br-HIPS without and with additives at the temperature of 500°C

The main purpose of using the additives was to remove the organobromine from the pyrolysis oils and obtained the clean oil products. The organobromine compounds in the oils were analyzed by the GC-ECD and the results are presented in Fig. 8. When the Br-HIPS sample was pyrolyzed with NZ, certain portion of the organobromine compounds still retained in the pyrolysis oil. Meanwhile, when RM and CL were applied, the intensity of ECD peaks apparently reduced dramatically. It implied that RM and CL could also effectively suppress the formation of organobrominated compounds and produce the valuable oil products.

3.3.3 Pyrolysis Residues

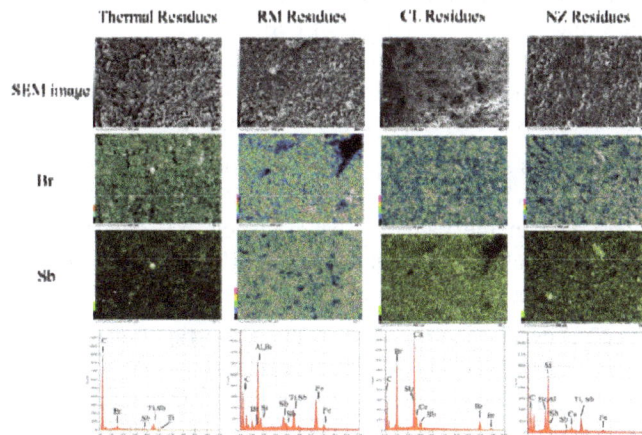

Fig. 9. SEM and EDX analysis of the residues derived from the pyrolysis of Br-HIPS sample without and with RM, CL and NZ, respectively

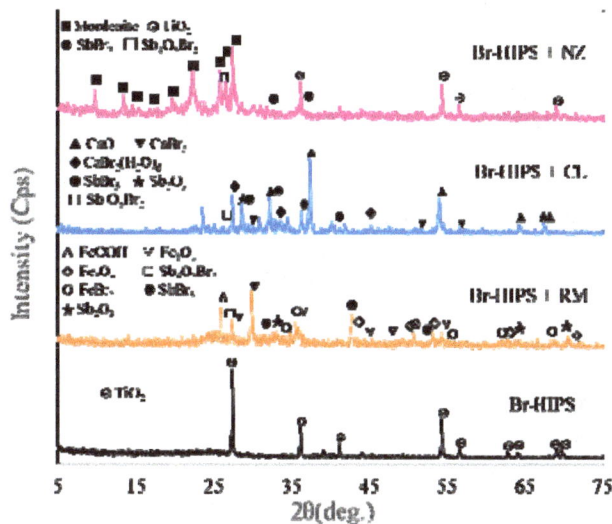

Fig. 10. XRD patterns of residues derived from the pyrolysis of Br-HIPS sample without and with RM, CL and NZ, respectively

When the Br-HIPS sample was pyrolyzed in the presence of catalysts, both bromine and antimony contents in the oils decreased dramatically. The bromine in the flame retardants is converted to not only HBr and the brominated organic compounds but also metal bromides left in the residues. However, because of very heterogeneous and intermingled nature of the residues, it was impossible to take representative sampling to analyze precise contents of Br and Sb in the residues (Hall et al, 2008). Therefore, the residues were qualitatively analyzed by the SEM-EDAX and XRD in order to determine and compare the additive change, residue compositions and bromine and antimony distributions in the residues. The results are presented in Fig. 9 and Fig. 10, respectively. In the EDAX pattern, when Br-HIPS was pyrolyzed in the absence of additives, the residue was found to contain carbon, titanium, bromine and antimony. The carbon can be caused by char formation. Only few of bromine and antimony was left in the residues. And from the XRD pattern, neither Sb_2O_3 nor $SbBr_3$ crystal was identified in the residues in Fig. 10, whilst only the TiO_2 crystal was observed, which is commonly employed as an additive in plastics (Hall et al, 2008). It was attributed to the reaction between Sb_2O_3 with HBr, derived from the BFRs decomposition, thereby generating the volatile $SbBr_3$ (Rzyman, Grabda, Oleszek-Kudlak, Shibata & Nakamura, 2010; Grause, Karakita, Kameda, Bhaskar & Yoshioka, 2012). Therefore, major parts of antimony and bromine were found in the oil rather the char when Br-HIPS was pyrolyzed (Jakab et al, 2003; Jung et al, 2012).

In the case of Br-HIPS and RM pyrolysis, there were lots of bromine and antimony elements were observed in the residues from the EDAX pattern. When compared with XRD pattern of RM in Fig. 2, the peaks of Fe_2O_3 and FeOOH disappeared or weakened while the additional peaks of $FeBr_2$, $Sb_4O_5Br_2$, $SbBr_3$ and Sb_2O_3 were detected. $FeBr_3$ was not identified in the residues, which owes to that $FeBr_3$ decomposes above 200°C to form Br_2 and $FeBr_2$ (Bhaskar et al, 2003; Terakado et al, 2011; Terakado et al, 2013). In addition, it was found that certain amount of Fe_2O_3 was reduced into Fe_3O_4 by the hydrogenation reaction.

When Br-HIPS was pyrolyzed with CL, as shown in Fig. 9, lots of bromine and antimony were also detected in the residues. From the XRD pattern, it was found that the peaks of CaO became weaker while the new peaks of $CaBr_2$ and $CaBr_2(H_2O)_6$ were observed in the pattern. The existence of $CaBr_2(H_2O)_6$ was attributed to the hygroscopic nature of $CaBr_2$ (Terakado et al, 2011; Terakado et al, 2013). In addition, the Sb_2O_3 were detected, which further confirmed the presence of lots of antimony existing in the residues.

As for the residues of Br-HIPS and NZ pyrolysis, no reflection characteristic of metal bromides was presented in the XRD pattern, except for $SbBr_3$ and $Sb_4O_5Br_2$. It was related to the fact that natural zeolite is mainly composed of SiO_2 and Al_2O_3, which would not react with evolved HBr. The amount of other metal oxides, such as CaO and Fe_2O_3, is few in NZ and then produced metal bromide species with too small particle sizes on the surface of the residue to be identified by the XRD method. As for NZ, the fixation of bromine and antimony was mainly associated with the physical adsorption of $SbBr_3$ and $Sb_4O_5Br_2$ on the surface of NZ (Hall et al, 2008). When compared with that of RM and CL, the bromine fixation ability of NZ was weaker and then the bromine and antimony content in the oils were relatively higher.

4. Conclusion

In this study, the pyrolysis of high impact polystyrene containing BFRs and Sb_2O_3 was carried out in the absence and presence of three additives (red mud, calcined limestone and natural zeolite) in a fixed bed reactor. In the pyrolysis of Br-HIPS without additives, with the increase of the pyrolysis temperature from 450°C to 550°C, the yield of the residue decreased while the yield of the gas increased slightly. The yield of the oil product was dependent upon the pyrolysis temperature, and the maximum oil yield (84.4 wt.%) was obtained at the pyrolysis temperature of 500°C. In the pyrolysis of Br-HIPS with the additives, it was found that red mud and natural zeolite could increase the yield of the gaseous product because of their zeolite property. In addition, the use of additives, especially red mud and calcined limestone, could increase the amount of styrene at the expense of ethylbenzene and 1,3-diphenylpropane. Furthermore, the additives could also upgrade the oil products by reducing the total bromine content, organobrominated compounds and antimony content in the oil products significantly. Red mud was the most effective additives used in this study. Because, on the one hand, red mud played a cracking catalyst effect on destroying the organobrominated compounds. On the other hand, it could work as a sorbent to fix HBr formed by the Br-HIPS degradation. The bromine fixation abilities of red mud and calcined limestone were similar with each other and better than that of natural zeolite, which implied that the reaction between bromine and metal oxides was the dominant mechanism of the bromine fixation. Additionally, from the XRD and SEM-EDAX analysis of the residues, it was found that there were lots of metal bromides compounds existing in the residues, which further confirmed the above bromine fixation mechanism.

Acknowledgement

The authors would like to thank Dr. Tohru Kamo of Advanced Industrial Science and Technology (AIST), Tsukuba, Japan for his valuable help and discussion.

References

Aguado, J., Serrano, D. P., Miguel, G. S., Castro, M. C., & Madrid, S. (2007). Feedstock recycling of polyethylene in a two-step thermo-catalytic reaction system. *Journal of Analytical and Applied Pyrolysis, 79*, 415-423. http://dx.doi.org/10.1016/j.jaap.2006.11.008

Ali, S., Garforth, A. A., Harris, D. H., Rawlence, D. J., & Uemichi, Y. (2002). Polymer waste recycling over "used" catalysts. *Catalysis Today, 75*, 247–255. http://dx.doi.10.1016/S0920-5861(02)00076-7

Álvarez, J., Ordóñez, S., Rosal, R., Sastre, H. V., & Díez, F. V. (1999). A new method for enhancing the performance of red mud as a hydrogenation catalyst. *Applied Catalysis A: General, 180*, 399-409. http://dx.doi.10.1016/S0926-860X(98)00373-1

Bhaskar, T., Matsui, T., Uddin, M., A., Kaneko, J., Mutoa, A., & Sakata, Y. (2003). Effect of Sb2O3 in brominated heating impact polystyrene (HIPS-Br) on thermal degradation and debromination by iron oxide carbon composite catalyst (Fe-C). *Applied Catalysis B: Environmental, 43*, 229-241. http://dx.doi.10.1016/S0926-3373(02)00306-5

Blazsó, M., & Czégény, Z. (2006). Catalytic destruction of brominated aromatic compounds studied in a catalyst microbed coupled to gas chromatography/mass spectrometry. *Journal of Chromatography A, 1130*, 91–96. http://dx.doi.10.1016/j.chroma.2006.05.009

Bozi, J., & Blazsó, M. (2009). Catalytic modification of pyrolysis products of nitrogen-containing polymers over Y zeolites. *Green Chemistry, 11*, 1638–1645. http://dx.doi.10.1039/b913894n

Cardona, S., C., & Corma, A. (2000). Tertiary recycling of polypropylene by catalytic cracking in a semibatch stirred reactor Use of spent equilibrium FCC commercial catalyst. *Applied Catalysis B: Environmental, 25*, 151–162. http://dx.doi.10.1016/S0926-3373(99)00127-7

Grause, G., Karakita, D., Kameda, T., Bhaskar, T., & Yoshioka, T. (2012). Effect of heating rate on the pyrolysis of high-impact polystyrene containing brominated flame retardants: Fate of brominated flame retardants. *Journal of Material Cycles and Waste Management, 14*, 259–265. http://dx.doi.10.1007/s10163-012-0067-8

Hall, W., & Williams, P. T. (2008). Removal of organobromine compounds from the pyrolysis oils of flame retarded plastics using zeolite catalysts. *Journal of Analytical and Applied Pyrolysis, 81*, 139-147. http://dx.doi.10.1016/j.jaap.2007.09.008

Hall, W., J., Miskolczi, N., Onwudili, J., & Williams, P., T. (2008). Thermal Processing of Toxic Flame-Retarded Polymers Using a Waste Fluidized Catalytic Cracker (FCC) Catalyst. *Energy & Fuels, 22*, 1691–1697. http://dx.doi.10.1021/ef800043g

Hinz, B., Hoffmockel, M., Pohlmann, K., Schädel, S., Schimmel, I., & Sinn, H. (1994). Dehalogenation of pyrolysis products. *Journal of Analytical and Applied Pyrolysis, 30*, 35-46. http://dx.doi.10.1016/0165-2370(94)00800-0

Jakab, E., Uddin, M. A., Bhaskar, T., & Sakata, Y. (2003). Thermal decomposition of flame-retarded high-impact polystyrene. *Journal of Analytical and Applied Pyrolysis, 68-69*, 83-89. http://dx.doi.10.1016/S0165-2370(03)00075-5

Jung, S. H., Kim, S. J., & Kim, J. S. (2012). Fast pyrolysis of a waste fraction of high impact polystyrene (HIPS) containing brominated flame retardants in a fluidized bed reactor: The effects of various Ca-based additives (CaO, Ca(OH)$_2$ and oyster shells) on the removal of bromine. *Fuel, 95*, 514-520. http://dx.doi.10.1016/j.fuel.2011.11.048

Koo, J. K., Kim, S. W., & Seo, Y. H. (1991). Characterization of aromatic hydrocarbon formation from pyrolysis of polyethylene-polystyrene mixtures. *Resources, Conservation and Recycling, 5*, 365-382. http://dx.doi.10.1016/0921-3449(91)90013-E

Lee, S. Y., Yoon, J. H., Kim, J. R., & Park, D. W. (2001). Catalytic degradation of polystyrene over natural clinoptilolite zeolite. *Polymer Degradation and Stability, 74*, 297–305. http://dx.doi.10.1016/S0141-3910(01)00162-8

Lee, S. Y., Yoon, J. H., Kim, J. R., & Park, D. W. (2002). Degradation of polystyrene using clinoptilolite catalysts. *Journal of Analytical and Applied Pyrolysis, 64*, 71–83. http://dx.doi.10.1016/S0165-2370(01)00171-1

Lópeza, A., Marco, I. D., Caballeroa, B. M., Laresgoiti, M. F., Adrados, A., & Aranzabal, A. (2011). Catalytic pyrolysis of plastic wastes with two different types of catalysts: ZSM-5 zeolite and Red Mud. *Applied Catalysis B: Environmental, 104*, 211–219. http://dx.doi.10.1016/j.apcatb.2011.03.030

Miskolczi, N., Hall, W. J., Angyal, A., Bartha, L., & Williams, P. T. (2008). Production of oil with low organobromine content from the pyrolysis of flame retarded HIPS and ABS plastics. *Journal of Analytical and Applied Pyrolysis, 83*, 115–123. http://dx.doi.10.1016/j.jaap.2008.06.010

Ni, M. J., Xiao, H. X., Chi, Y., Yan, J. H., Buekens, A., Jin, Y. Q., & Lu, S. Y. (2012). Combustion and inorganic bromine emission of waste printed circuit boards in a high temperature furnace. *Waste Management, 32*, 568–574. http://dx.doi.10.1016/j.wasman.2011.10.016

Ongondo, F. O., Williams, I. D., & Cherrett, T. J. (2011). How are WEEE doing? A global review of the management of electrical and electronic wastes. *Waste Management, 31*, 714–730. http://dx.doi.10.1016/j.wasman.2010.10.023

Ordóñez, S., Sastre, H., & Deíz, F. V. (2001). Characterisation and deactivation studies of sulfided red mud used as catalyst for the hydrodechlorination of tetrachloroethylene. *Applied Catalysis B: Environmental, 29*, 263–273. http://dx.doi.org/10.1016/S0926-3373(00)00207-1

Park, E. S., Kang, B. S., & Kim, J. S. (2008). Recovery of Oils with High Caloric Value and Low Contaminant Content by Pyrolysis of Digested and Dried Sewage Sludge Containing Polymer Flocculants. *Energy & Fuels, 22*, 1335–1340. http://dx.doi.10.1021/ef700586d

Rzyman, M., Grabda, M., Oleszek-Kudlak, S., Shibata, E., & Nakamura, T. (2010). Studies on bromination and evaporation of antimony oxide during thermal treatment of tetrabromobisphenol A (TBBPA). *Journal of Analytical and Applied Pyrolysis, 88*, 14–21. http://dx.doi.10.1016/j.jaap.2010.02.004

Sushil, S., & Batra, V. S. (2008). Catalytic applications of red mud, an aluminium industry waste: A review. *Applied Catalysis B: Environmental, 81*, 64–77. http://dx.doi.10.1016/j.apcatb.2007.12.002

Syamsiro, M., Wu, H., Komoto, S., Cheng, S., Noviasri, P., Prawisudha, P., & Yoshikawa, K. (2013). Co-Production of Liquid and Gaseous Fuels from Polyethylene and Polystyrene in a Continuous Sequential Pyrolysis and Catalytic Reforming System. *Energy and Environment Research, 3*, 2. http://dx.doi.org/10.5539/eer.v3n2p90

Terakado, O., Ohhashi, R., & Hirasawa, M. (2011). Thermal degradation study of tetrabromobisphenol A under the presence metal oxide: Comparison of bromine fixation ability. *Journal of Analytical and Applied Pyrolysis, 91*, 303–309. http://dx.doi.10.1016/j.jaap.2011.03.006

Terakado, O., Ohhashi, R., & Hirasawa, M. (2013). Bromine fixation by metal oxide in pyrolysis of printed circuit board containing brominated flame retardant. *Journal of Analytical and Applied Pyrolysis, 103*, 216–221. http://dx.doi.org/10.1016/j.jaap.2012.10.022

United Nations Environment Programme [UNEP]. (2009). Retrieved from http://www.unep.org/yearbook/2009/

Vasile, C., Brebu, M., A., Totolin, M., Yanik, J., Karayildirim, T., & Darie, H. (2008). Feedstock Recycling from the Printed Circuit Boards of Used Computers. *Energy & Fuels, 22*, 1658–1665. http://dx.doi.10.1021/ef700659t

Wang, S. B., Ang, H. M., & Tadé, M. O. (2008). Novel applications of red mud as coagulant, adsorbent and catalyst for environmentally benign processes. *Chemosphere, 72*, 1621–1635. http://dx.doi.10.1016/j.chemosphere.2008.05.013

Yang, X. N., Sun, L. S., Xiang, J., Hu, H., & Su, S. (2013). Pyrolysis and dehalogenation of plastics from waste electrical and electronic equipment (WEEE): A review. *Waste Management, 33*, 462–473. http://dx.doi.org/10.1016/j.wasman.2012.07.025

Yanik, J., Uddin, M. A., Ikeuchi, K., & Sakata, Y. (2001). The catalytic effect of Red Mud on the degradation of poly (vinyl chloride) containing polymer mixture into fuel oil. *Polymer Degradation and Stability, 73*, 335–346. http://dx.doi.10.1016/S0141-3910(01)00095-7

Yathavan, B. K., & Agblevor, F. A. (2013). Catalytic Pyrolysis of Pinyon–Juniper Using Red Mud and HZSM-5. *Energy & Fuels, 27*, 6858-6865. http://dx.doi.org/10.1021/ef401853a

Zhang, L., Zhang, B., Yang, Z. Q., & Yan, Y. F. (2014). Pyrolysis behavior of biomass with different Ca-based additives. *RSC Advances, 4*, 39145–391555. http://dx.doi.10.1039/c4ra04865b

14

Impact of Building and Refrigeration System's Parameters on Energy Consumption in the Potato Cold Storage

Ramkishore Singh[1], S P Singh[2] & I J Lazarus[1]

[1] Department of Physics, Durban University of Technology, Durban 4000, South Africa

[2] School of Energy & Environmental Studies, Takashashila Campus, Khandwa Road, Devi Ahilya University, Indore- 452017, India

Correspondence: Ramkishore Singh, Department of Physics, Durban University of Technology, Durban 4000, South Africa.E-mail: singh.ramkishore@gmail.com

Abstract

Cold storage is a key element for long term preservation and distribution of perishable items. A large quantity of perishable items is gone wastage every year due to inadequate storage facility in the countries like India. Their energy extensive refrigeration process makes them accountable for substantial share of total energy consumption in the country. Moreover, energy inefficiency in operation and building designs in the existing cold storages inhibits expansion of desirable storage facility. This study aims to explore the building and refrigeration system's parameters that may influence the overall energy consumption in the storage. A survey was conducted in ten Indian potato cold storages and detail data of their energy consumption, building construction and refrigeration systems was collected. The collected data was analyzed quantitatively and qualitatively. Trend analysis technique was also used to establish relationship between specific energy consumption and identified parameters. Results show that the utilization factor has major impact on the energy consumption in the storages and the existing cold storages are not utilizing their storage capacities fully. Comparatively higher energy consumption was observed in the cold storages that have low utilization factors. Building aspect ratios and effective U-value of the building envelop were also identified one of the crucial parameters that can play a major role to improve the energy performance of the cold storages. As expected, a negligible variation in the specific energy consumption of the storages with the specific installed refrigeration systems and AHUs was observed.

Keywords: potato cold storage, energy efficient building, refrigeration, capacity utilization factor, effective U-value

1. Introduction

Energy crises, global warming and food security are major concerns for the whole world and need to be given immediate attention. Cold storages are essential for food security as well as for quality of perishable items, however, operation of the storages is an energy extensive process and consumes   substantial amount of the total energy consumption in the countries like India. As of 2012, India had approximately 6,300 cold storages, with a total storage capacity of 30.11 million metric tons. Recently, it was estimated that the cold storage capacity in India needs to doubled, to a total of 61.13 million metric tons, to accommodate the yearly production of the perishable food items in the country (Sridar, 2013). Despite that more than 80% of the existing storage facility is being used for potato, fruits and vegitables storage only(Singh, Singh, & Lazarus, 2014), a sizable quantity (approximately 10-15%) of the total potato production is gone wastage every year due to lack of sufficient storage facility in the country (Agriwatch, 2012). Also, the Indian cold storages are mainly run on the grid electricity and their energy expenses account approximately 28% of the total costs(TERI, 2012). Therefore, the energy performance of the existing cold storage facility needs to be assessed and discussed properly because the poor performance of the storage facilities not only increase the energy consumption but also responsible for low quality of refrigerated product (Gottschalk, K., Nagy, L., 2003; Nourian, F., Ramaswamy, H.S., 2003). In recent past, few researchers have emphasized to improve the energy efficiency and operation in the potato storages in different parts of the world including India (Chourasia & Goswami, 2009; Hasse, H., Becker, M., Grossmann, K., Maurer, 1996; Marchant, A.N. and Davies, 1994; Marchant, A.N., Lidstone, P.H. and Davies, 1994; USAID, 2009; Xie, J., Qu, X.H., Shi ,J.Y. and Sun, 2006). Chourasia and Goswami (Chourasia & Goswami, 2009) discussed few important factors that can influence the energy consumption in the potato cold storage. They also suggested some possible solutions to improve the energy performance of the storage. Hasse et al. (Hasse, H., Becker, M., Grossmann, K., Maurer, 1996) accessed the potential for energy savings by improving automation and control in the refrigeration system. Devres and Bishop (Devres, Y. O. and Bishop, 1995) studied the energy consumption, using a theoretical model, in a real potato cold storage in the UK climate. Xie et al. (Xie, J., Qu, X. H., Shi, J. Y. and Sun, 2006) analysed several design parameters such as corner baffle, the stack mode of foodstuffs for a minitype cold storage and analysed the parameters that affect the flow field in the storage. Marchant and coworkers (Marchant, A.N. and Davies, 1994; Marchant, A.N., Lidstone, P.H. and Davies, 1994) designed a part of an expert system controller by using artificial intelligent techniques to improve the indoor environment as well as the energy performance of the cold storage. Moreover, few researchers also suggested the refrigeration units, which can run on solar energy (Grenier, Ph., Guilleminot, J.J., Meunier and Pons, 1988; Oertel, K. and Fischer, 1998) and biomass energy (Anbazhaghan, Saravanan, & Renganarayanan, 2005).

Despite numerous efforts that were made in this area, effect of building and refrigeration system's parameters on the energy consumption in the potato storage has hardly been explored. This study aims to identify and figure out the parameters those can influence energy consumption in existing potato cold storages and can be helpful in further development of new energy efficient cold storages. In this study, particularly, the impact of aspect ratio and effective overall heat transfer coefficient on the energy consumption in the potato cold storages were discussed using the trend analysis. Also, the variation in energy consumption with capacity utilization factor was analyzed. Results of this study could be important reference materials for the cold storage designers, architects, consultants and owners for assessing the energy consumption patterns in the existing storages and selecting a more accurate approach for future development of the energy efficient storages. However, study is limited for a small cluster in the composite climate and possibly results may vary in other parts of the same climate or in other climatic zones. Therefore, similar studies in other climatic zones are suggested, which could lead to more generalized results.

2. Methodology

2.1 Study Region

A survey, for energy consumption, building and refrigeration systems' parameters, was conducted in a few potato cold storages in and around Indore City. The city is located in the mid-western part of India, South of the Tropic of Cancer (22.71°N, 75.91°E), and has been selected one of the special economic zones (SEZs) in India ("Indore Special Economic Zone," 2003). The city has maximum number of cold storages (approximately 52) in the region (Singh et al., 2014), which play a vital role in regulating the demand of perishable products in the region. More than 50% of the existing storage facility is being used for potato storage only (Singh et al., 2014). The climate of the Indore is described as composite climate having three distinct seasons summer, monsoon and winter.

2.2 Data Collection and Analysis

Present study intends to determine the energy consumption pattern in the potato cold storages through field survey and experimental procedures. The energy consumption and other detail were collected from a few cold storages in the selected region using sample survey technique. A detailed questionnaire was prepared and supplied randomly to the potato cold storages. The surveyed cold storages were a mix of private and cooperative categories. Responsible people e.g. owners, managers and operators at each of the cold storages were asked to answer all the questions stated in the questionnaire. The information was asked in the questionnaire specifically to obtain data about their electrical energy consumptions, storage patterns, construction pattern, design features, aspect ratios and orientation of the storage building as well as types of refrigeration system, frequency of use and their capacities which were being used. Also, the surveyed cold storages were visited physically a number of times for validating the information obtained through the questionnaire and to enable the respondents answer correctly to the questions in the questionnaire. Moreover, a few selected persons for example owner, manager and operator were interviewed and information, on technical, economic and social challenges in running the storages, were collected. In this article, only relevant information is presented. The monthly electricity bills were also collected directly from the responsible energy supplying agency in the region to justify and maintain the accuracy and completeness of the energy consumption data of the surveyed cold storages, All of the received questionnaires were examined carefully and only ten questionnaires, which had most of the desirable information and covered complete range of storage capacities, were analyzed quantitatively. The storage capacities of the selected cold storages vary between 6000 tons and 15000 tons (Singh, R., Singh, S.P., 2010).

2.2.1 Energy Consumption and Storage Patterns

The energy consumption and stored potato data were collected for the selected cold storages for four consecutive years. Yearly storage pattern of potato in the storages and their respective specific energy consumptions (SEC) are presented in Figure 1 and Table 1 respectively.

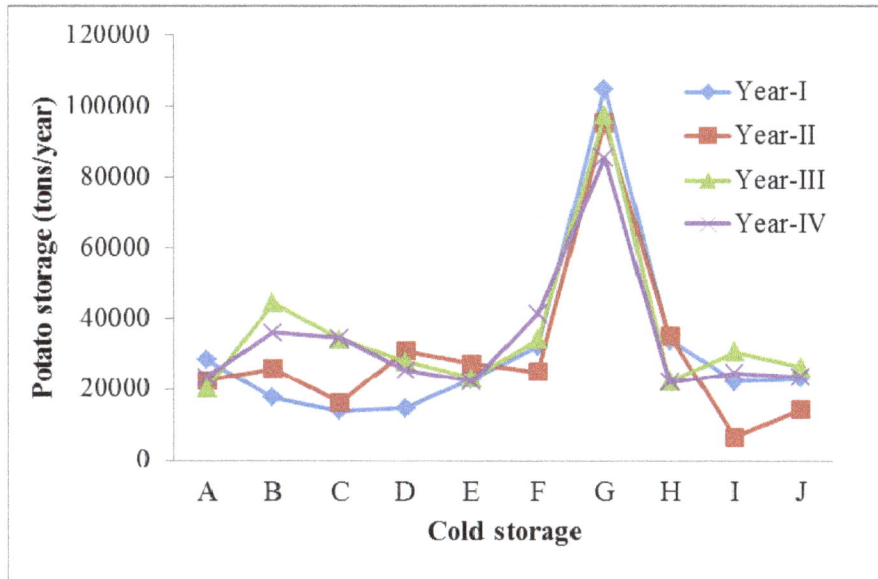

Figure 1. Annual potato storage patterns for years I to IV

Table 1. Annual Specific Energy Consumption (SEC) of Potato cold storages

Cold Storage	Annual specific energy consumption(kWh/ton)				
	Year-IV	Year-III	Year-II	Year-I	Average
A	10.07	10.86	10.77	10.49	10.5
B	10.01	9.92	12.33	13.69	11.5
C	9.00	9.90	12.79	13.00	11.2
D	14.63	14.96	13.67	21.47	16.2
E	12.79	13.38	11.30	12.58	12.5
F	9.47	9.92	10.17	9.66	9.8
G	9.70	10.84	10.38	10.61	10.4
H	15.56	16.82	9.64	10.80	13.2
I	11.35	12.05	26.08	13.81	15.8
J	12.15	8.90	11.93	12.38	11.3

2.2.2 Building Construction Pattern

The buildings were found to be constructed with a slanted type roof made of a corrugated sheet of tin or asbestos cement sheets. The buildings also have a false flat ceiling under the slanted roof, which keeps outdoor and indoor environment separate. Section between the slanted roof and the false ceiling also has a few windows, which allow continuous exchange of the air in the section with the ambient air. This air exchange helps to remove the heat effect created by the solar radiation falling directly on the slanted roof section. The air gap between the slanted part and the flat ceiling also created an extra insulation, which reduces the heat transfer between storage chamber and the outdoor ambient environment. In all of the cold storages, similar construction pattern was observed. The construction materials used in the construction of the storage buildings are mainly bricks, thermocole insulation, wooden rafter, wood powder, plaster and asphalt sheet. The buildings also have a few columns and beams made of RCC, however, effect of column and beams were not taken into consideration in this study. It can be observed from the data that the construction materials used in the construction of the cold storages' buildings are same, however, thicknesses of some construction materials vary. The thickness and arrangements of construction materials observed in the construction of walls and roof in the selected storage buildings are given in Tables 2 and 3.

Table 2. Variation in thickness of the construction materials used in walls in different cold storage

Cold Storage	Thickness (m)					
	Plaster	Brick	Plaster	Thermocole	Iron net	Plaster
A	0.0254	0.3560	0.0254	0.0762	0.003	0.0380
B	0.0254	0.2286	0.0254	0.0762	0.003	0.0380
C	0.0254	0.2286	0.0254	0.1016	0.003	0.0380
D	0.0254	0.2286	0.0254	0.0762	0.003	0.0127
E	0.0381	0.2286	0.0254	0.1016	0.003	0.0380
F	0.0381	0.2286	0.0127	0.0809	0.003	0.0380
G	0.0254	0.2286	0.0127	0.0800	0.003	0.0254
H	0.0254	0.2286	0.0191	0.0635	0.003	0.0127
I	0.0254	0.2286	0.0254	0.0762	0.003	0.0127
J	0.0254	0.2286	0.0254	0.0800	0.003	0.0188

Note: Actual measurement was taken in inches.

Table 3. Variation in thickness of construction materials used in ceiling in different cold storage

Cold Storage	Thickness (m)					
	Wood Powder	Asphalt sheet	Thermocole	Iron	Plaster	Wood
A	0.2032	--	0.1016	0.003	0.038	0.0380
B	0.0508	0.004	0.1016	0.003	0.038	0.0635
C	0.1524	0.004	0.1016	--	--	0.0254
D	0.0762	0.004	0.0762	--	--	0.0254
E	0.0508	0.012	0.0762	--	--	0.1016
F	0.0381	0.004	0.1016	--	--	0.0254
G	--	0.004	0.1300	--	--	0.0254
H	0.0508	0.004	0.0762	--	--	0.0254
I	0.0508	0.004	0.0762	--	--	0.0254
J	0.1524	--	0.0254	--	--	0.0762

Note: Actual measurement was taken in inches.

A representative arrangement for the wall and the flat ceiling constructions is shown in Figure 2. Thickness of each layer in actual construction varies from one building to other (See Tables 2 & 3).

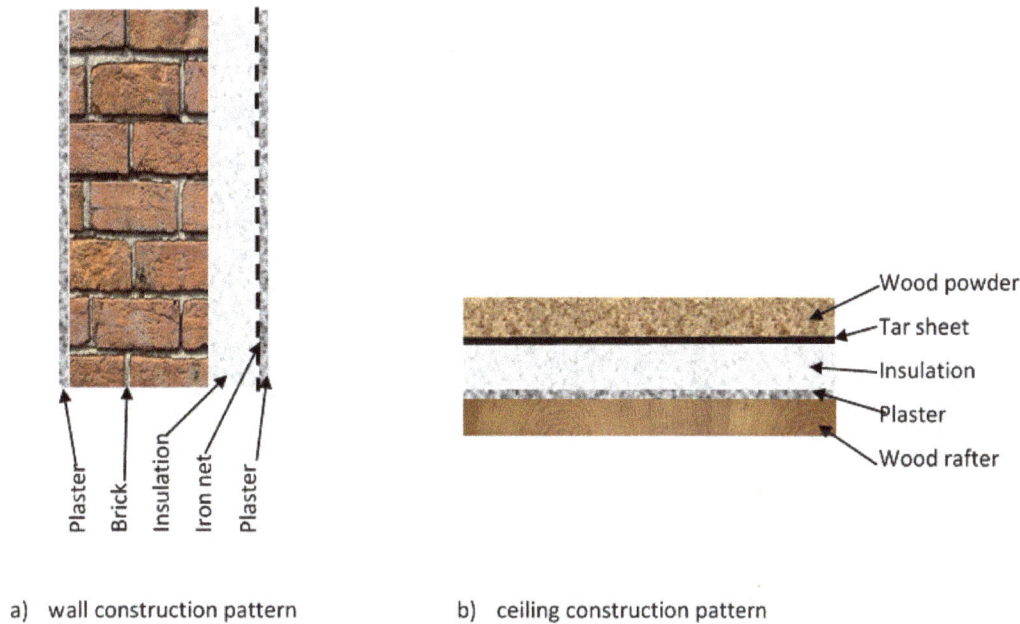

a) wall construction pattern   b) ceiling construction pattern

Figure 2. Construction patterns of wall and roof

2.2.3 Calculation of Ueff

In this study, U_{eff} represents effective overall heat transfer coefficient for complete building envelop and is estimated by following equation:

$$U_{eff}=(U_r \times A_r + U_w \times A_w)/(A_r + A_w) \qquad (1)$$

Where U_r, U_w, A_r, A_w represent overall heat transfer coefficients and area of roof and walls respectively.

The overall heat transfer coefficient is calculated using the following general formula:

$$1/U = R = 1/h_i + l_1/k_1 + l_2/k_2 \ldots \ldots l_n/k_n + 1/h_o \qquad (2)$$

Where h_i, h_o are heat transfer coefficients for inner and outer layer of air films, $l_1, l_2 \ldots l_n$ and $k_1, k_2, \ldots k_n$ are thicknesses and thermal conductivities for the construction materials used in layer 1, layer 2....layer n respectively.

The effects of convection and radiation on the inner and outer surfaces of building components (e.g. walls and roof) are usually included in the combined heat transfer coefficients h_i and h_o, respectively. The conductivities of each the construction materials used in the calculation is listed in Table 4. The thicknesses of the construction layers used in the calculation of overall U-factor of the buildings' components and eventuially U_{eff} of the building envelops were taken from the Tables 2 and 3. Surface area of each surface was calculated using the building dimensions given in Table 5. The U_{eff} has been correlated with the energy consumption later in the results and discussion section.

Table 4. Thermal conductivity of construction materials

	Plaster	Brick	Thermocole	Iron Net	Wood rafter/ Wood powder	Asphalt
Thermal conductivity(W/m-K)	0.72	0.84	0.035	53.6	0.13/0.1	0.5

Table 5. Dimensions of the cold storages

Cold storage	Chamber	Dimensions in (m)		
		Length	Width	Height
A	1	58.87	25.91	5.94
	2	24.38	25.91	18.29
B	1	47.52	29.7	20.46
	2	39.62	30.48	18.29
C		31.32	31.32	20.1
D		75.41	30.48	15.24
E		33.53	30.48	19.81
F	1	33.53	22.86	9.14
	2	30.48	30.48	12.8
G	1	68.53	12.19	5.94
	2	44.73	19.05	10.67
	3	25.44	15.21	11.25
	4	58.82	38.41	14.63
H		83.12	30.48	17.76
I		45.72	36.58	15.24
J		45.72	36.58	15.24

2.2.4 Refrigeration System

During the survey, it was observed that in all cold storages, a conventional vapour compression refrigeration (VCR) system is being used to produce desirable indoor storage environmental conditions. The temperature and humidity in all the surveyed storages were found to be maintained at 4°C and between 80 and 95% respectively. Compressors, condensers and air handling units are main components of refrigeration unit. It was noted that reciprocating heavy duty compressors are being used in all the selected storages. The pictorial view of the compressor room in the refrigeration system used in one of the surveyed cold storages is shown in Figure 3.

Figure 3. Compressor's room of a refrigeration unit in a cold storage

Two types of air handling units, i) bunker type and ii) fan coil units (FCUs) are being used for air circulation and maintain the temperature in the refrigerated chambers(Figure 4). About 80 percent of the cold storages were found to be using bunker type air handling units and other 20% are using fan coil unit. The air handling units were found to be installed in the upper most section of the cold storages chambers and are being used to re-circulate the indoor air continuously. It was also informed by responsible persons that the highly CO_2 concentrated indoor air of the storage chambers is completely replaced with fresh outdoor air once every day. The process of air replacement is generally taken place in the early morning hours(between 2 am and 4 am) or when ambient air is at minimum temperature level of the day, so that the risk of high cooling load due to higher ambient temperature could be avoided.

Figure 4. Air handling unit a) Banker type b) Fan coil unit

3. Results and Discussions

The energy consumption pattern of the cold storages is represented by the specific energy consumption, which is could be a suitable parameter to compare the energy performance of the storages. The table 1 show that the yearly specific energy consumption varies between 9.00 and 15.56kWh/ton, 8.90 and 16.82kWh/ton, 9.64 and 26.08kWh/ton and 9.66 and 21.47kWh/ton for years IV, III, II and I respectively. An average specific energy consumption varies between 9.8 and 16.2 kWh/ton/year. This large variation could be due to variation in operation pattern and difference in refrigeration system's performance. The variation also indicates the possibilities of a significant energy savings in some of the storages by implementing appropriate energy saving measures. Some of the possible measures to improve the energy efficiency in the cold storages were discussed in our previous study (Singh, R., Singh, S.P., 2010). The variation in specific energy consumption could also be influenced by the building and refrigeration systems' parameters (e.g. capacity utilization factor (CUF), overall heat transfer coefficient (U_{eff}), aspect ratios of buildings, capacities of refrigeration units and air handling units). To assess the possible relationship between these parameters and the energy consumptions, trend analysis were used. The trend analysis is particularly helpful for identifying the general behavior of the interpreter and also examines changes in its behavior across the whole spectrum of its numerical values.

3.1 Effect of Capacity Utilization

During the survey, it was observed that only a small fraction of total storage capacities of the cold storages were utilizing. The capacity utilization may depend on the crop yield in the respective year and demand of the produce in the market. If crop yield is high, a higher storage capacity will require to accommodate the maximum produce, while higher market demand may lead to lower storage requirement. The cold storage capacity utilization factor is determined as follows:

$$\text{Capacity utilization factor} = \frac{\text{Used capacity of cold storage(tons)}}{\text{Full capacity of cold storage(tons)}} \quad (3)$$

Figure 5. shows the relationship between the average specific energy consumption as a function of capacity utilization factor.

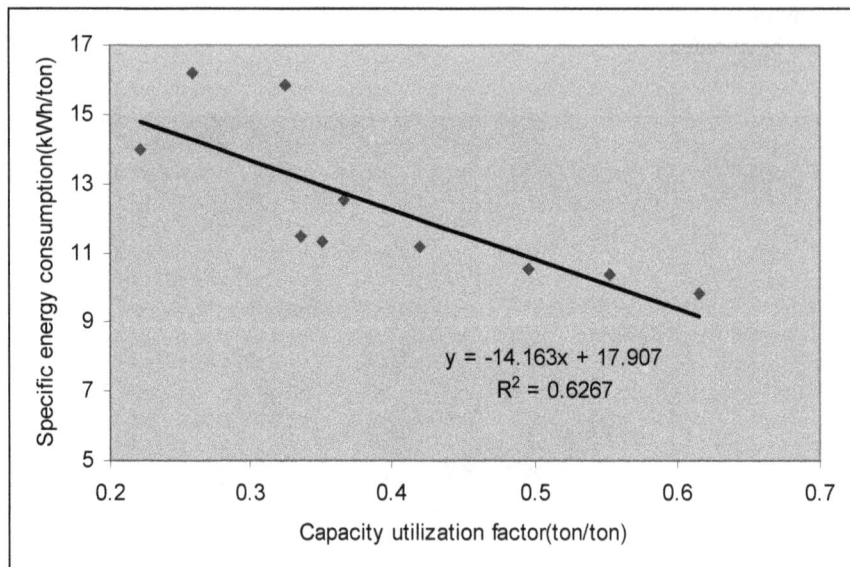

Figure 5. SEC vs. capacity utilization

It can clearly be observed from Fig.5 that the capacity utilization factor influences the specific energy consumption significantly. An average negative slope of the SEC with changes in CUF was estimated 14.163kWh/ton that means the increase in CUF by 0.1 decreases the energy consumption by 1.46kWh/ton. The negative slope of the trend line indicates higher energy performance for higher utilization of the storage, which is obvious. It would also be worth to mention here that none of the storage showed the capacity utilization factor 1. The low range of CUF varies between 0.2 and 0.6 is an indication of poor utilization of the storage facility and huge potential of maximizing energy efficiency of the potato storages. If the storages utilize their storage capacities at maximum strength, the current minimum energy consumption rate, 9kWh/ton/year, could be reduced further by a significant factor. Therefore, it is recommended that the maximum utilization of the storage capacities should be ensured. The maximum utilization can be done by dividing the larger storage chambers into many smaller storage chambers connecting with a separate smaller refrigeration unit. A separate assessment is required to optimize the size of the storage chamber, which is not part of this study and can be done separately.

3.2 Effect of Aspect Ratio of Building

Shape of the building envelop, determined by aspect ratios, plays a crucial role in the energy performance of the storage because most part of the external surface area of the building expose directly to sun (Singh, S.P., Singh, R.K., Sodha, 2009). Compactness is expressed by the aspect ratios of building envelop for a specified volume. As heat losses are proportional to the building surface area, therefore, higher compactness results in lower heat transfer and improves the energy efficiency. Figure 6 shows the effect of aspect ratios of building envelop on the average specific energy consumption. In this figure, data from only single chamber cold storages (C,D,E,H,I and J) is presented due to unavailability of separate energy data for each buildings attached with the same refrigeration system. For example, cold storage A has two buildings with different dimensions and orientation and operated using a combined refrigeration system consists a few compressors, a condenser and separate air handling units.

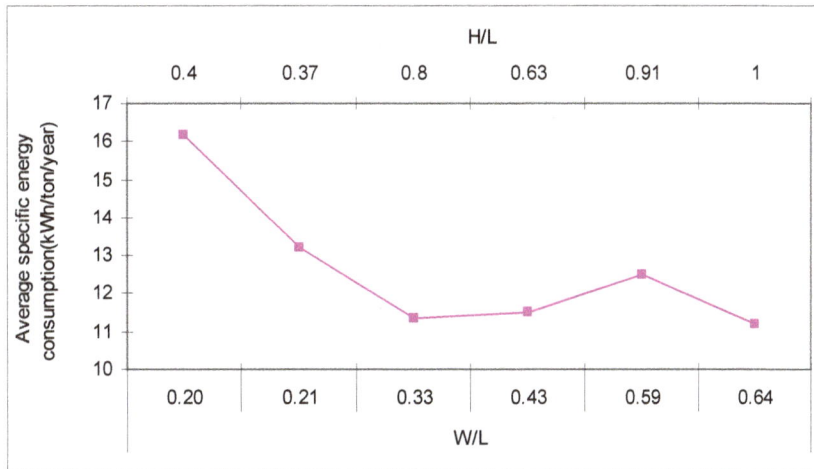

Figure 6. SEC vs. aspect ratios

It can be observed from the fig.6 that the average specific energy consumption varies with the aspect ratios of the cold storages. The building of the aspect ratios W/L=0.64, H/L=1 was found to be the most energy efficient one. The buildings of the aspect ratios of W/L=0.33, H/L=0.8 and W/L=0.43, H/L=0.63 demand a higher energy compare to the best one but the difference is not significant. Best option in terms of energy efficiency might change with location and climatic condition and, therefore, similar studies for other locations in same climatic zone and/or other climatic zones are suggested for generalizing the results. Moreover, a separate analysis is required for the cold storages having multiple buildings served by the one refrigeration system.

3.3 Effect of U Value

As discussed previously in section 2.2.3, the overall heat transfer coefficient of building envelop is another important parameter that affects the thermal performance of the building critically. The U-value entirely depends on the materials used in the construction and their thicknesses. Figure 7 shows a liner relationship between the overall effective heat transfer coefficient and the specific energy consumption of the cold storages. Though the relationship is not very strong but a higher specific energy demand can be observed for higher effective heat transfer coefficient as expected. A separate study for an optimization of effective overall heat transfer coefficient values, for minimum energy demand, is required for the storage buildings in different climates including composite climate.

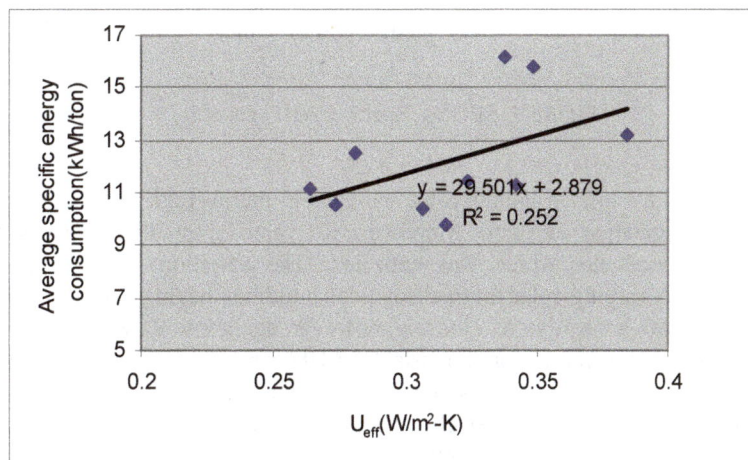

Figure 7. SEC vs Ueff

3.4 Effect of Size of Refrigeration and Air Handling Units

In this section, an effect of refrigeration and air handling units on the energy consumption in the cold storages was also discussed. Figures 8 & 9 show the variation in specific energy consumption with installed specific refrigeration and AHU capacities. It should be worth to mention that, the installed specific capacities were calculated based on the designed storage capacities of the storages and not for the stored produced at the time of survey.

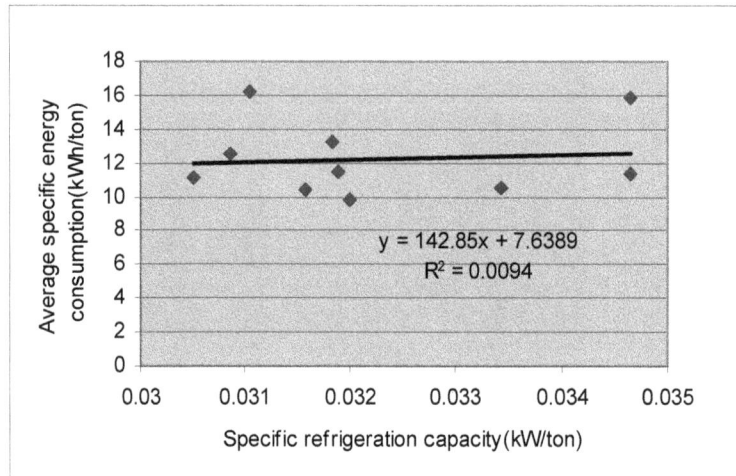

Figure 8. SEC vs. Specific refrigeration capacity

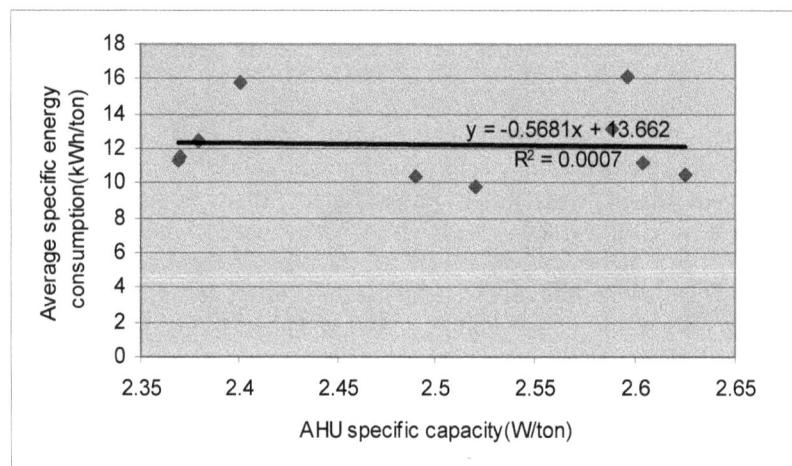

Figure 9. SEC vs. Specific AHU capacity

The results presented in Fig. 8 show an unnoticeable increase in the average specific energy consumption the with specific installed refrigeration capacity. Moreover, R^2 value show that there is not a considerable relationship between two parameter which was expected. The variation in the average specific energy consumption could be due to varying solar thermal loads of buildings having different aspect ratios for equal storage capacities. Fig. 9 shows a negative relationship, however, again low R^2 value indicate the independency of the average specific energy consumption on the installed AHU capacities. The variation in the energy demand could be due to similar reasons explained for installed refrigeration capacities. A further similar analysis can be done with large number of sample size in other climates and locations to develop common and more effective design guidelines for developing new and energy efficient cold storages.

4. Conclusion

A survey was conducted, in ten potato cold storages, for their energy consumptions, storage patterns, construction patterns of buildings and the installed refrigeration systems. Based on the collected that impact of various building parameters i.e. aspect ratios and U_{eff}, and utilization capacity was analyzed and discussed. It was observed that the specific energy consumption varies between 9kWh/ton/year and 26kWh/ton/year among the surveyed cold storages. This huge variation indicate a substantial potential to improve the energy efficiency in the cold storages. Moreover, a poor utilization factor observed between 0.2 and 0.6, which indicates partial utilization of the available storage capacity. The energy consumption in the cold storages with low utilization factor was observed higher which means operating the storage in partially filled condition is not a good practice from energy efficiency point of view. The capacity utilization factor could be increased by creating a number of smaller storage chambers, instead of one bigger chamber, connected with a separate refrigeration unit. The aspect ratios of building envelops and effective U- value also influence the energy consumption significantly. Therefore, dimensions of the storage buildings, construction materials and their thicknesses should be decided wisely beforehand the final construction. Results also reveal that the specific energy consumption found almost invariant with the installed specific refrigeration and AHU capacities as expected.

References

Agriwatch. (2012). *Market intelligence system: Baseline data for potato and onion.* Retrieved from http://sfacindia.com/PDFs/Onion & Potato Baseline Report.pdf

Anbazhaghan, N., Saravanan, R., & Renganarayanan, S. (2005). Biomass Based Sorption Cooling Systems for Cold Storage Applications. *International Journal of Green Energy*, 2(4), 325–335. http://dx.doi.org/10.1080/01971520500287891

Chourasia, M. K., & Goswami, T. K. (2009). Efficient design , operation , maintenance and management of cold storage compressor, 1(1), 70–93.

Devres, Y. O., & Bishop, C. F. H. (1995). Computer model for weight loss and energy conservation in a fresh produce Refrigerated Store. *Applied Energy*, 50, 97–117. http://dx.doi.org/10.1016/0306-2619(95)92627-6

Gottschalk, K., Nagy, L., & F. I. (2003). Improved climate control for potato stores by fuzzy controllers. *Computers and Electronics in Agriculture*, 40(1–3), 127–140. http://dx.doi.org/10.1016/S0168-1699(03)00016-4

Grenier, Ph., Guilleminot, J. J., Meunier, & Pons, F. M. (1988). Solar powered solid absorption cold store. *Journal of Solar Energy Trans. ASME*, 1b(3), 192–197. http://dx.doi.org/10.1115/1.3268256

Hasse, H., Becker, M., Grossmann, K., & Maurer, G. (1996). Top-down model for dynamic simulation of cold-storage plants. *International Journal of Refrigeration*, 19(1), 10–18. http://dx.doi.org/10.1016/0140-7007(95)00077-1

Indore Special Economic Zone. (2003). *http://www.indoresez.gov.in/.* Retrieved January 22, 2013, from http://www.indoresez.nic.in/mp.htm

Marchant, A. N., & Davies, T. W. (1994). Refrigerated storage: real-time control using intelligent parameter passing. *International Journal of Refrigeration*, 17(2), 109–116. http://dx.doi.org/10.1016/0140-7007(94)90051-5

Marchant, A. N., Lidstone, P. H., & Davies, T. W. (1994). Artificial Intelligence Techniques for the Control of Refrigerated Potato Stores Part 2: Heat and Mass Transfer Simulation. *Journal of Agricultural Engineering Research*, 58(1), 27–36. http://dx.doi.org/10.1006/jaer.1994.1032

Nourian, F., Ramaswamy, H. S., & K. A. C. (2003). Kinetics of quality change associated with potatoes stored at different temperatures. *LWT - Food Science and Technology*, 36(1), 49–65. http://dx.doi.org/10.1016/S0023-6438(02)00174-3

Oertel, K., & Fischer, M. (1998). Adsorption cooling system for cold storage using methanol/silicagel. *Applied Thermal Engineering*, 18, 773–786. http://dx.doi.org/10.1016/S1359-4311(97)00107-5

Singh, R., Singh, S. P., & Lazarus, I. J. (2014). Energy Performance Improvements in Indian Potato Cold Storages. *Journal of Basic and Applied Scientific Research*, 4(7), 44–50. Retrieved from http://www.textroad.com/ArticleView.aspx?articleId=JBASR-2874-2

Singh, R., Singh, S. P., & S. R. N. (2010). Energy saving potential in Indian cold storages. In *2ndBharatiya Vigyan Sammelan*. Indore.

Singh, S. P., Singh, R. K., & Sodha, M. S. (2009). Empirical Relations for Orientation of Optimum Size Solar Efficient Cold Storage Building. *SESI Journal*, 19(1&2), 32–39.

Sridar, N. (2013). *The food wastage & cold storage developing realistic solutions Infrastructure relationship in India.*

TERI. (2012). Retrieved from http://www.teriin.org/index.php?option=com_ongoing&task=about_project& pcode=2009RT01

USAID. (2009). *Private Sector Competitiveness Enhancement program. Cold chain and storage action plan.*

Xie, J., Qu, X. H., Shi, J. Y., & Sun, D. W. (2006). Effects of design parameters on flow and temperature fields of a cold store by CFD simulation. *Journal of Food Engineering*, *77*(2), 355–363. http://dx.doi.org/10.1016/j.jfoodeng.2005.06.044

15

Environmental Impact Analysis of the Emission from Petroleum Refineries in Nigeria

Oladimeji T. E.[1], Sonibare J. A.[2], Odunfa K. M.[3] & Oresegun O. R.[1]

[1] Chemical Engineering Department, Covenant University, Ota, Ogun State Nigeria

[2] Chemical Engineering Department, Obafemi Awolowo University, Ile-Ife, Nigeria

[3] Mechanical Engineering Department, University of Ibadan, Ibadan, Nigeria

Correspondence: Oladimeji Temitayo Elizabeth, College of Engineering, Covenant University. Ota, Ogun State. Nigeria. Email: temitayo.fatoki@covenantuniversity.edu.ng

Abstract

Health and environmental hazards, a thing of global concern have been the major characteristics of the petroleum refinery areas worldwide, Nigeria inclusive. This is as a result of the emissions from petroleum refineries which resulted into air quality degradation of the host environment. This problem which has equally affected the climatic conditions of the petroleum producing areas is more pronounced in Nigeria due to lack of implementing adequate policies to protect the host environment. This study is carried out to investigate the atmospheric conditions of the petroleum refineries and identify the environmental impact of emissions of criteria pollutants from the proposed project in the area of influence. Emission inventory of criteria pollutants was carried out on the four existing and twenty-three proposed petroleum refineries in Nigeria. Using no control-measure option, the estimated annual criteria air pollutants emissions from point sources in the existing refineries are 1,217 tons/annum for PM_{10}, 45,124 tons/annum for SO_2, 167,570 tons/annum for NO_x, 3,842 tons/annum for VOC and 242,469 tons/annum for CO. An additional 1,082 tons/annum of PM_{10}, 168,944 tons/annum of SO_2, 688,687 tons/annum of NO_x, 9,122 tons/annum of VOC and 569,975 tons/annum of CO were predicted to be added into the Nigeria airshed by the proposed petroleum refineries. The highest pollutant emitting state was predicted to be Rivers State with the highest number of refineries while the least pollutant emitting states were predicted to be Kaduna, Edo, Lagos and Anambra States with only one refinery in each of the state. The ability to adopt appropriate control measures will determine the rate of emission of criteria pollutants released into the country's airshed.

Keywords: criteria air pollutants, emission factor, emission inventory, Nigeria, refinery

1. Introduction

Nigeria is the highest producing nation in Afica with production rate of 2.35 million barrel per day (bbl/dy) (Klieman, 2012). International Monetary Fund (IMF) establishes that oil and natural gas export revenue account for 96% of total revenue in 2012, also it has been estimated that demand and consumption of petroleum in Nigeria grows at the rate of 12.8% annually, however, petroleum products are unavailable and expensive because a significant percentage of required petroleum products are imported while limited quantity is supplied to the nation due to instability in price and scarcity of the products. In view of this, it is necessary for the country to develop her refining potential (Sonibare *et al.*, 2007). One of the strategies developed in Nigeria to increase her refining potential includes approval of more refineries to complement the existing ailing four state-owned refineries, which despite having a combine production capacity of 445,000 barrels per day (bpd), remain unable to meet the nation's petroleum products demand. Nigeria has four existing petroleum refineries (Table 1), while twenty-three proposed refineries (Table 2) have been licensed and are at various stages of completion (DPR, 2004; DPR, 2010). The more the refineries, the more the emissions generated, and the more dangerous it becomes on health and the environment which may result in degraded air quality of host environment, hence the need to control the rate of emissions in the refineries.

As petroleum refinery and petrochemical industries are most desirable for national development and improved quality of life, the unwholesome and environmentally unacceptable pollution effects of the waste from these industries are of major concern (Reinermann and Golightly, 2005). This is because, in the process of converting crude oil into petroleum and petrochemical products, wastes of different kinds are generated. These wastes are released to the environment in form of gases, particles and liquid effluent which becomes hazardous to the environment and to human health. Air emissions can come from number of sources within a petroleum refinery which include; equipment leaks (from valves, flanges, pump seals, drains and compressor seals), high-temperature combustion processes in the burning of fuels, the heating of steam and process fluids and the transfer of products. Numerous pollutants are emitted into the environment through normal emissions, fugitive releases, accidental releases or plants upsets from various refining operations such as separation, conversion and treatment processes (US DOE, 2007). About 900 cases of accidents in oil refining processes and exploration was reported in Nigeria in the year 2000 which is as a result of equipment failure and malfunctioning, deterioration and ageing of pipelines etc. Major components of these emissions are criteria air pollutants, which include particulate matter (PM), oxides of nitrogen (NO$_x$), sulphur dioxide (SO$_2$), carbon monoxide (CO) and volatile organic compounds (VOCs).

From previous studies, the environmental and health implications of these pollutants cannot be overemphasized. During the past decades, criteria pollutants caused some health problems and damages to plants such as DNA breakdown in bone marrow cells, respiratory diseases and reduced lung function in humans (Zhongua et al., 2003), reduction of reproductive processes (Cape, 2003) in plants and formation of acid rain which deteriorates water quality and affect aquatic habitat (Johnson et al., 2003). About 2.5 million deaths per year was found to result from indoor exposures to particulate matter in rural and urban areas in developing countries, representing 4-5% of the 50-60 million global deaths that occur annually (WHO, 2002). The effects may not be confined to the local area of production but extend to regional as well as global scales due to possibility of long transportation (Mitra and Sharma, 2002).

This study aims at collection and collation of detailed information concerning the air pollutant emissions in existing and proposed petroleum refineries in Nigeria (Figure 1) using emission factor approach. This is for the purpose of determining the potential impacts of these emissions on air quality in Nigeria's airshed.

1.1 Petroleum Refineries and Air Emission Sources

A typical modern refinery takes in crude as feed through the heat exchangers (for temperature increase) and desalters (for salt water removal) into the atmospheric distillation column, where distillation process takes place under atmospheric pressure. In the distillation process, fractions get separated in increasing order of boiling points. The top products (highly volatile) get condensed in a reflux condenser with some portion of condensed fractions going back as reflux (Sonibare et al., 2007). Several other processes are required for the crude oil to be finally converted into the final products of different fractions. Several operations such as separation, conversion, treatments, feedstock and product handling involved in the petroleum refining process are potential sources of criteria air pollutants (James et al., 2001). The most significant point sources of emission in the petroleum refineries are fluid catalytic cracking units, steam boilers, process heaters, and flares. These sources emit various pollutants through catalytic cracking of hydrocarbons, combustions and equipment leaks of volatile compounds.

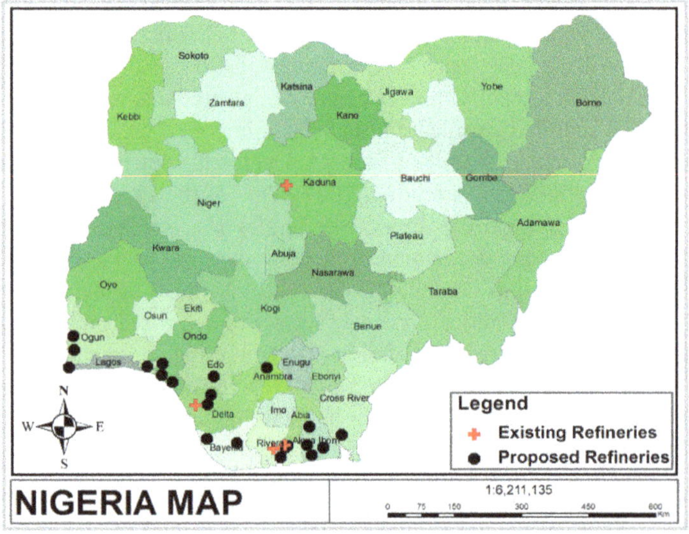

Figure 1. Existing and proposed petroleum refineries in Nigeria

Table 1. Existing refineries in Nigeria

S/N	Name of Refinery	Capacity (bbl/dy)	Year of Production	Location		
				Town	L.G.A	State
1	Kaduna Refining & Petrochemical Company	110,000	1980	Kaduna	Chikun	Kaduna
2	Port Harcourt Refining Company I	60,000	1965	Port Harcourt	Eleme	Rivers
3	Port Harcourt Refining Company II	150,000	1989	Port Harcourt	Eleme	Rivers
4	Warri Refining & Petrochemical Company	125,000	1978	Warri	Effunrun	Delta

Source: DPR (2001)

Table 2. Proposed Refineries in Nigeria

S/N	Name of Refinery	Capacity (bbl/dy)	Date of issue	Location		
				Town	L.G.A	State
1	Amakpe International Refinery	12,000	2004	Ikot Ekpene	Eket	Akwa Ibom
2	Amexum Corporation	100,000	2009	Ikang	Akpabuyo	Cross River
3	Antonio Oil	27,000	2009	Iwopin	Iwopin	Ogun
4	Chase Wood Consortium Nigeria Limited	70,000	2004	Eket	Eket	Akwa Ibom
5	Clean Water Refinery	60,000	2004	Onne	Eleme	Rivers
6	Gasoline Associate & International Limited Refinery	100,000	2009	Ipokia	Ipokia	Ogun
7	Ilaje Refinery & Petrochemicals	100,000	2004	Ilaje	Ilaje	Ondo
8	Niger Delta Refinery & Petrochemical Limited	100,000	2004	Niger Delta	Warri South	Delta
9	NSP Refineries and Oil Services Limited	120,000	2004	Ohali Ogba	Andoni	Rivers
10	Ode Aye Refinery Limited	100,000	2004	Ode Aye	Okiti Pupa	Ondo
11	Ologbo Refinery Company Nigeria Limited	12,000	2010	Ologbo	Ologbo	Edo
12	Orient Petroleum Resources Limited	55,000	2004	Aguleri Out	Anambra West	Anambra
13	Owena Oil & Gas Limited	60,000	2004	Ilaje	Ilaje	Ondo
14	Qua Petroleum Refinery Limited	100,000	2004	Ibeno	Ibeno	Akwa Ibom
15	Rehoboth Natural Resources Limited	12,000	2008	Immingiri	Yenekoa	Bayelsa
16	Resources Refinery & Petrochemical Limited	100,000	2004	Ikot Abasi	Ikot Abasi	Akwa Ibom
17	Rivgas Petroleum & Energy Limited	30,000	2004	Port Harcourt	Eleme	Rivers
18	Sapele Refinery Limited	100,000	2004	Okpe-Sobo	Sapele	Delta
19	South West Refinery & Petrochemical Company	100,000	2004	Ogun-water side	Iwopin	Ogun
20	Starrex Petroleum	100,000	2004	Onne	Onne	Rivers

	Refinery					
21	Tonwei Oil Refinery	200,000	2002	Ekeremor	Ekeremor	Bayelsa
22	Total Support Limited	12,000	2004	Free trade zone	Calabar	Cross River
23	Union Atlantic Petroleum Ref.	100,000	2002	Badagry	Badagry	Lagos

Source: DPR (2004); NNPC (2008); DPR (2010)

2. Methodology

This study involved preparation of detailed emission inventory with estimation of emissions of criteria air pollutants from point sources of existing and proposed refineries in Nigeria. An emission factor approach was used to determine the criteria air pollutant emissions in Nigeria's petroleum refineries and its contribution to the country's airshed.

Detailed information on process unit capacity from the existing and proposed refineries in Nigeria was obtained from previous research reports, company's website and Department of Petroleum Resources (NNPC, 2008; EIA, 2000; US DOE, 2007; DPR 2001; DPR, 2004). The emission factor of criteria pollutants for process units in the petroleum refinery in AP-42 of the United States Environmental protection Agency (US EPA, 1989) was used (Table 3). The process unit capacity was combined with the emission factor of pollutants for each process unit which is calculated as follows;

Table 3. Emission factor of pollutants for process units in petroleum refinery

S/N	Process Name	Pollutants					
		PM_{10}	SO_x	NO_x	VOC	CO	Unit
1	Process Heaters	7.4S	158.6S	55.0	0.3	5.0	1000 gallons burned
2	Fluid catalytic cracking unit	-	493	71	220	13700	1000 barrels processed
3	Flares	-	-	0.027	5.6	14.8	Valve in operation
4	Boilers	-	159.3S	67	-	5	1000 gallons burned

"S" means sulphur content "-" means no emission factor for the pollutant.

Source: US EPA (1989)

$$E = A \times EF_1 \left[1 - \left(\frac{D}{100} \right) \right] \tag{1}$$

where

E = Emission Estimate for the Process Unit (tons/yr)

A = Activity Rate (bbl/yr)

EF_1 = Controlled Emission Factor (lbs /bbl)

D = % Control Efficiency

Due to lack of information on the level of efficiency of the control device, control efficiency was assumed to be zero for "worst case" scenarios and equation 1 became:

$$E = A \times EF_2 \tag{2}$$

where

EF_2 = Uncontrolled emission factor (lbs/bbl)

Total emissions of the criteria pollutants in tons/yr from point sources of existing and proposed refineries were determined using equation (2) and the result is shown in Table 4-5.

3. Results and Discussion

The total estimated annual criteria air pollutants from point sources in the existing refineries are 1,217 tons/annum for PM_{10}, 45,124 tons/annum for SO_2, 167,570 tons/annum for NO_x, 3,842 tons/annum for VOC and 242,469 tons/annum for CO (Table 4). An additional 1,082 tons/annum of PM_{10}, 168,944 tons/annum of SO_2, 688,687 tons/annum of NO_x, 9,122 tons/annum of VOC and 569,975 tons/annum of CO (Table 5) were predicted to be added into the Nigeria airshed by the proposed petroleum refineries. Since emission rates are directly proportional to process unit capacity/ activity rate from equation (2), Tonwei Oil Refinery proposed to operate at 200,000 bbl/dy will release the maximum emission of air pollutant while Amakpe Refinery, Total Support Refinery, Ologbo Refinery and Rehoboth Refinery proposed to operate at the same capacity of 12,000 bbl/dy will release the minimum emission of air pollutants. Also, the highest emitting state is Rivers state with the highest number of refineries (2 existing and 4 proposed refineries) while the least emitting states are Kaduna, Edo, Lagos and Anambra States (1 existing and 3 proposed refineries respectively) with only 1 refinery in each of the state. The significant point sources of these emissions are fluid catalytic cracking units (FCCU), process heaters, steam boilers and flares. From the criteria pollutants emitted, CO emission is the highest due to higher rate of incomplete combustion of fuel in the refinery units while particulate matter, which has to do with solid and liquid droplets suspended in air is the least emitted.

Criteria Air Pollutants (CAPs) includes the common air pollutants all over the world which could endanger public health and environment and it originates from various sources in the refinery such as the fluid catalytic cracking units, steam boilers, process heaters and flares e.t.c. (James et al., 2001). The CAPs on its environment indicated the potential for petroleum refinery to create an elevation in ambient concentrations - both indoor and outdoor in Nigeria, though composition could be impacted by local emissions (Zielinska and Fujita, 2003). Na et al. (2001) established a strong link between downtown hydrocarbon levels and nearby petrochemical complex while Kajihara et al. (2003) also noticed a decrease in ambient benzene level with a reduction in benzene concentration discharged into the atmosphere. Possible impacts of anticipated emissions of criteria air pollutants on human (Guo et al., 2004), vegetation (Rao et al., 2001) soil and water (Kim et al., 2001; Johnson et al., 2003) are strong enough for necessary control measures to be implemented in both operating and proposed Nigerian petroleum refineries. The ability to adopt appropriate control measures will determine how much emission of criteria air pollutants will be released into the country's airshed. Both technology and policy control options are recommended in order to reduce the criteria air pollutants in the Nigeria's airshed.

Table 4. Air   pollutant emissions from point sources of existing petroleum refinery units

S/N	Refinery	Process Unit	Criteria pollutant emissions (tons/yr)				
			PM$_{10}$	SO$_2$	NO$_x$	VOC	CO
1	Kaduna Refining & Petrochemical Company (110,000 bpsd)	Fluid Catalytic Cracking	-	1,894.52	274.62	8,457.99	52,504.38
		Process Heaters	73.00	1,567.77	6,020.76	34.76	549.24
		Flares	-	-	0.04	0.11	0.18
		Steam Boilers	-	105,771.09	49,365.38	-	3,684.76
2	Port Harcourt Refining Company I (60,000 bpsd)	Fluid Catalytic Cracking	-	434.52	69.52	194.67	12086.71
		Process Heaters	20.85	465.81	1793.72	10.43	163.38
		Flares	-	-	0,011	0.03	0.07
		Steam Boilers	-	2071.81	9688.14	-	723.04
3	Port Harcourt Refining Company II (150,000 bpsd)	Fluid Catalytic Cracking	-	3,597.85	517.96	1,609.48	100009.99
		Process Heaters	128.62	2,787.90	10,710.15	59.09	973.34
		Flares	-	-	0.07	0.14	0.05
		Steam Boilers	-	9,309.24	43,476.71	-	3,246.76
4	Warri Refining & Petrochemical Company (125,000 bpsd)	Fluid Catalytic Cracking	-	2,339.47	337.18	1,042.86	65,008.25
		Process	99.44	2,130.90	8,179.48	45.19	743.91

	Heaters					
	Flares	-	-	0.04	0.11	0.24
	Steam Boilers	-	7,953.52	37,361.46	-	277.40
	Total Emissions	1,216.66	45,124.43	167,569.95	3,841.56	242,468.55

"-" means no emission for the pollutant.

Table 5. Air pollutant emissions from point sources of proposed petroleum refinery units

S/N	Refinery	Process Unit	[a]Criteria pollutant emissions (tons/yr)				
			PM_{10}	SO_2	NO_x	VOC	CO
1	Amakpe International Refinery (12,000 bpsd)	Fluid Catalytic Cracking	-	13.70	2.43	10.43	483.19
		Process Heaters	1.05	1.05	73.00	0.35	0.69
		Flares	-	-	0.0004	0.001	0.003
		Steam Boilers	-	560.05	2,602.24	-	209.24
2	Amexum Corporation (100,000 bpsd)	Fluid Catalytic Cracking	-	1206.24	173.81	542.29	33,576.52
		Process Heaters	59.09	1,296.62	4,977.90	27.81	451.90
		Flares	-	-	0.04	0.07	0.14
		Steam Boilers	-	5,760.05	26,902.24	-	2009.24
3	Antonio Oil (27,000 bpsd)	Fluid Catalytic Cracking	-	10.43	1.74	6.95	333.72
		Process Heaters	3.48	93.85	361.53	1.99	34.76
		Flares	-	-	0.003	0.00007	0.011
		Steam Boilers	-	420.62	1,964.0	-	14.62
4	Chase Wood Consortium Nigeria Limited (70,000 bpsd)	Fluid Catalytic Cracking	-	590.95	83.42	264.19	13.90
		Process Heaters	31.28	636.14	2,436.81	13.90	222.48
		Flares	-	-	0.01	0.03	0.07
		Steam Boilers	-	2,930.43	13,678.8	-	983.8
5	Clean Water Refinery (60,000 bpsd)	Fluid Catalytic Cracking	-	434.52	69.52	194.67	12086.71
		Process Heaters	20.85	465.81	1793.72	10.43	163.38
		Flares	-	-	0,011	0.03	0.07
		Steam Boilers	-	2071.81	9688.14	-	723.04
6	Gasoline Associate & International Limited Refinery (100,000 bpsd)	Fluid Catalytic Cracking	-	1206.24	173.81	542.29	33,576.52
		Process Heaters	59.09	1,296.62	4,977.90	27.81	451.90
		Flares	-	-	0.04	0.07	0.14
		Steam Boilers	-	5,760.05	26,902.24	-	2009.24
7	Ilaje Refinery & Petrochemicals (100,000 bpsd)	Fluid Catalytic Cracking	-	1206.24	173.81	542.29	33,576.52
		Process Heaters	59.09	1,296.62	4,977.90	27.81	451.90
		Flares	-	-	0.04	0.07	0.14
		Steam Boilers	-	5,760.05	26,902.24	-	2009.24
8	Niger Delta Refinery & Petrochemical Limited (100,000 bpsd)	Fluid Catalytic Cracking	-	1206.24	173.81	542.29	33,576.52
		Process Heaters	59.09	1,296.62	4,977.90	27.81	451.90
		Flares	-	-	0.04	0.07	0.14
		Steam Boilers	-	5,760.05	26,902.24	-	2009.24
9	NSP Refineries and	Fluid Catalytic	-	1,741.58	246.81	775.19	48,350.34

		Cracking					
	Oil Services Limited	Cracking					
	(120,000 bpsd)	Process Heaters	86.9	1,866.71	7,164.43	38.24	650.05
		Flares	-	-	0.04	0.11	0.20
		Steam Boilers	-	8,294.19	38,738.66	-	2,892.20
10	Ode Aye Refinery	Fluid Catalytic	-	1206.24	173.81	542.29	33,576.52
	Limited	Cracking					
	(100,000 bpsd)	Process Heaters	59.09	1,296.62	4,977.90	27.81	451.90
		Flares	-	-	0.04	0.07	0.14
		Steam Boilers	-	5,760.05	26,902.24	-	2009.24
11	Ologbo Refinery	Fluid Catalytic	-	13.70	2.43	10.43	483.19
	Company Nigeria	Cracking					
	Limited	Process Heaters	1.05	1.05	73.00	0.35	0.69
	(12,000 bpsd)	Flares	-	-	0.0004	0.001	0.003
		Steam Boilers	-	560.05	2,602.24	-	209.24
12	Orient Petroleum	Fluid Catalytic	-	368.48	55.62	163.38	10,167.86
	Resources Limited	Cracking					
	(55,000 bpsd)	Process Heaters	17.38	392.81	1,505.19	10.43	139.05
		Flares	-	-	0.001	0.03	0.04
		Steam Boilers	-	1,741.58	8,137.76	-	608.33
13	Owena Oil & Gas	Fluid Catalytic	-	434.52	69.52	194.67	12086.71
	Limited	Cracking					
	(60,000 bpsd)	Process Heaters	20.85	465.81	1793.72	10.43	163.38
		Flares	-	-	0,011	0.03	0.07
		Steam Boilers	-	2071.81	9688.14	-	723.04
14	Qua Petroleum	Fluid catalytic	-	1206.24	173.81	542.29	33,576.52
	Refinery Limited	cracking					
	(100,000 bpsd)	Process Heaters	59.09	1,296.62	4,977.90	27.81	451.90
		Flares	-	-	0.04	0.07	0.14
		Steam Boilers	-	5,760.05	26,902.24	-	2009.24
15	Rehoboth Natural	Fluid catalytic	-	13.70	2.43	10.43	483.19
	Resources Limited	cracking					
	(12,000 bpsd)	Process Heaters	1.05	1.05	73.00	0.35	0.69
		Flares	-	-	0.0004	0.001	0.003
		Steam Boilers	-	560.05	2,602.24	-	209.24
16	Resources Refinery	Fluid catalytic	-	1206.24	173.81	542.29	33,576.52
	& Petrochemical	cracking					
	Limited	Process Heaters	59.09	1,296.62	4,977.90	27.81	451.90
	(100,000 bpsd)	Flares	-	-	0.04	0.07	0.14
		Steam Boilers	-	5,760.05	26,902.24	-	2009.24
17	RIVGAS Petroleum	Fluid catalytic	-	107.77	13.90	48.66	3,024.29
	& Energy Limited	cracking					
	(30,000 bpsd)	Process Heaters	6.95	118.18	448.40	2.43	3,024.29
		Flares	-	-	0.003	0.001	0.01
		Steam Boilers	-	517.96	2,419.43	-	180.76
18	Sapele Refinery	Fluid catalytic	-	1206.24	173.81	542.29	33,576.52
	Limited	cracking					
	(100,000 bpsd)	Process Heaters	59.09	1,296.62	4,977.90	27.81	451.90
		Flares	-	-	0.04	0.07	0.14
		Steam Boilers	-	5,760.05	26,902.24	-	2009.24
19	South West Refinery	Fluid catalytic	-	1206.24	173.81	542.29	33,576.52
	& Petrochemical	cracking					
	Company	Process Heaters	59.09	1,296.62	4,977.90	27.81	451.90
	(100,000 bpsd)	Flares	-	-	0.04	0.07	0.14
		Steam Boilers	-	5,760.05	26,902.24	-	2009.24
20	Starrex Petroleum	Fluid catalytic	-	1206.24	173.81	542.29	33,576.52
	Refinery	cracking					

	(100,000 bpsd)	Process Heaters	59.09	1,296.62	4,977.90	27.81	451.90
		Flares	-	-	0.04	0.07	0.14
		Steam Boilers	-	5,760.05	26,902.24	-	2009.24
21	Towei Oil Refinery (200,000 bpsd)	Fluid catalytic cracking	-	2,158.71	312.85	962.91	60,006.0
		Process Heaters	229.86	5,183.0	19,904.66	107.77	1,811.1
		Flares	-	-	0.11	0.20	0.70
		Steam Boilers	-	23,040.20	107,605.47	-	8,033.48
22	Total Support Limited (12,000 bpsd)	Fluid catalytic crack	-	13.70	2.43	10.43	483.19
		Process Heaters	1.05	1.05	73.00	0.35	0.69
		Flares	-	-	0.0004	0.001	0.003
		Steam Boilers	-	560.05	2,602.24	-	209.24
23	Union Atlantic Petroleum Refinery (100,000 bpsd)	Fluid catalytic crack	-	1206.24	173.81	542.29	33,576.52
		Process Heaters	59.09	1,296.62	4,977.90	27.81	451.90
		Flares	-	-	0.04	0.07	0.14
		Steam Boilers	-	5,760.05	26,902.24	-	2009.24
	Total Emission (tons/yr)		1,081.8	168,943.69	688,687.1	9,121.6	569,975.4

[a]Calculated by the authors   "-"means no emission factor for the pollutant.

4. Conclusion and Recommendations

From the emission inventory carried out in Nigerian petroleum refineries, results revealed that petroleum refinery is one of the major sources of air pollution in Nigeria. The environmental and health impacts of these pollutants indicate that adequate control as to be put place, as more refineries are established, to ensure reduction in the emissions.

Boilers, process heaters, and other process equipment are responsible for the emission of particulates, carbon monoxide, nitrogen oxides (NO_x), sulphur oxides (SO_2), and carbon dioxide (CO). Fluid catalytic cracking units and other units release particulate matters (PM) while Volatile organic compounds (VOCs) are released from storage, product loading and   handling facilities and fugitive emissions such as leaks from flanges, valves, seals and drains. These air emissions can be controlled in the refineries by taking the following measures;

i. NO_x emission reduction can be achieved by replacing furnace burners with low-NO_x burners and substitution of conventional liquid (nitrogen-containing) fuels with gaseous fuel (nitrogen-free).

ii. SO_2 emission reduction can be achieved by the introduction of Clean Fuels (low sulfur) specifications for petrol and diesel to lower sulphur dioxide emissions.   Also through desulfurization of fuels   or by directing the use of high-sulfur fuels to units equipped with SOx emissions controls

iii.   CO reduction is implemented by controlling the oxygen levels in the CO burners which is the main source of CO emissions.

iv.   VOC reduction involves the systematic leak detection and repair of any fittings that emit fugitive VOCs by replacing with more reliable equipment, including the environmentally friendly valve design, which tests proved to provide the best seal to prevent VOC fugitive leaks. Also losses from storage tanks and product transfer areas can be reduced by vapour recovery systems and double seals.

v.   PM reduction is achieved by installing cyclones on the catalytic cracking unit stacks, this cyclones operate by creating a centrifugal force that physically separates the particulates from the rest of the gas, particulates that would previously have been emitted are now collected from the cyclones and are then available for reuse.

References

Cape, J. N. (2003). Effects of airborne volatile organic compounds on plants. *Environmental Pollution, 122*, 145-157.

[DPR] Department of Petroleum Resource. (2001). Nigeria Oil Industry Statistical Bulletin. Department of Petroleum Resources, Ministry of Petroleum Resources, Lagos, Nigeria.

[DPR] Department of Petroleum Resource. (2004). Nigeria Oil Industry Statistical Bulletin. Department of

Petroleum Resources, Ministry of Petroleum Resources, Lagos, Nigeria.

[DPR] Department of Petroleum Resource. (2010). Private refineries and petrochemical plants status. Nigeria Oil Industry Statistical Bulletin. Department of Petroleum Resources, Ministry of Petroleum Resources, Lagos, Nigeria.

[EIA] Energy Information Administration. (2000). Refinery Capacity Report. Energy Information Administration, Washington, DC.

Gou, H., Lee, S. C., Chan, L. Y., & Li, W. M. (2004). Risk assessment of exposure to volatile organic compounds in different indoor environments. *Environ. Res., 94*, 57-66.

James, H., Gary, G., Glenn, E., & Handwerk, H. (2001) Petroleum Refining: Technology and Economics (4th ed.). CRC Press.

Johnson, R. L., Thomas, T. B., & Zogorski, J. J. (2003). Effects of daily precipitation and evapotranspiration patterns on flow and VOC transport to groundwater along a watershed flow path. *Environmental Science Technology, 37*, 4944-4954.

Kajihara, H., Fushimi, A. & Nakanishi, J. (2003). Verification of the effect on risk due to reduction of benzene discharge. Chemosphere, 53: 285-290.

Kim, H., Annabel, M. D., & Rao, P. S. (2001). Gaseous transport of volatile organic chemicals in unsaturated porous media: Effect of water partitioning and air-water interfacial adsorption. *Environ. Sci. Technol., 35*, 4457-4462.

Klieman, K. A. (2012). "U.S. Oil Companies, the Nigerian Civil War, and the Origins of Opacity in the Nigerian Oil Industry", 1(1), 155–165.

Mitra, A. P., & Sharma, C. (2002). Indian aerosols: present status. *Chemosphere, 49*, 1175 -1190

Na, K., Moon, K. C., & Fung, K. (2001). Concentrations of volatile organic compounds in an area of Korea. *Atmospheric Eniron, 35*, 2747-2756.

[NNPC] Nigeria National Petroleum Corporation. (2008). Warri Refining and Petrochemical Company Limited. *Technical Report, 4*, 74-76. Retrieved from http://www.osha.gov/dts/osta/otm/otm_iv/otm_iv_html. Retreived February 20, 2012.

Rao, P. S., Gavane, A. G., Ankam, P., Ansari, M. F., Pandit, V. I., & Nema, P. (2004). Performance evaluation of a green belt in a petroleum refinery: *A case study. Ecol. Eng., 23*, 77-84

Reinermann, P., & Golightly, R. (2005). Emission data management. Petroleum quarterly, www.eptq.com. Assessed December 12, 2012.

Sonibare, J. A., Akeredolu, F. A.,Obanijesu E. O., & Adebiyi, F. M. (2007). Contribution of Volatile Organic Compounds to Nigeria's airshed by Petroleum Refineries. *Petroleum Science and Technology, 25*, 503-516.

[U.S. DOE] United State Department of Energy. (2007). Energy and Environmental profile of the U.S. Petroleum Refining Industry. Energetics, Inc., Columbia, MD.

[USEPA] United States Environmental Protection Agency. (1989). Airs Facility Subsystem Source Codes and Emission Factor Listing for Criteria Air Pollutants. Research Triangle Park, N.C. U.S. Environmental Protection Agency.

WHO. (2002).The Health Effects of Indoor Air Pollution Exposure in Developing Countries. Publication of the World Health Organization, Geneva, Switzerland. WHO/SDE/OEH/02.05

Zielinska, B., & Fujita, E. M. (2003). Characterization of ambient volatile organic compounds at the western boundary of the SCOS97-NARSTO Modelling Domain. *Atmospheric Environ., 37*, 171-180.

Zhonghua, L., Dong, W., & Sheng, Z. Y. (2003). DNA damage and changes of antioxidative enzymes in chronic benzene poisoning mice. *Bin Za Zhi, 21*, 423 -425.

16

Assessment of Investment in Small Hydropower Plants

Perica Ilak[1] & Slavko Krajcar[1]

[1] Faculty of Electrical Engineering and Computing, University of Zagreb, Croatia

Correspondence: Perica Ilak, Faculty of Electrical Engineering and Computing, University of Zagreb, Zagreb, HR-10000, Unska 3, Croatia. E-mail: perica.ilak@fer.hr

Abstract

In this study an assessment of investments in a hydropower plants is considered. The objective of this work is to maximize the net present value from selling energy. Because of the stochastic nature of river flows, flow duration curve is constructed to take into account hydropower plant capacity factor, important for the assessment of investment. Proposed mixed-integer linear programming model is flexible and accounts for nonlinear three-dimensional (3-D) relationship between the produced power, the discharged water, and the head of the associated reservoir. Continuous chain of imaginary run-of-river plants is considered on the Sava river stretch (Croatia) from border with Slovenia to city of Sisak.

Keywords: assessment of investment, optimization, mathematical model, small hydropower plant, run-of-the-river, mixed-integer linear programming, Sava river, City of Zagreb

Nomenclature

B	Set of indices of the blocks of the piecewise linearization of the unit performance curve $B=\{1,2\}$, $blok \in B$.
$c(P_{max}(i))$	Specific investment cost function [€/kW].
$d_k(i,t)$	0/1 variable used for discretisation of performance curves $k \in \{1,2\}$.
$E0_l(i)$	Minimal possible energy output of plant i for performance curve l, $l \in \{1,2\}$, [MWh].
$E_e(i,t)$	Electricity produced for energy market by plant i in time period t [MWh].
$H_{max}(i)$	Maximal possible head of a pondage i [m].
$H_r(i)$	Rated head of a plant i [m].
$H(i,t)$	Head of pondage i in time period t [m].
H	Conversion factor equal to 3600 [m³s/m³h].
I	Set of indices of the reservoirs/plants, $I=\{'Podsused','Prečko','Zagreb1','Zagreb2','Zagreb3','Zagreb4'\}$, $i \in I$.
$I_{>10MW}$	Subset of plants with capacity above 10 MW, $I_{>10MW} \subset I$.
$I_{\leq 10MW}$	Subset of plants with capacity of 10 MW and under, $I_{\leq 10MW} \subset I$.
i_x	Inflation index.
$I(t)$	Investment cost of a cascade in time period t [€].
J	Set of indices of the perf. curves $J=\{1$-high lvl., 2-middle lvl., 3-low lvl.$\}$, $j \in J$.
n	Capacity factor.
$N_{I \times B}$	Number of units scenario matrix of dimension IxB.
$NPV_{A \times B \times C}$	Depository matrix of $Q_{*,a}, Q_{*,b}, N_{*,c}$ combinations net present values of dimension $AxBxC$.
$om(P_{max}(i))$	Specific operating and maintenance cost function [%].
$on(i,t)$	0/1 variable which is equal to 1 if plant i is on-line in time step t.

$O\&M(t)$	Operating and maintenance cost of a cascade in time period t [€].
$P_{max}(i)$	Capacity of plant i [MW].
$Q_{min}(i)$	Minimum water discharge of plant i [m³/s].
$Q_{max}(i)$	Maximum water discharge of plant i [m³/s].
$Q_{res}(i)$	Residual flow of plant i [m³/s].
$Q_{I \times A}$	Rated flow scenario matrix of dimensions IxA.
$Q_{I \times A}$	Rated head scenario matrix of dimension IxC.
$q_v(i,t,blok)$	Variable water discharge of a plant i in time period t of block $blok$ [m³/s].
$q(i,t)$	Total water discharge of plant i in time step t [m³/s].
$R(t)$	Revenue of a cascade in time period t [€].
r	Discount rate.
$s(i,t)$	Spillage of the reservoir i in time step t [m³].
T	Set of indices of the steps of the optimization period, $T=\{1, 2, \ldots, T_{max}\}$, $t \in T$, $T_{max} \in \{20,25,30\}$.
U_i	Set of upstream reservoirs of plant i.
$V_{min}(i)$	Minimal possible utilizable volume of a plant i [m³].
$V_{max}(i)$	Maximal possible utilizable volume of a plant i [m³].
$V(i,t)$	Utilizable volume of a plant i in time interval t [m³].
$W(i,t)$	Forecasted natural water inflow of the reservoir i in time step t [m³].
$w(i,t,blok)$	0/1 variable which is equal to 1 if water discharged by plant i has exceeded block $blok$ in time step t.
$weight(t)$	Distribution of investment cost along y time intervals, $t \in y \subset T$.
$X(i,t)$	Water content of the reservoir i in time step t [m³].
$X_{avg}(i,t)$	Average water content of the reservoir i in time step t [m³].
$X_{max}(i)$	Maximal content of the reservoir i [m³].
$X_l(i)$	l-th discrete level of the content of the reservoir i, $l \in \{1,2\}$ [m³].
$X_{min}(i)$	Minimal content of the reservoir i [m³].
$X(i,0)$	Initial water content of the reservoir i [m³].
$X(i,T_{max})$	Final water content of the reservoir i [m³].
Y	Number of hours in one year, 8760 [h].
$\pi_{fid}(t)$	Feed-in-tariff in time step t [€/MWh].
$\pi_e(t)$	Forecasted price of electricity in time step t [€/MWh].
$\rho_j(i)$	Slope of the performance curve j of plant i [MWh/m³].
$\rho_v(blok)$	Slope of the block $blok$ of the utilizable volume function [m³/m³/s].
$\Delta Q_{blok}(i)$	Maximum water discharge of block $blok$ of plant i B$=\{1,2\}$, $blok \in B$ [m³/s].

1. Introduction

Harnessing a stream for hydroelectric power is a major undertaking and careful planning is a necessary if a profitable hydropower plant (HPP) is to result. At the same time country water laws and environmental concerns must be accounted (WVU, 1978) for detailed insight of environmental impacts of small HPP and how to they compare to other renewables is given in (Kucukali & Baris, 2009).The general procedure in hydrologic study is to establish how much water is available to divert through turbine and the hydraulic head associated with this flow.

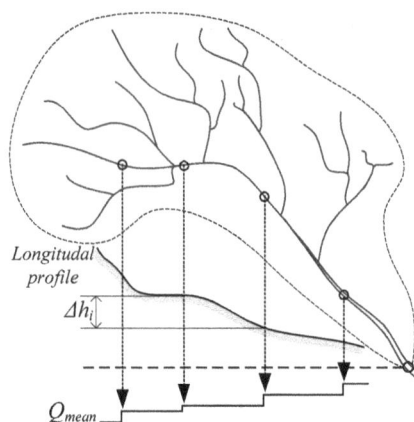

Figure 1. Line potential = function of mean annual flow and head difference Δh_i of a reach

Hence a river line potential is observed. The line potential denotes the theoretical potential of streams and rivers which could be harnessed through a continuous chain of imaginary run-of-river plants (a scheme operating with no appreciable water storage). The relevant hydro potential is obtained by subdividing a stream or river into reaches along which discharge and longitudinal slope are approximately uniform as illustrated in Figure 1. It is normally defined on the basis of mean annual discharge and the difference in elevation between beginning and end of each reach. In this study tributary line potential is not considered. More details on how to evaluate river energy potential in (Weiss & Faeh, 1990; Korkmaz, 2007). When river line potential is established utilizable potential of each river reach is calculated. The utilizable potential is a function of utilizable volume of water and performance curves.

Scope of this work is limited to a basic understanding of modeling hydropower systems from financial point of view with easy to implement linear programming model, check (IRENA, 2012) for more on financial aspect of hydropower systems. This work will give interesting insight for those considering investing in small distributed hydro production. For those interested in technical aspect of small scale HPP check all volumes of (USACE, 1979). Also short insight how to evaluate potential location for HPP is given in (WVU, 1978) and more detailed insight in (ESHA, 2004). A review of different types of models developed to evaluate the cost of the small hydropower projects is given in (Mishra et al., 2011). In (Kucukali, 2011) new risk metric is implemented in hydro model for assessment of investment in river-type hydropower plant projects. To create dynamic models which seek to answer when to invest in small scale hydropower plants consult (Knutsen et al., 2010). Reconnaissance level of detail is warranted and some basic requirements when considering investing in hydropower plants will be discussed.

In Section III problem description and mathematical formulation of a model will be given, in Section IV case study is given of a potential investment in cascaded run-of-the-river HPPs on the Sava river and results are shortly discussed, annual energy production of cascade is estimated and if estimate appear to have a value exceeding the cost of building cascade then a feasibility level of study will be warranted which can be refined and prepared in greater detail that consider construction and maintenance (WVU, 1978).

2. Problem Description and Formulation

2.1 Flow-Duration Curve

Utilizable volume V of water is defined by flow duration curve (FDC) as illustrated on Figure 2 and pondage dimensions as illustrated on Figure 3. FDCs of observed river reaches are based on daily data since FDC developed from monthly data generally become increasingly less reliable if power storage is relatively small or nonexistent and will overestimate average energy production by as much as 15 to 50 percent, depending on the day-to-day variability of flow (USACE, 1979).

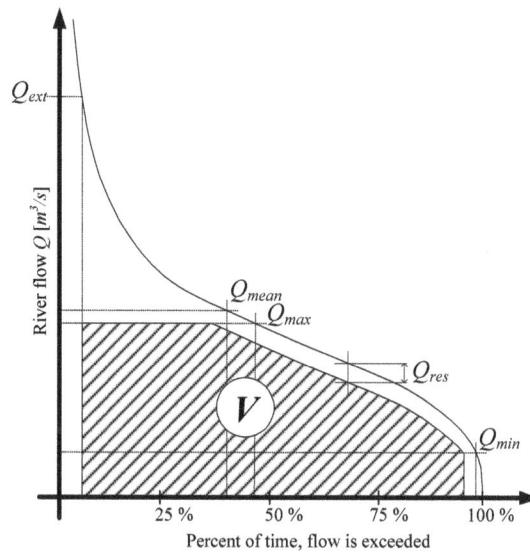

Figure 2. FDC of a river reach

For extremely high flows Q_{ext} the tailwater will rise so high that the net power head will become too small for the power plant to function. Unless there is available storage to regulate flows to more favorable discharge rates or a sluice to divert extreme flows from main riverbed then HPP will be inoperable in extremely high flows. Residual flow Q_{res} and HPP minimum turbine discharge Q_{min} are taken into account to correctly evaluate utilizable volume.

2.2 Pondage

Multipurpose projects usually allocate the total available storage to the various purposes proportional to some cost and benefit relationship or to achieve prescribed objectives. Often these objectives have conflicting demands on storage.

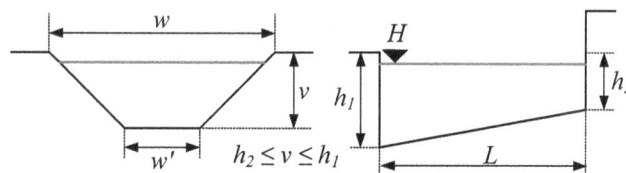

Figure 3. Pondage dimensions

Improved riverbed (Figure 3) of each river reach in this study is considered as pondage of associated HPP. Purpose of run-of-river scheme in this study is to maximize profit form electricity production and to stabilize head H of each river reach and consequently prevent flooding of nearby area. Pondage parameters w, w', v, L, h_1 and h_2 are determined by algorithm shown in Figure 8.

Pondage continuity equation is formulated as (1)-(2). For high exactness Y should strive to 0.

$$X(i,t) = X(i,t-1) + \frac{1}{Y} \cdot W(i,t) + \frac{1}{Y} \cdot \sum_{j \in U}[V(j,t) + s(j,t)] - \frac{1}{Y} \cdot [V(i,t) + s(i,t)], \forall i \in I, \forall t \in T \quad (1)$$

$$X(i,t) \leq X_{max}(i),$$

$$X(i,t) \geq X_{min}(i), \forall i \in I, \forall t \in T \quad (2)$$

The time delay for water transportation is not considered.

2.3 Head Dependent Energy Production

In order to obtain reasonable estimate of the annual power production performance curves are used (Figure 4). Defined performance curves account for efficiency characteristics and operating range limitation consistent with the turbine type likely to be installed (Perez-Diaz et al., 2010). More on performance curves in (Conejo et al., 2002; USACE, 1979).

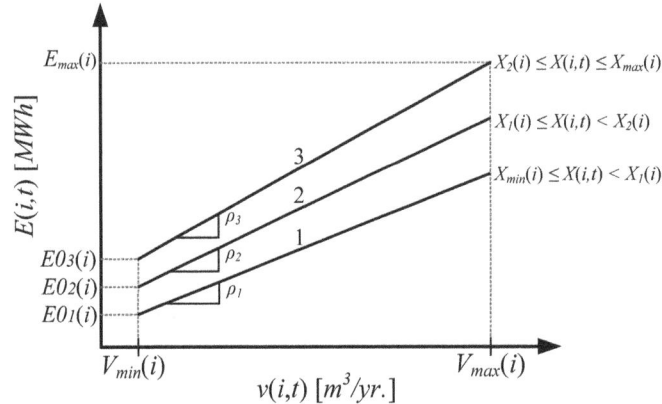

Figure 4. Three-dimensional unit performance curves used for energy evaluation

Generally to assure that best choice of rated head H_r, rated turbine discharge Q_{max} and number of units N is selected, all combinations of scenarios H_r, Q_{max} and N should be calculated, procedure for determining optimal rated flow, number of units and head is shown in Figure 8.

General procedure is to calculate and compare energy produced of turbines having a higher and a lower rated flow. Following establishment of the rated flow it needs to be checked if power is being lost because turbine discharge is consistently below lower boundaries, then HPP maximal capacity P_{max} is lowered and more units are added. If energy production $E(i,t)$ increase is substantial, cost of the alternatives may be determined from the HPP specific investment cost function (Figure 6) and O&M specific costs function (Figure 7). Also first selection of the number of turbines needs to be compared with the lesser number of units.

The rated head of the turbine can be further refined by optimization in a similar manner. The annual power production is computed for higher and lower heads with the same capacity rating. The rated head yielding the highest annual output should be used.

In this paper performance curves (Figure 4) have been modeled through a piecewise linear formulation of Hill chart (Guan et al., 1999). Figure 4 shows linear performance curves with its associated slope ρ which is defined by HPP conversion capabilities [MW/m^3].

There are 3 performance curves associated for 3 water contents (Figure 4).

Activation of corresponding performance curve is done by approach presented in (Conejo et al., 2002), with some improvements presented in (Rajšl & Ilak, 2012), and is shown in (3)-(9).

$$X_{avg}(i,t) = \frac{X(i,t)+X(i,t-1)}{2}, \forall i \in I, \ \forall t \in T \tag{3}$$

$$X_{avg}(i,t) \geq X_1(i) \cdot [\, d_1(i,t) - d_2(i,t)\,] + X_2(i) \cdot d_2(i,t), \ \forall i \in I, \forall t \in T \tag{4}$$

$$X_{avg}(i,t) \leq X_{max}(i) \cdot d_2(i,t) + X_1(i) \cdot \big[\, 1 - d_1(i,t)\big] + X_2(i) \cdot [\, d_1(i,t) - d_2(i,t)], \forall i \in I, \forall t \in T \tag{5}$$

$$d_1(i,t) \geq d_2(i,t),$$
$$d_2(i,t) \geq d_3(i,t),$$
$$d_3(i,t) \geq d_4(i,t), \forall i \in I, \forall t \in T \tag{6}$$

Energy production performance curves (7)-(12).

$$E(i,t) - E0_1(i) \cdot on(i,t) - v(i,t) \cdot \rho(i) - E_{max}(i) \cdot [d_1(i,t) + d_2(i,t)] \leq 0$$
$$E(i,t) - E0_1(i) \cdot on(i,t) - v(i,t) \cdot \rho(i) + E_{max}(i) \cdot [d_1(i,t) + d_2(i,t)] \geq 0, \forall i \in I, \ \forall t \in T \qquad (7)$$

$$E(i,t) - E0_2(i) \cdot on(i,t) - v(i,t) \cdot \rho(i) - E_{max}(i) \cdot [1 - d_1(i,t) + d_2(i,t)] \leq 0$$
$$E(i,t) - E0_2(i) \cdot on(i,t) - v(i,t) \cdot \rho(i) + E_{max}(i) \cdot [1 - d_1(i,t) + d_2(i,t)] \geq 0, \forall i \in I, \forall t \in T \qquad (8)$$

$$E(i,t) - E0_3(i) \cdot on(i,t) - v(i,t) \cdot \rho(i) - E_{max}(i) \cdot [2 - d_1(i,t) - d_2(i,t)] \leq 0$$
$$E(i,t) - E0_3(i) \cdot on(i,t) - v(i,t) \cdot \rho(i) + E_{max}(i) \cdot [2 - d_1(i,t) - d_2(i,t)] \geq 0, \forall i \in I, \forall t \in T \qquad (9)$$

To correctly calculate annual energy production it is necessary to calculate annual utilizable volume $V(i,t)$ correctly (Figure 5). To do that $V(i,t)$ needs to be expressed as a function of HPP water discharge $q(i,t)$ as it is shown in (11).

Turbine water discharge of plant i in time step t (10).

$$q(i,t) = Q_{min}(i) + \sum_{blok=1}^{2} q_v(i,t,blok), \forall i \in I, \ \forall t \in T \qquad (10)$$

Utilizable volume of plant i in time step t (11).

$$V(i,t) = f(q(i,t)) = V_{min}(i) + v(i,t), \forall i \in I, \forall t \in T \qquad (11)$$

Where $V_{min}(i)$ is minimum possible utilizable volume of plant i (12) which can be determined from Figure 10 and $v(i,t)$ is variable utilizable volume of plant i in time step t (13).

$$V_{min}(i) = g(Q_{min}(i)), \ \forall i \in I, \ \forall t \in T \qquad (12)$$

$$v(i,t) = H \cdot Y \cdot \sum_{blok=1}^{2} \rho_v(blok) \cdot q_v(i,t,blok), \forall i \in I, \ \forall t \in T \qquad (13)$$

Function $V(i,t)$ illustrated in Figure 5 is modeled through use of binary variables (14).

$$q_v(i,t,1) \leq \Delta Q_1(i) \cdot on(i,t)$$
$$q_v(i,t,1) \geq \Delta Q_1(i) \cdot w(i,t,1)$$
$$q_v(i,t,2) \leq \Delta Q_2(i) \cdot w(i,t,1)$$
$$q_v(i,t,2) \geq \Delta Q_2(i) \cdot w(i,t,2), \ \ \forall i \in I, \ \forall t \in T \qquad (14)$$

To prevent model to calculate with discharge above turbine maximal discharge $Q_{max}(i)$ in time period t constraint (15) is defined.

$$q(i,t) \leq Q_{max}(i), \ \forall i \in I, \ \forall t \in T \qquad (15)$$

Important characteristic of function (11) is its concave nature important for assessment of investment (Figure 5). The greater the chosen value of the maximal turbine discharge, the smaller proportion the year that the system will be operating on full power, i.e. it will have a lower capacity factor n as it is depicted on Figure 5 (ESHA, 2004; BHA, 2005).

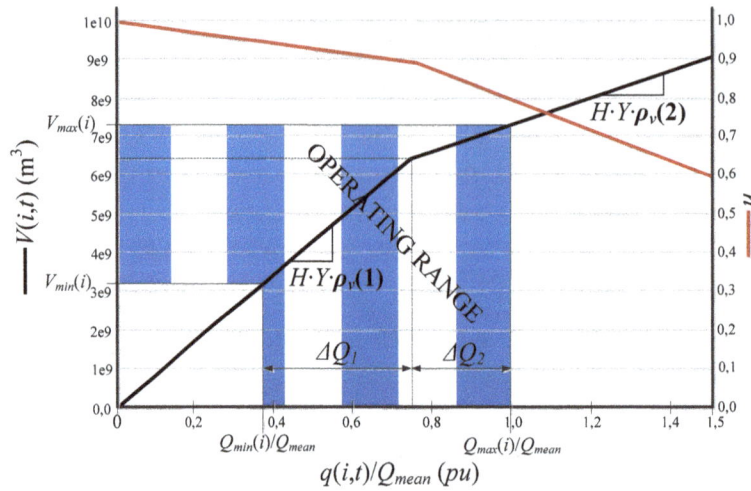

Figure 5. Utilized volume and capacity factor as a function of turbine flow

2.4 Revenues and Expenses

The largest share in investment cost for large hydropower plant is typically taken up by civil works for the construction of the hydropower plant (such as dam, tunnels, canal and construction of powerhouse, etc.). Electrical and mechanical equipment usually contributes less to the cost. However, for hydropower projects where the installed capacity is less than 5 MW, the costs of electro-mechanical equipment may dominate total cost due to high specific cost of small-scale equipment (IRENA, 2012). Specific cost of investment in HPP is depicted on Figure 6 and is piecewise linear function.

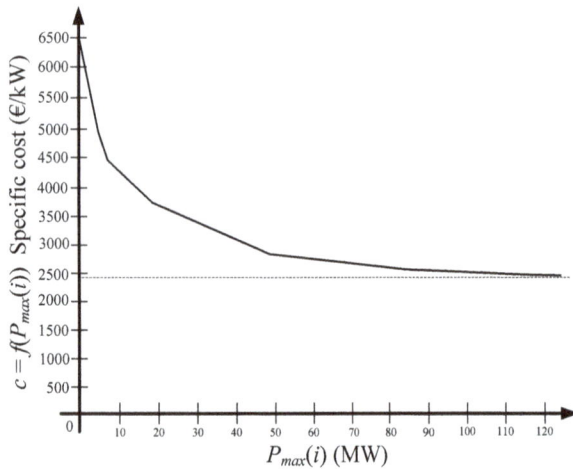

Figure 6. Specific investment cost of a HPP as a function of installed capacity

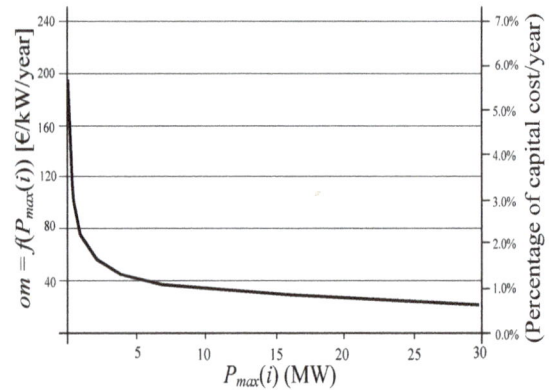

Figure 7. Specific O&M costs as a function of installed capacity

HPP usually require little maintenance, and operation costs will be low. When in cascade along a river, centralized control can reduce O&M costs to very low levels. In this study O&M specific costs are depicted on Figure 7 according to (IRENA, 2012).

2.5 Objective Function and Algorithm

Maximize:

$$NPV = \sum_{t=y+1}^{t=T} \frac{(1+i_x)^t (R(t) - O\&M(t))}{(1+r)^t} - \sum_{t=1}^{t=y} \frac{(1+i_x)^t I(t)}{(1+r)^t} \qquad (16)$$

Expression (16) assumes that the project will be developed in y years (time intervals). At the end of y-th year the whole development is finished and paid. The electricity revenues and O&M costs are made effective at the end of each year and begin at the end of the y-th year (ESHA. 2004).

Revenue part of (13):

$$R(t) = \pi_e(t) \cdot \sum_{i \in I_{>10MW}} E(i,t) + \pi_{fid} \cdot \sum_{i \in I_{\leq 10MW}} E(i,t) \ \forall i \in I, \forall t \in T \tag{17}$$

Operating and maintenance part:

$$O\&M(t) = \frac{\sum_{i \in I} om(P_{max}(i))}{100} \cdot \sum_{t=1}^{t=y} \frac{(1+i)^t I(t)}{(1+r)^t} \ \forall i \in I, \forall t \in T \tag{18}$$

Investment cost:

$$I(t) = weight(t) \cdot \sum_{i \in I} c(P_{max}(i)) \cdot P_{max}(i) \ \forall i \in I, \ \forall t \in T \tag{19}$$

Expression (20) defines how investment cost is distributed over y years.

$$\sum_{t=1}^{y} weight(t) = 1 \tag{20}$$

Simple heuristic algorithm for reconnaissance detail of study for cascaded run-of-river HPPs.

Start

$A \in \mathbb{N}, B \in \mathbb{N}, C \in \mathbb{N}, i \in I$

$Matrices \ Q_{I \times A}, N_{I \times B}, H_{I \times C}, NPV_{A \times B \times C};$
$Colum \ vectors \ Q_{max_{I \times 1}}, N_{I \times 1}, H_{r_{I \times 1}};$

$a = 1,2,3,\ldots,A; \ b = 1,2,3,\ldots,B; \ c = 1,2,3,\ldots,C;$
$while \ a \leq A$
$\{$
$Q_{max} = Q_{*,a}$
$\quad while \ b \leq B$
$\quad \{$
$N = N_{*,b}$
$\quad\quad while \ c \leq C$
$\quad\quad \{$
$\quad\quad\quad H_r = H_{*,c}$
$\quad\quad\quad\quad\quad\quad\quad ***MILP***: \max\{Net \ Present \ Value\};$
$\quad\quad\quad NPV_{a,b,c} = Net \ Present \ Value;$
$\quad\quad\quad c++;$
$\quad\quad \}$
$\quad b++;$
$\quad \}$
$a++;$
$\}$
$Matrix \ maximal \ value: \max_{a,b,c}\{NPV_{A \times B \times C}\} = NPV_{a^*,b^*,c^*}$
$Optimal \ vectors: Q_{*,a^*}, N_{*,b^*}, H_{*,c^*}$

Stop

Figure 8. Heuristic algorithm for assessment of investment in cascaded run-of-river HPPs

Model is defined as a mixed-integer linear program using indexed assignments (Rosenthal, 2012). The presented results were obtained on 3.4 GHz based processor with 8 GB RAM using CPLEX under General Algebraic Modeling System (GAMS).

3. Case Study and Results

Virtual hydropower system (HPS) Sava (Croatia) is modelled as illustrated on Figure 9. Observed 6 river reaches consists of: 6 pondages, 6 run-of-river HPP and a sluice.

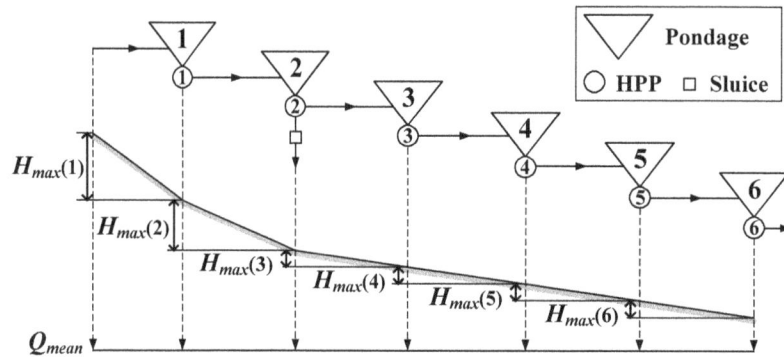

Figure 9. HPS sava

Because of computational effectiveness, it should be noted that time periods of 1 year are considered.

Since there is a sluice in pondage 2, extreme water flows will not reduce utilizable volume by extreme tailwater rise.

The conventional propeller and very low head (VLH) Kaplan turbines are considered which are operated at power outputs with flows from 40 to 100 percent of rated flow $Q_{max}(i)$. Performance curves (Figure 4) and pondage (Figure 9) parameters are shown in Table 1.

Table 1. HPP performance curves and associated pondage parameters

HPP	ρ_1	ρ_2	ρ_3	X_{min}	X_1	X_2	X_{max}	H_{max}
1	2,207E-5	2,207E-5	2,207E-5	3E6	3.3E6	3.66E6	4.078E6	9
2	1,839 E-5	1,839E-5	1,839E-5	4.6E6	5E6	5.4E6	6.75E6	7,5
3	9,197E-6	9,197E-6	9,197E-6	1.2E6	1.4E6	1.7E6	1.903E6	3,75
4	9,197E-6	9,197E-6	9,197E-6	1.2E6	1.4E6	1.7E6	1.9406E6	3,75
5	9,197E-6	9,197E-6	9,197E-6	0.4E6	0.5E6	0.6E6	0.7875E6	3,75
6	9,197E-6	9,197E-6	9,197E-6	1.7E6	2.0E6	2.3E6	2.718E6	3,75

Since there is no important tributary in observed river reaches, one FDC is constructed and is assigned to all 6 river reaches. FDC used in this paper is piecewise linear function with 3 linear parts and is an arithmetic mean of four FDC based on daily flows of periods from time intervals: 1997 to 1987, 1988 to 1993, 1988 to 1998 and 1994 to 1999. Resulting FDC is linearized and is depicted on Figure 10.

Mean annual discharge Q_{mean} of each river reach is 320 m³/s. Residual water flow Q_{res} is 20 m³/s. Maximal flow Q_{ext} of each river reach is 800 m³/s. Electric energy price π_e is 43.6 €/MWh and is average price of base load power at EPEX Spot (EEX, 2012) for 2003 to 2012 period. Feed-in-tariff π_{fit} is 56 €/MWh. Discount rate r is 8.2 %. Inflation index i is 2 %, number of investment years y is 1 and utilizable volume function slopes $\rho_v(1)$ and $\rho_v(2)$ are 0.8616 and 0.3618 respectively. Number of units is set to $N(i) = 1$ for all i and since stabilizing head $H(i,t)$ of each river reach is one of the major concerns in this study and maximal possible head $H_{max}(i)$ is predetermined by geographical and urbanization constraints thus rated head $H_r(i)$ won't be optimized.

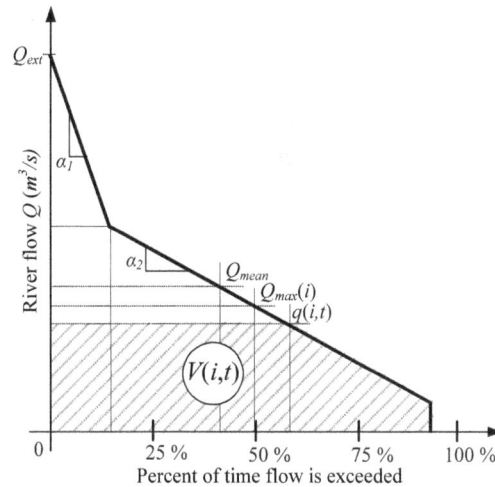

Figure 10. Linearized FDC less residual flow and extreme flows

Consequently results are obtained for set of three observed periods $T_{max} \in \{20, 25, 30\}$ with scenario matrix $Q_{I \times A} = [1.5, 1.25, 1.0, 0.75, 0.5, 0.4, 0.33]^T [pu]$ with $A = 1$, since one scenario of rated flows is used.

Table 2. Output data for time horizon of 20 years

Q_{max}/Q_{mean}	1,5	1,25	1	0,75	0,5	0,4	0,33
n	0,61171	0,69375	0,7787	0,8616	0,9025	0,9259	0,9875
NPV [mil. €]	-164	-120	-43,6	-29,3	-37,6	-27,9	-33,5
NPV [%]	-32,929	-27,396	-12,035	-10,135	-16,686	-14,686	-19,606
IRR [%]	3,128	4,12	6,4842	6,731	5,64011	5,99	5,09
W [GWh]	646,78	638,38	565,78	461,22	314,92	253,55	215,36
$O\&M$ [mil. €]	39,5	34,6	30,9	30,1	27,2	23,8	26,2
I [mil. €]	459	402	331	259	198	166	145
$P_{max}(1)$ [MW]	35,785	29,799	23,839	17,879	11,919	9,536	7,867
$P_{max}(2)$ [MW]	29,798	24,832	19,866	14,899	9,933	7,946	6,556
$P_{max}(3)$ [MW]	14,899	12,416	9,933	7,45	4,966	3,973	3,278
$P_{max}(4)$ [MW]	14,899	12,416	9,933	7,45	4,966	3,973	3,278
$P_{max}(5)$ [MW]	14,899	12,416	9,933	7,45	4,966	3,973	3,278
$P_{max}(6)$ [MW]	14,899	12,416	9,933	7,45	4,966	3,973	3,278

Table 3. Output data for time horizon of 25 years

Q_{max}/Q_{mean}	1,5	1,25	1	0,75	0,5	0,4	0,33
n	0,61171	0,69375	0,7787	0,8616	0,9025	0,9259	0,9875
NPV [mil. €]	-130	-87	-11,0	-3,25	-19,4	-12,2	-20,9
NPV [%]	-25,798	-19,889	-3,007	-1,122	-8,498	-6,349	-12,025
IRR [%]	4,8552	5,7157	7,83875	8,06	7,12	7,39	6,593
W [GWh]	666,74	638,38	5,66E+07	461,22	314,92	253,55	215,36
$O\&M$ [mil. €]	44,0	38,6	34,5	33,6	30,3	26,5	29,2
I [mil. €]	459	402	331	259	198	166	145
$P_{max}(1)$ [MW]	35,785	29,799	23,839	17,879	11,919	9,536	7,867
$P_{max}(2)$ [MW]	29,798	24,832	19,866	14,899	9,933	7,946	6,556
$P_{max}(3)$ [MW]	14,899	12,416	9,933	7,45	4,966	3,973	3,278
$P_{max}(4)$ [MW]	14,899	12,416	9,933	7,45	4,966	3,973	3,278
$P_{max}(5)$ [MW]	14,899	12,416	9,933	7,45	4,966	3,973	3,278
$P_{max}(6)$ [MW]	14,899	12,416	9,933	7,45	4,966	3,973	3,278

Table 4. Output data for time horizon of 30 years

Q_{max}/Q_{mean}	1,5	1,25	1	0,75	0,5	0,4	0,33
n	0,61171	0,69375	0,7787	0,8616	0,9025	0,9259	0,9875
NPV [mil. €]	-104	-63,8	14,2	16,1	-5,87	-56,8	-11,5
NPV [%]	-20,553	-14,387	3,603	5,452	-2,544	-0,292	-6,547
IRR [%]	5,868	6,6225	8,58	8,7891	7,91	8,165	7,425
W [GWh]	679,33	638,38	565,78	461,22	314,92	253,55	215,36
$O\&M$ [mil. €]	47,4	41,5	37,1	36,1	32,6	28,5	31,4
I [mil. €]	459	402	330	259	198	166	145
$P_{max}(1)$ [MW]	35,785	29,799	23,839	17,879	11,919	9,536	7,867
$P_{max}(2)$ [MW]	29,798	24,832	19,866	14,899	9,933	7,946	6,556
$P_{max}(3)$ [MW]	14,899	12,416	9,933	7,45	4,966	3,973	3,278
$P_{max}(4)$ [MW]	14,899	12,416	9,933	7,45	4,966	3,973	3,278
$P_{max}(5)$ [MW]	14,899	12,416	9,933	7,45	4,966	3,973	3,278
$P_{max}(6)$ [MW]	14,899	12,416	9,933	7,45	4,966	3,973	3,278

4. Discussion

Simulation shows that it is unlikely that schemes using significantly more than the mean river flow (Q_{mean}) will be economically attractive nor will they be environmentally acceptable (Tables 1-3). Therefore the turbine design flow for a run-of river scheme (a scheme operating with no appreciable water storage) will not normally be greater than Q_{mean}. The exception would be a scheme specifically designed to capture very high winter flows, which is not the case here since pondage parameters are defined by geographical and urbanization constraints (ESHA, 2004; BHA, 2005).

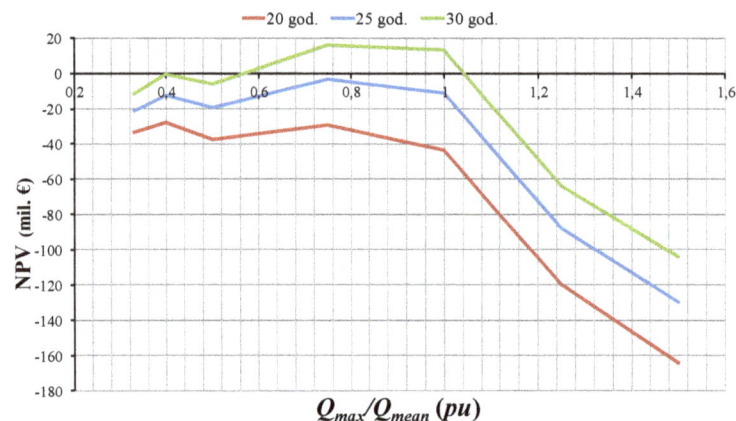

Figure 11. NPV value for different HPP rated flows

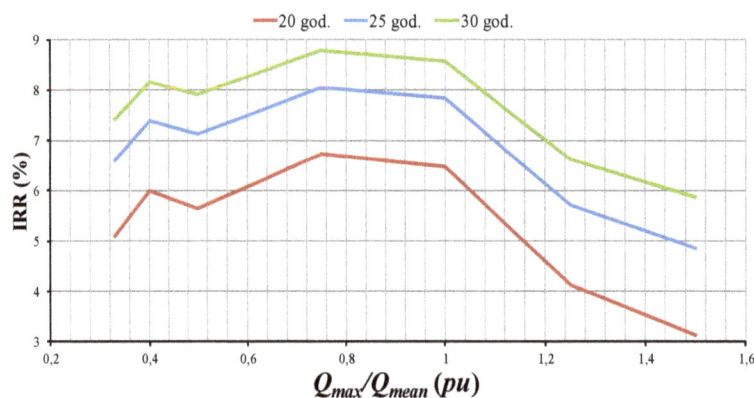

Figure 12. IRR value for different HPP rated flows

Simulation showed that for observed period of 30 years positive NPV (Figure 11) and highest IRR (Figure 12) were obtained. Model of virtual HPS Sava presented here is flexible and allows simulation for wide range of input data. Project was evaluated to be economically favorable and more detailed energy evaluations needs to be conducted using accurately forecasted data along with different scenarios and shorter time interval which in the case of energy production estimates during feasibility studies should not exceed one month.

Adjusting model to desired accuracy and detail will result in computational intensive simulation and will provide valuable data on run-of-river cascade long-term schedule and its economic justification.

Acknowledgements

The authors would like to express their sincere thanks to the Department of Energy and Power Systems of University of Zagreb Faculty of Electrical Engineering and Computing.

References

BHA. (2005). A Guide To UK Mini-Hydro Developments.

Conejo, A., Arroyo, J., Contreras, J., & Villamor, F. (2002). Self-Scheduling of a Hydro Producer in a Pool-Based Electricty Market. *IEEE Transactions On Power Systems, 17*(4). http://dx.doi.org/10.1109/TPWRS.2002.804951

EEX. (2012). *European Energy Exchange AG*. Retrieved January 25, 2012, from http://www.eex.com

ESHA. (2004). Layman's Guidebook on how Develop a Small Hydro Site.

Guan, X., Svoboda, A., & Li, C. A. (1999). Scheduling hydro power systems with restricted operating zones and discharge ramping constraints. *IEEE Trans. Power Syst., 14*, 126-131. http://dx.doi.org/10.1109/59.744500

IRENA. (2012). Hydropower Renewable Energy Technologies: Cost Analysis Series, vol. 1.

Knutsen, L. H., & Holand, R. (2010). Investment Strategy for Small Hydropower Generation Plants in Norway.

Korkamz, O. (2007). A Case Study on Feasibility Assessment of Small Hydropower Scheme.

Kucukali, S. (2011). Risk assessment of river-type hydropower plants using fuzzy logic approach. *Energy Policy, 39*(10), 6683-6688. http://dx.doi.org/10.1016/j.enpol.2011.06.067

Kucukali, S., & Baris, K. (2009). Assessment of small hydropower (SHP) development in Turkey: Laws, regulations and EU policy perspective. *Energy Policy, 37*(10), 3872-3879. http://dx.doi.org/10.1016/j.enpol.2009.06.023

Mishra, S., Singal, S. K., & Khatod, D. K. (2011). Optimal installation of small hydropower plant. *A review, 15*(8), 3862-3869. http://dx.doi.org/10.1016/j.rser.2011.07.008

Pérez-Díaz, J. I., Wilhelmi, J. R., & Arévalo, L. A. (2010). Optimal short-term operation schedule of a hydropower plant in a competitive electricity market. *Energy Conversion and Management, 51*(12), 2955-2966. http://dx.doi.org/10.1016/j.enconman.2010.06.038

Rajšl, I., Ilak, P., Delimar, M., & Krajcar, S. (2012). Dispatch Method for Independently Owned Hydropower Plants in the Same River Flow. *Energies, 5*, 3674-3690. http://dx.doi.org/10.3390/en5093674

Rosenthal, R. (2012). GAMS-A User's Guide; GAMS Development Corporation: Washington, DC, USA, 2012. Retrieved January 25, 2012, from http://www.gams.com/dd/docs/bigdocs/GAMSUsersGuide.pdf

USACE. (1979). U.S. Army corps of enginners. Feasibility Studies for Small Scale Hydropower Additions: Hydrologic Studies Volume III.

Weiss, H. W., & Faeh, A. O. (1990). Methods for evaluating hydro potential. *Proceedings of two Lausanne Symposia*.

WVU. (1978). A cross-section, Small Hydroelectric plants. *West Virginia University*, FS-13.

17

A Laboratory Investigation of the Effects of Saturated Steam Properties on the Interfacial Tension of Heavy-Oil/Steam System Using Pendant Drop Method

Madi Abdullah Naser[1,2], Asep Kurnia Permadi [1], Wisup Bae [2], Wonsun Ryoo [3] & Septoratno Siregar [1]

[1] Department of Petroleum Engineering, Bandung Institute of Technology, Indonesia.

[2] Department of Energy and Mineral Resources Engineering, Sejong University, South Korea

[3] Department of Chemical Engineering, Hongik University, South Korea

Correspondence: Wisup Bae, Department of Energy and Mineral Resources Engineering, Sejong University, 98 Gunja-dong, Gwangjin-gu, Seoul 143-747, South Korea. E-mail: wsbae@sejong.ac.kr

The research is financed by Department of Energy and Mineral Resources Engineering, Sejong University.

Abstract

For about a century, steam injection has been widely used as the most popular thermal recovery method for heavy-oil in sandstone reservoirs. In order to achieve higher recovery efficiency, which corresponds to the lowest possible value of residual oil saturation and economic success of steam injection projects, an accurate laboratory measurement of the interfacial tension between steam and heavy-oil is essential. However, laboratory investigation and visualization of the effects of steam injection on the interfacial tension between heavy-oil and steam as a function of saturation temperature and pressure is not well documented in the literature.

The objective of this study is to investigate the influences of the two main factors which affect the interfacial tension of heavy-oil and steam namely saturation pressure and temperature. An optical cell, which was fitted with a goniometer system and a procedure to generate steam for the measurement of interfacial tension have been used. The difference between the density of heavy-oil and steam, which was used for pendant drop measurements, was calculated at specific temperature and pressure conditions using Katz's method. Meanwhile, the density of steam was obtained from an international steam table.

The interfacial tension of heavy-oil/steam was measured in small intervals, ranging from 115 to 181 Celsius and 25 to 150 pounds per square inch. The results show that the interfacial tension decreases when the saturation temperature and pressure increases. This finding might be useful as an important reference for understanding and visualization the mechanism of interfacial tension during steam injection.

Keywords: pendant drop method, interfacial tension, saturation pressure, saturation temperature, heavy-oil, steam generation

1. Introduction

Injection of steam into a heavy-oil sandstone reservoir has become an important and successful oil recovery process in the last few decades because this method is capable of enhancing the oil production by reducing its viscosity and residual oil saturation in the swept zones of the reservoirs. This lower oil saturation, in turn, is related to the interfacial tension between the displacing fluid and the oil, which approaches zero. As a result, high recovery from the swept zones can be achieved (Huygens *et al.,* 1995). Interfacial tension exists when there are two immiscible fluids (gas-liquid or liquid-liquid), which are in contact with a few large-diameter molecules. It is normally measured in dynes/cm (Donaldson & Alam, 2008). The interfacial tension between gas and crude oil ranges from zero to approximately 34 dynes/cm and it is a function of temperature, pressure, and compositions of both the hydrocarbon and aqueous phases (PetroWiki, 2013). The interfacial tension between the displacing and the displaced fluids in the reservoir is of importance for understanding the oil recovery mechanism (Guo & Schechter, 1997), and it might affect the efficiency of the oil recovery process.

A number of laboratory and theoretical studies have been conducted to investigate the mechanisms of the effects of temperature, pressure, and the composition of each phase on the interfacial tension of crude-oil/brine systems (McCaffery, 1972; Wang & Gupta, 1995; Hjelmeland & Larrondo, 1986), water/bitumen systems (Rajayi & Kantzas, 2011), and gas/water systems (Rushing, et al., 2008; Okasha & Al-Shiwaish, 2010). However, experimental study of the effects of high saturation temperature and pressure on heavy-oil/steam interfacial tension was not well documented in the literature. There was only one study which was related to the interfacial tension measurement of light-oil (n-decane and n-tetradecane)/steam systems. Huygens, et al. 1995 measured the interfacial tensions between n-decane, n-tetradecane, mixtures (60-40 weight %) of n-decane and n-tetradecane and saturation steam, using image processing techniques at the temperature (ranging from 100°C to 135°C in 5 °C intervals) and saturation pressure conditions. First, the experiment was performed with increasing temperature, and second, it was carried out with decreasing temperature. However, the mechanisms and procedures of steam generation were not clearly explained in their study. They observed that the interfacial tension of oil/steam systems tends to decrease with increasing temperature at saturation pressure conditions. Also recently, Yaser, et al. 2012 measured the interfacial tension between steam and Athabasca oil bitumen empirically using pendant drop method with two types of different bitumen. They measured the interfacial tension at several temperatures, ranging from 120°C to 220°C in 20 °C intervals, and saturation pressure conditions. They observed that the interfacial tension between Athabasca oil bitumen and steam also decreased when temperature and pressure increased.

1.1 Measurement of Interfacial Tension in Fluid-Fluid Systems Methods

Several methods have been utilized to analyze and measure the interfacial tension between two immiscible fluids at different conditions. Those methods were reviewed in detail by Drelich, et al. 2012. In general, those techniques, such as Wilhelmy plate and Du Nouy ring methods, can be used for direct measurement of the interfacial tension by using a microbalance. Maximum bubble pressure and growing drop techniques can be used to measure the maximum pressure to force a gas bubble into a liquid while pressure is continuously changed and monitored when the bubble is growing. Capillary rise and drop volume techniques are used to analyze the balance between capillary rise and drop volume or weight. Pendant drop and sessile drop techniques are used to analyze the drop profile sitting on a solid surface or hanging profile.

1.2 Pendant Drop Technique

The pendant drop technique is an old-fashioned and quick method in which the interfacial tension is measured between two liquids (Worthington. 1881). Applying this technique for interfacial tension measurement requires a camera with high-magnification lens to record the shape of the drop. In this method, the maximum diameter and the ratio between that parameter and the diameter at a point where the distance starting from that point to the drop apex is of maximum diameter have been used to evaluate the size and shape parameters as shown in Figure.1. In order to determine the interfacial tension between steam and heavy-oil from drop profile, two parameters should be experimentally determined which are the radius of curvature at the drop apex R_o and the shape factor β, then the interfacial tension can be calculated by using Equation 1 to Equation 4. (Herd, et al. 1992).

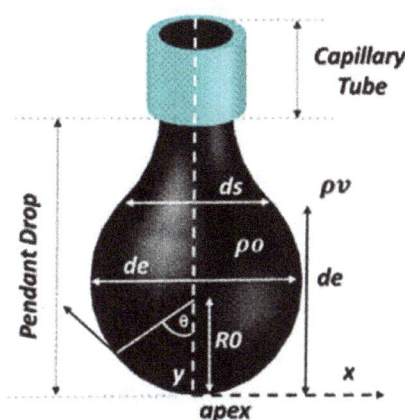

Figure 1. A close-up 3D view of pendant drop measurement. (After Naser *et al.*, 2015)

$$\gamma = \frac{\Delta_\rho \, g \, R_o^2}{\beta}$$

(1)

$$S = \frac{ds}{de}$$

(2)

$$\beta = 0.12836 - 0.7577\,S + 1.7713\,S^2 - 0.5426\,S^3$$

(3)

$$R_o = \frac{de}{2\,(\,0.9987 + 0.1971\,\beta - 0.073\,\beta^2 + 0.34708\,\beta^3\,)}$$

(4)

1.3 Objectives

The main objectives of this study are to investigate and visualize the influences of the two main factors affecting the interfacial tension of heavy-oil and steam and its parameters which are high saturation pressure and temperature. To achieve that, an experimental technique and corresponding procedures are developed to generate steam within a small optical cell. Steam is generated in this equipment by supplying discontinuous heat energy to water. In this case, the water is continuously injected into the closed optical cell at high saturation temperature conditions. After steam is generated, drop image processing technique, using a Goniometer/Tensiometer with DROP_image Advanced-p/n 290-U1, is applied to measure the interfacial tension of the heavy-oil/steam system at elevated saturation temperature and pressure using the pendant drop method.

2. Experimental Materials

All interfacial tension measurement experiments are conducted using one specific type of dead heavy-oil at high saturation temperature and pressure condition. This heavy oil comes from a sandstone reservoir in Duri Field located in Indonesia. The density and viscosity of the heavy oil are 20 °API and 486.584 cp, respectively, at atmospheric pressure and 15 °C. The true boiling point (TBP) distillation data according to the ASTM D-2892 Standard was used for characterizing the heavy oil composition. Table 1 summarizes the composition and density of the heavy-oil resulted from the experiment that was done in this study. The asphaltene content is determined according to the modified ASTM method by Wang and Buckley 2002. The ASTM procedure specifies that a volume of n-pentane that is 40 times the volume of the aliquot of oil should be added. The resulted asphaltene content of this heavy oil is about 1.44 % wt. The experimental saturation pressure and saturation temperature ranges from 25 psia to 150 psia and from 115 °C to 181 °C, respectively, for both steam generation and interfacial tension measurements.

Table 1. Composition of the heavy oil based on the true boiling point distillation data

Composition	Temperature Cut Point (°C)	Density (g/cc)	Average Molecular Weight	Weight %	Cumulative Weight %
Light Naphtha	80	0.68	104	0.2	0.2
Medium aphtha	150	0.77	116	1.5	1.7
Heavy Naphtha	200	0.83	142	2.6	4.3
Kerosene	260	0.87	172	4.8	9.1
Atmosphere Gas Oil	340	0.91	212	12.5	21.6
Light Vacuum Gas Oil*	450	0.93	324	18.1	39.7
heavy Vacuum Gas Oil*	570	0.93	476	16.5	56.2
Vacuum Residue	-	0.97	1020	43.8	100

* Vacuum distillation was conducted under 2.66 mbar.   Density of the used heavy oil was measured as 0.94 g/cc.

3. Experimental Apparatus

The apparatus consists of a small optical cell used for steam generation and an image analyzing system (Goniometer/Tensiometer with DROP_image Advanced -p/n 290-U1) used for interfacial tension measurement. Two heater guns are used to prevent dew formed on glass window and high-pressure syringe pump is used to inject oil and water into the cell. Upper and lower cartridge heaters are used to heat the cell to the required saturation temperature. Temperature sensors, attached at outside and inside the cell, and controlling box are used for controlling the cell temperature. There are also pressure sensors, a light source, and a video camera. Figures 2 and 3 show a schematic diagram of the equipment's real appearance used for the steam generation and interfacial tension measurement in this study.

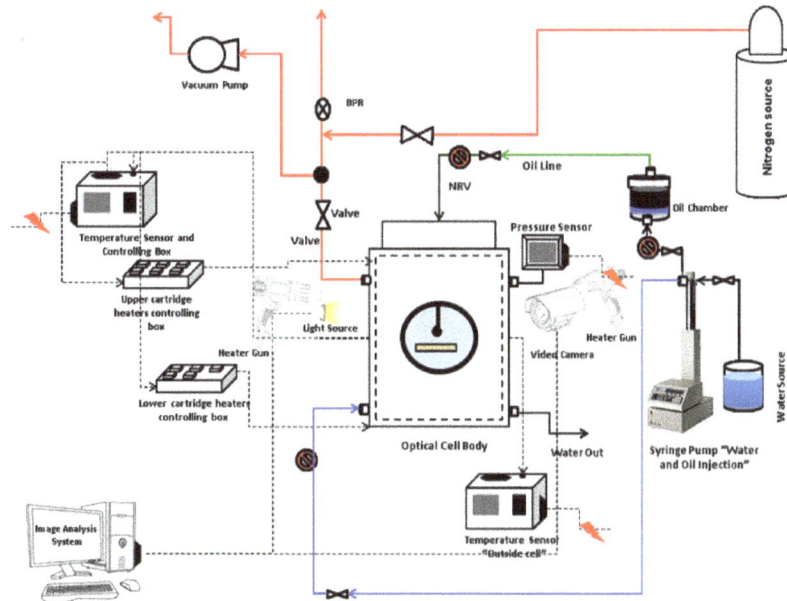

Figure 2. Schematic diagram of high-pressure high-temperature optical cell experimental setup. (After Naser et al., 2015)

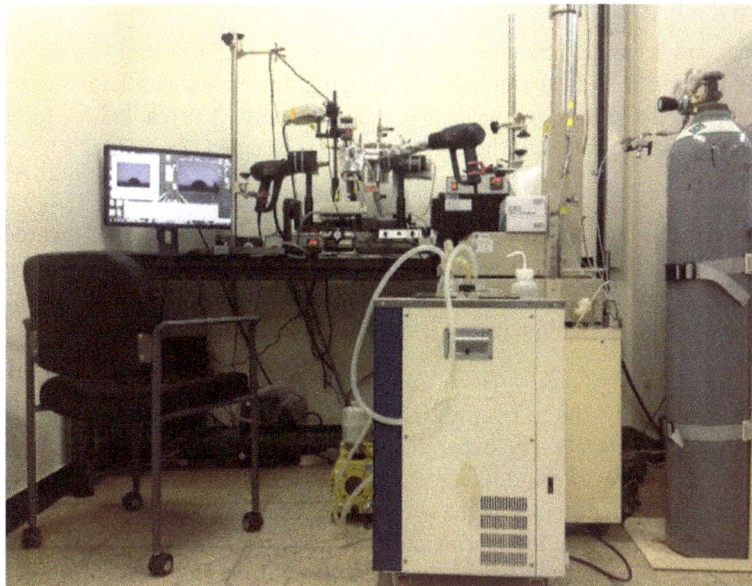

Figure 3. High-pressure high-temperature optical cell's apparatus. (After Naser et al., 2015)

4. Procedure of Experiment

The following seven steps are used in the procedure for preparing the equipment, which is utilized for steam generation as well as interfacial tension measurement. They are shown in Figure 4 and described in detail as follows.

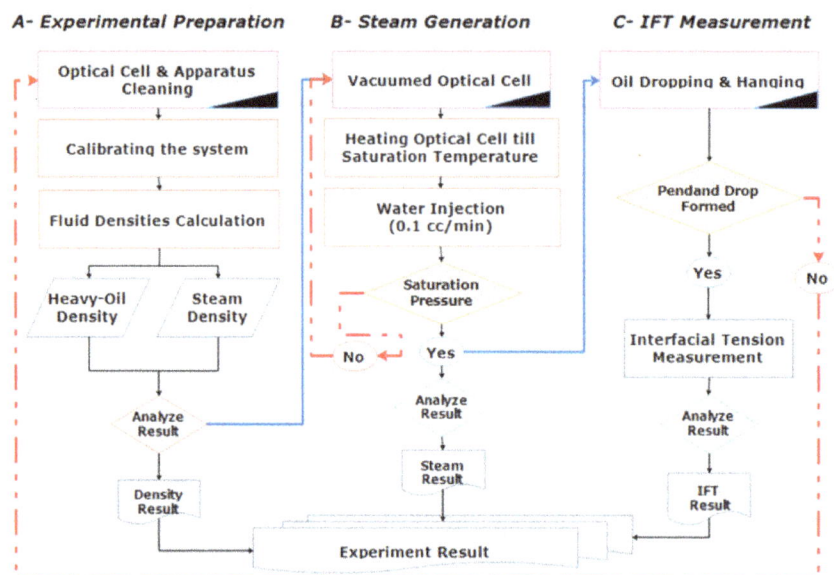

Figure 4. Experimental flow chart procedures of experiment preparation, steam generation, and interfacial tension used in this study

4.1 The first step

In order to generate steam, measure and record the accurate values of interfacial tension, cleaning must be done before each experiment is performed. The optical cell, glass windows, connection tubes, and valves in the system are cleaned with toluene and flushed by hot water to remove all oil contaminants which may contact with glass windows and cell surface or plug in the tube.

4.2 The second step

In order to calculate correct and accurate interfacial tension values, the magnifications of the video and picture capturing devices must be known. In the calibration step, the program measures a known-size object, and the vertical and horizontal size of the pixels are calculated automatically. The calibration procedure must be performed every time when the magnification of the optical device is changed (Finn. 2012).

4.3 The third step

In order to measure the interfacial tension of heavy-oil and steam using pendant drop method, the difference between density of heavy-oil and steam is required. In-house software is utilized for heavy-oil density calculation at specific temperature and pressure conditions using Katz's method (Ahmed. 1989). The density of steam is obtained from the steam tables (Wolfgang, et al. 2008).

4.4 The fourth step

In order to generate pure steam (nearly 100 % quality) within the optical cell, the cell and its connections must be vacuumed by a vacuum pump before every experiment. This procedure is used to reduce any gases that possibly affect the saturation steam pressure and temperature. After all systems have been vacuumed, the cell is directly heated to the required saturation temperature while the temperature outside the cell is kept to be $1^{\circ}C$ larger than that inside the cell to reduce heat loss during steam generation. Also, it is to keep the process running as close as possible to a desired set point of saturation steam temperature.

4.5 The fifth step

In order to clearly monitor and to prevent quick condensation, the water is injected into the cell at the rate of 0.1ml/min by ISCO pump. At the beginning, the cell's pressure and ISCO pump's pressure will gradually increase with the continuous water injection until both pressures stabilize. In this procedure, both pressures need to reach a certain value and they need to be stabilized. In addition to that, the inside and outside temperatures of the cell also need to be stabilized. At that point, the heavy oil dropping and hanging procedure can be commenced. If one of those requirements is not satisfied, then one needs to repeat the steam generation procedure all over again as shown in Figure. 4.

4.6 The sixth step

In order to minimize the volume and maximize the size of the oil pendant drop during dropping and hanging procedure, the oil must be injected inside the cell using a needle which has optimum outside and inside diameters. After the steam is successfully generated, the heavy-oil is injected and held at the bottom of the needle tip. The volume of the droplet is carefully controlled by a valve and syringe pump. If the volume cannot be controlled and held, it is needed to repeat the experiment started from the apparatus cleaning procedure, as shown in Figure 4. Once the volume of pendant drop has been controlled and held successfully, the interfacial tension measurement procedure begins.

4.7 The seventh step

In order to more realistically simulate the conditions of the reservoir and to achieve reliable interfacial tension results, the oil droplet should be immersed in the steam for several minutes before the interfacial tension measurement began. This time allows the surface between the heavy-oil and the steam to reach an equilibrium condition. Then, the measurement is commenced after the droplet has been in equilibrium and no more light components of the heavy oil are evaporated.

5. Experiment Results and Discussions

Since the pendant drop method is the most adaptable method not only for high pressure but also for a wide range of high temperatures, it is selected for measuring the heavy-oil/steam interfacial tension in this study. In addition, it is an absolute, simple, and accurate method compared with other methods or techniques, which are empirically derived or need further correction to the measured data.

In this section, the results of heavy oil density, steam density, density difference, steam generation, and measurements of heavy-oil/steam interfacial tension at saturation pressure of 25, 50, 100, and 150 psia and temperature of 115, 138, 164, and 181°C conditions is presented. In addition, it should be noted that these experiment results are only preliminary results, because all the experiments have not been completed yet (not until all required saturation temperatures and saturation pressures can be achieved) due to technical problems and increases in heat losses at the evaluated saturation temperatures. However, the findings demonstrated by these results may provide a good indicator of the relationships between heavy-oil/steam interfacial tension, saturation pressure, and saturation temperature under reservoir conditions.

5.1 Results of Heavy-Oil and Steam Density Calculation

The calculations of the fluid densities, which are required for interfacial tension measurement, are similar to that in many of previous studies in which empirical correlations are used to estimate the fluid density. Table 2 shows the heavy-oil density, steam density, and density difference as a function of saturation temperature and saturation pressure. From Figure 5 it can be seen that the density of steam increases with temperature and pressure. Since the steam is in the form of gas phase, pressure has a greater effect than temperature on its density. Meanwhile, the density of heavy-oil decreases with temperature and pressure. Since it is liquid then temperature has a greater effect than pressure on its density as shown in Figure 6.

Table 2. Steam density, heavy-oil density, and density difference as a function of saturation temperature and saturation pressure

Saturation Steam Pressure (psia)	Saturation Steam Temperature (°C)	Heavy Oil Density (g/cm3)	Steam Density (g/cm3)	Difference Density (g/cm3)
25	115	0.8080	0.0010	0.8070
50	138	0.8023	0.0019	0.8004
75	153	0.7972	0.0028	0.7944
100	164	0.7931	0.0036	0.7895
150	181	0.7854	0.0053	0.7801

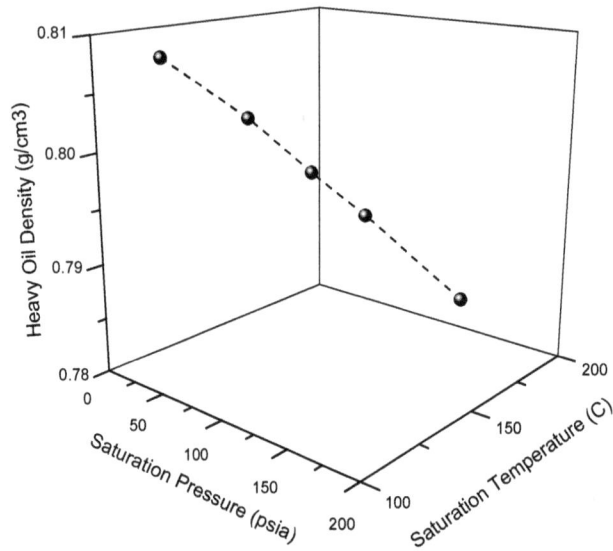

Figure 5. Heavy oil density as a function of saturation temperature and saturation pressure

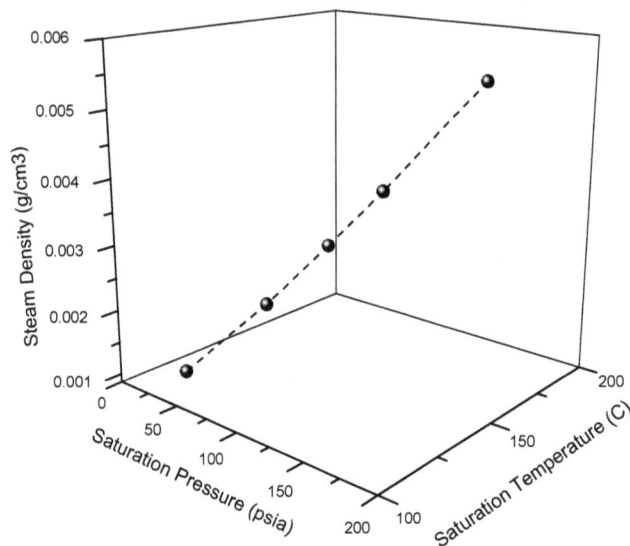

Figure 6. Steam density as a function of saturation temperature and saturation pressure

5.2 Result of Steam Generation

In the fifth procedure, it was pointed out that the cell's pressure or steam pressure in all the steam generation experiments were increased gradually by continuous water injection until they reached and are stabilized at 25, 50, 100, and 150 psia as shown in Figure 7. Those constant pressures are called saturated steam pressure and the steam quality at the beginning of the stabilized line is nearly 100 %. In those experiments, when the saturated steam pressures reached certain values and are stabilized, the interfacial tension measurement begins. From Figure 8 and Table 3, it can be observed that the values of the saturated steam pressure, which corresponds to a saturated steam temperature, obtained from this experiment are exactly the same with the values provided in the international steam tables. In this experiment, the increase in heat loss during steam generation is largely the cause of not completing the experiment until the required saturation temperature is achieved. The other reason is some technical problems.

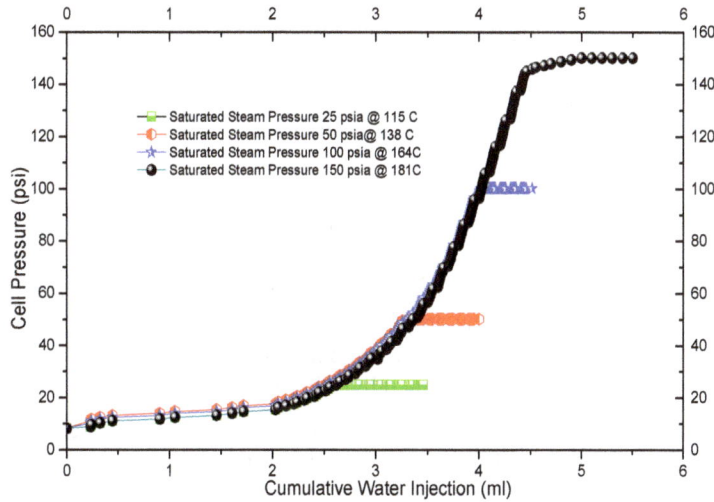

Figure 7. Saturation steam pressure results at different saturation temperature and saturation pressure

Table 3. Saturation pressure vs. saturation temperature of this study and international steam tables

Temperature (C)	Saturation Pressure (psia)	
	This Study	Steam Table
115	24.9	24.94
138	50.2	49.53
153	75.6	74.78
164	100.2	99.15
173	125.6	124.45
181	150.2	149.49
188	175.6	174.18
194	200.2	198.98

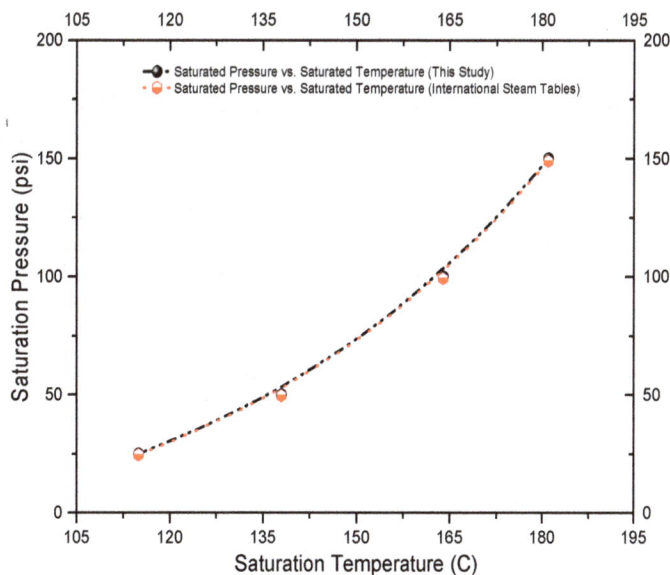

Figure 8. Comparison of saturation pressure vs. saturation temperature of this study and international steam tables

5.3 Results of Interfacial Tension Experiments

In this section, the outcomes of measuring heavy-oil/steam interfacial tension by the pendant drop method are shown. Before starting the experiment, the system is calibrated to prove the validity of the experimental system and measured parameters. The surface tension measurement of distilled water/air at standard condition (25 °C and atmospheric pressure) in this study by the pendant drop method is 71.95 mN/m. In addition, similar calibration result of the surface tension of distilled water/air at standard condition was also presented by Vargaftik, et al. (1983) in air/water systems, which is 71.99 mN/m. In order to minimize the errors of interfacial tension measurement, each experiment has been repeated three times which is denoted as Test#1, Test#2, and Test#3). Each test has a maximum number of measuring points, which is around 3500, with the time interval between two consecutive points is one second. The interfacial tension, the shape factor, the radius of curvature at the heavy oil drop's apex, the drop surface area, the drop volume, the contact angle at the drop limit (horizontal) hairline, the total height measured from hairline to apex, the maximum width, the maximum of optimizations performed, and errors or other parameters may be obtained by using the method.

In order to achieve an accurate interfacial tension measurement, equilibrating the system is one of the most important steps in the procedure. Once the oil droplet is successfully formed, its volume should be minimum and the size at the tip of the needle should be maximum. After that, it is suspended inside the optical cell for several minutes before the measurement begins. In order to get equilibrium interfacial tension value, all the measurements are conducted at different aging time during which the oil drop will be exposed to steam. The equilibrium interfacial tension is a static value, which reaches and stabilizes at a certain point. This value will be different at different periods of aging time ranging from 33 to 50 min after the oil drop has been exposed to steam. During the aging time, the steam will dissolve in the heavy oil or vice versa. Figure 9 through Figure 12 show the results of just one of the three tests at the saturation pressures of 25, 50, 100, and 150 psia and temperatures of 115, 138, 164, and 181°C. In general, it can be observed that the steam injection affects the heavy-oil/steam system interfaces.

Figures 9 (a), 10 (a), 11 (a), and 12 (a) show the shape factors for all of the systems studied. They show that the steam and aging time do have effects on shape factors of the oil drop. It can be seen that the shape factor increases when the aging time increases in all the systems. This effect on the drop shape is reflected by the ratio of ds/de of the oil drop. The maximum oil drop diameter (de) decreases due to the steam (the pendant drops tend to be smaller), while the diameter at the distance de from the drop apex (ds) remains relatively the same. As a result, the shape factor (β) increases as it is clearly shown in the figures.

Figures 9 (b), 10 (b), 11 (b), and 12 (b) show the radius of curvature at the oil drop apex (R_o) for all of the systems studied. It can be seen that it decreases when the aging time increases. This phenomenon is reflected by an increase in the value of the shape factor (β) and a decrease in the maximum drop diameter of the oil drop. As the volume of the oil drop tends to be smaller, the radius of curvature at the oil drop apex decreases.

From the results, the interfacial tension decreases when the aging time increases for all of the systems. This can be illustrated in Figures 9 (c), 10 (c), 11 (c), and 12 (c). However, the effect of aging time on the equilibrium interfacial tension is a slightly different. The interfacial tension decreases drastically during the first 8.33 min of exposure to the steam. From 8.33 min to 25 min, the interfacial tension decreases moderately and after that it begins to stabilize.

Figures 9 (d), 10 (d), 11 (d), and 12 (d) show a digital photograph of the droplet for the heavy-oil/steam systems. There are some important phenomena which have been observed during the equilibrium and aging time periods. Sometimes, a pendant drop volume is larger than required. Because of that, it tends to be pulled downward. Inversely, when it is smaller, the oil drop moves up and down along the needle tube until it stabilizes at the bottom of the needle tip. After stabilized condition is achieved, most of the pendant drops can stay at the bottom of the needle tip for several hours but some of them stay for lower period of time. The end result of this is uncompleted-shape pendant drop as clearly shown in Figure 13. This same phenomenon was also observed by Daoyong & Yongan 2004 for crude-oil/CO_2 interactions. Using the capillary rise method, which is the oldest surface tension method, they explained this effect by the balanced forces between the tension force (heavy oil with needle) and gravity force of the oil drop as well as the wettability.

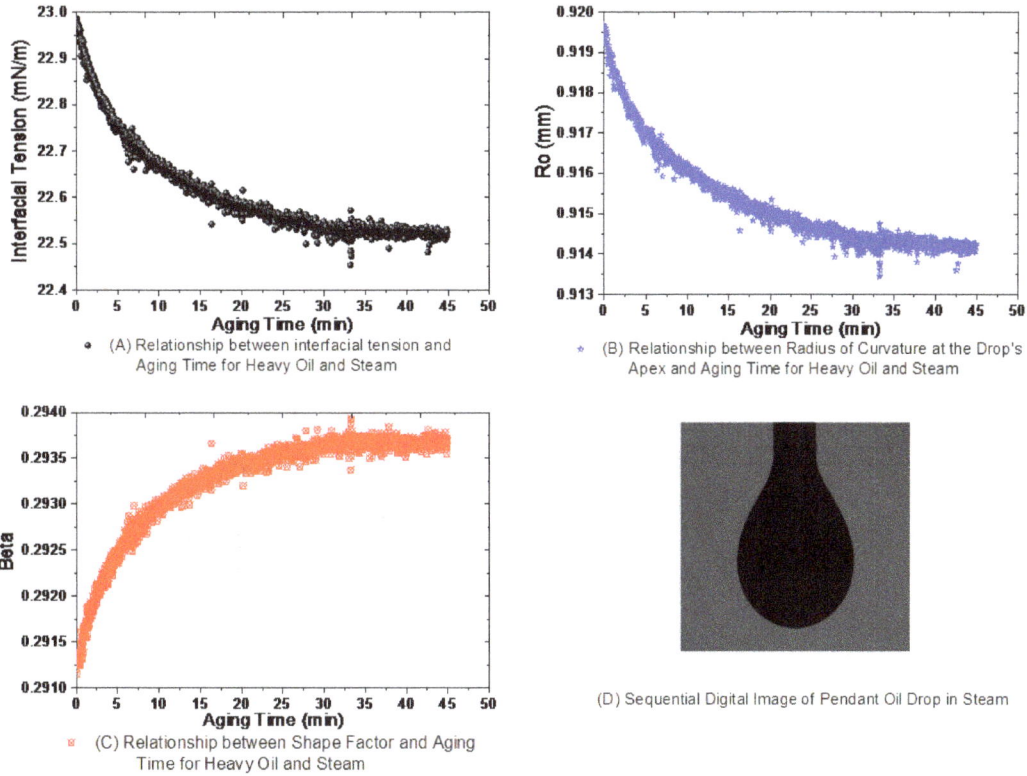

(A) Relationship between interfacial tension and Aging Time for Heavy Oil and Steam

(B) Relationship between Radius of Curvature at the Drop's Apex and Aging Time for Heavy Oil and Steam

(C) Relationship between Shape Factor and Aging Time for Heavy Oil and Steam

(D) Sequential Digital Image of Pendant Oil Drop in Steam

Figure 9. Interfacial tension measurements and its parameters of heavy-oil/steam system at 115°C and 25 psia

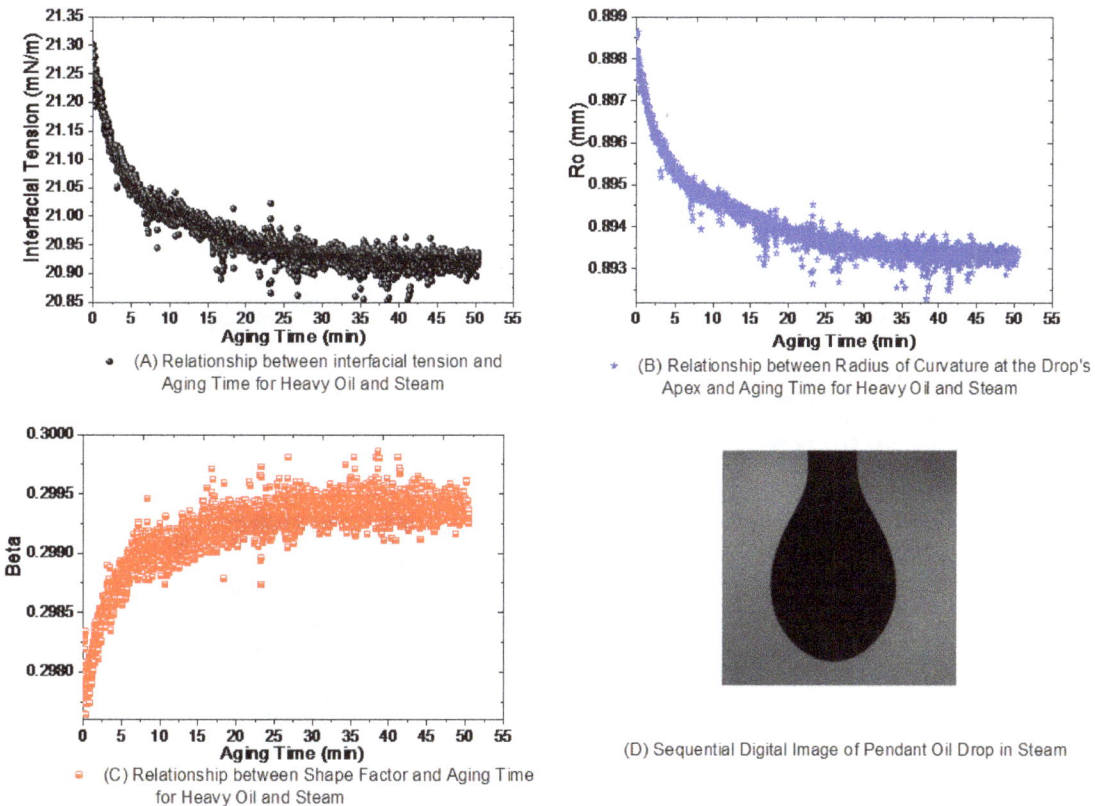

(A) Relationship between interfacial tension and Aging Time for Heavy Oil and Steam

(B) Relationship between Radius of Curvature at the Drop's Apex and Aging Time for Heavy Oil and Steam

(C) Relationship between Shape Factor and Aging Time for Heavy Oil and Steam

(D) Sequential Digital Image of Pendant Oil Drop in Steam

Figure 10. Interfacial tension measurements and its parameters of heavy-oil/steam system at 138°C and 50 psia

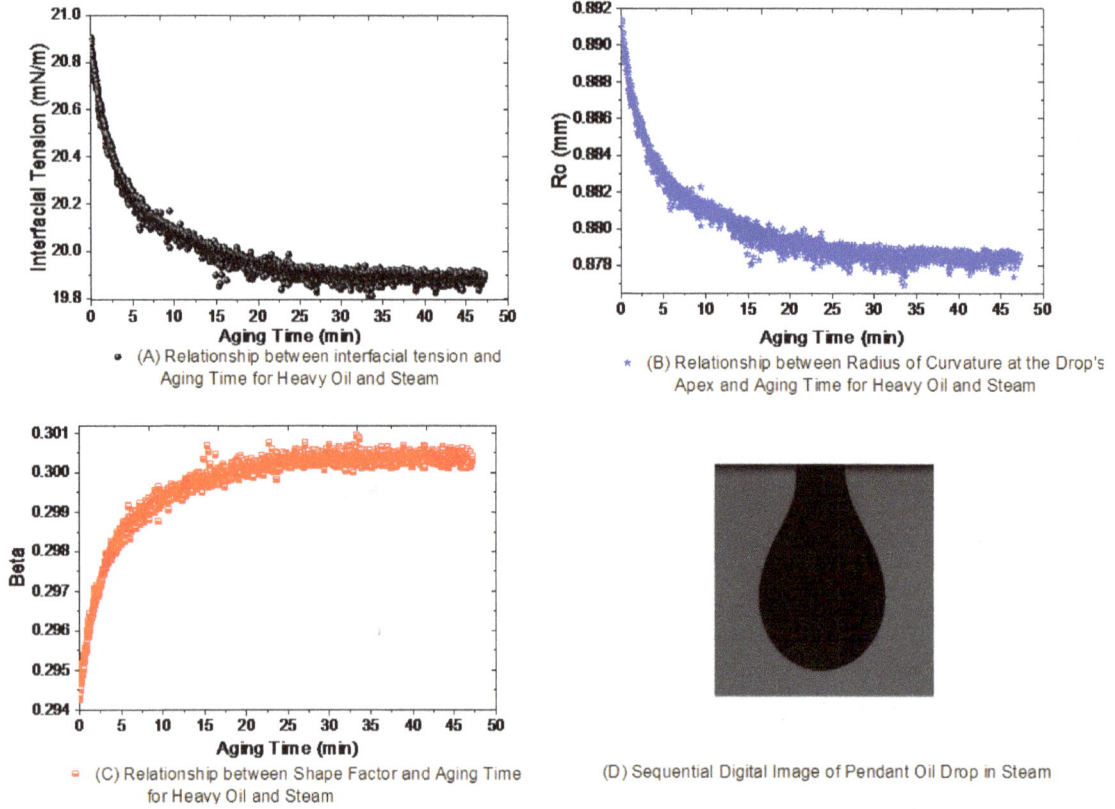

(A) Relationship between interfacial tension and Aging Time for Heavy Oil and Steam

(B) Relationship between Radius of Curvature at the Drop's Apex and Aging Time for Heavy Oil and Steam

(C) Relationship between Shape Factor and Aging Time for Heavy Oil and Steam

(D) Sequential Digital Image of Pendant Oil Drop in Steam

Figure 11. Interfacial tension measurements and its parameters of heavy-oil/steam system at 164°C and 100 psia

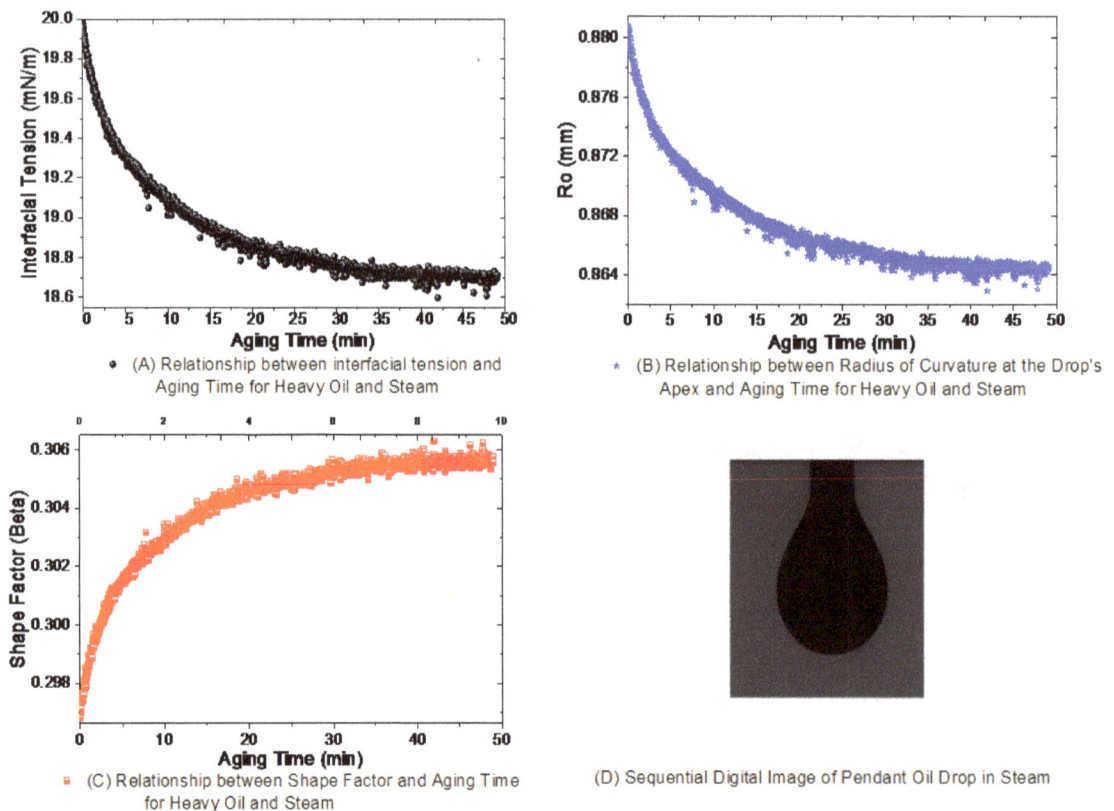

(A) Relationship between interfacial tension and Aging Time for Heavy Oil and Steam

(B) Relationship between Radius of Curvature at the Drop's Apex and Aging Time for Heavy Oil and Steam

(C) Relationship between Shape Factor and Aging Time for Heavy Oil and Steam

(D) Sequential Digital Image of Pendant Oil Drop in Steam

Figure 12. Interfacial tension measurements and its parameters of heavy-oil/steam system at 181°C and 150 psia

The same phenomena found in this study were also observed by Shariat, *et al*. 2012 in gas/water systems in which the light intensity affected the accuracy of interfacial tension results (results not shown). Because of these phenomena, the camera or the software cannot read and digitize the droplet correctly. These phenomena can be explained as the size or the shape of the droplet looks like being decreased and sometimes the edge of the droplet is not clear because of a decrease in the light intensity (i.e. too dark). On the other hand, by increasing the light intensity too much (i.e. too bright), the size or the shape of the droplet looks like being increased and sometimes the edge of the droplet is not clear enough compared with the background.

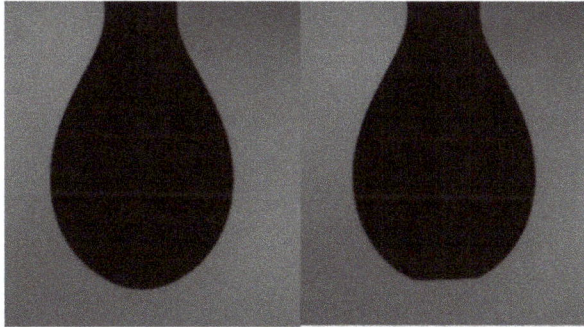

Figure 13. An example of sequential digital image of oil pendant drop in steam with upward and downward movement and wettability alteration

Figure 14 shows the 3D scatter plot comparing the three-test results (Test#1, Test#2, and Test#3) of interfacial tension measurements with its parameters at different saturation temperatures and pressures. It shows that, in general, the interfacial tension decreases with elevating saturation pressure and saturation temperature in all the systems studied as it is clearly shown in Figure 14. As can be seen from the data in the figure, the equilibrium interfacial tension values of all the tests are not very different. For example, the interfacial tension values at saturation pressure of 25 psia and temperature of 115 $^{\circ}$C in Test#1, #2, and #3 are 22.41, 22.44, and 22.52 mN/m, respectively. It is also determined from Figure 15 that the radius of the curve at the oil drop apex decreases with elevating saturation pressure and saturation temperature for all the systems studied with values that are all quite similar. Figure 16 shows the shape factor at different saturation temperatures and pressures in all the three tests.

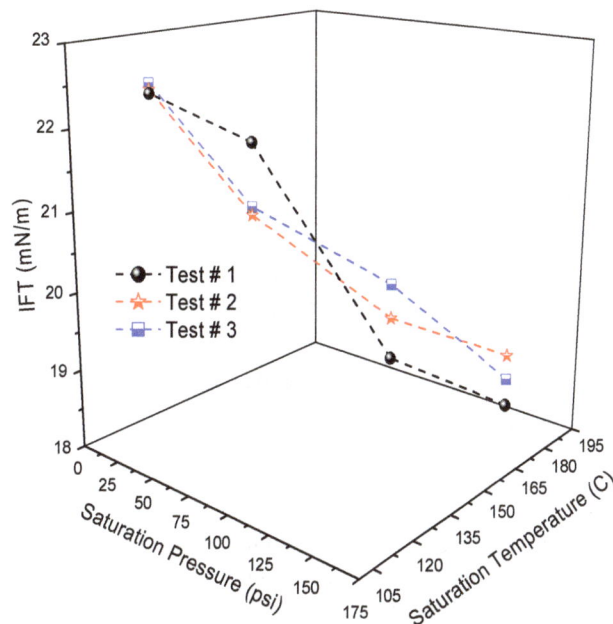

Figure 14. 3-D scatter plot of comparison of three tests of interfacial tension measurements of heavy-oil/steam at different saturation temperature and pressure, (A): Test#1, (B): Test#2, (C): Test#3

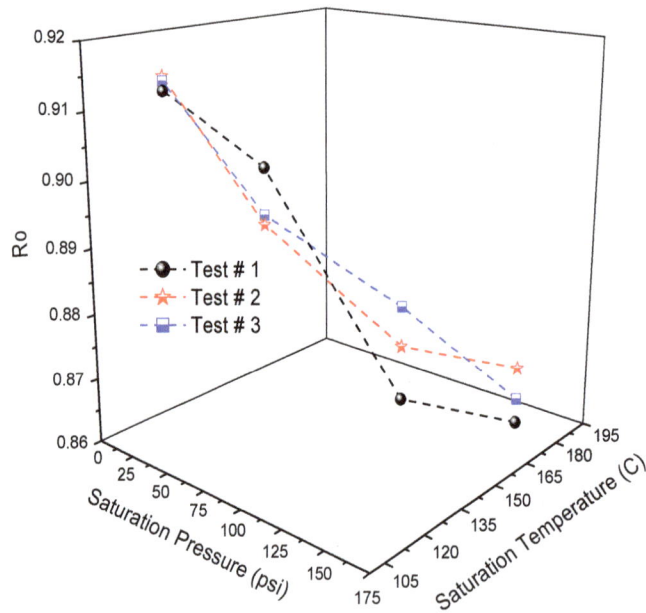

Figure 15. 3-D scatter plot of comparison of three tests of the radius of curvature at the drop apex of heavy-oil/steam at different saturation temperature and pressure, (A): Test#1, (B): Test#2, (C): Test#3

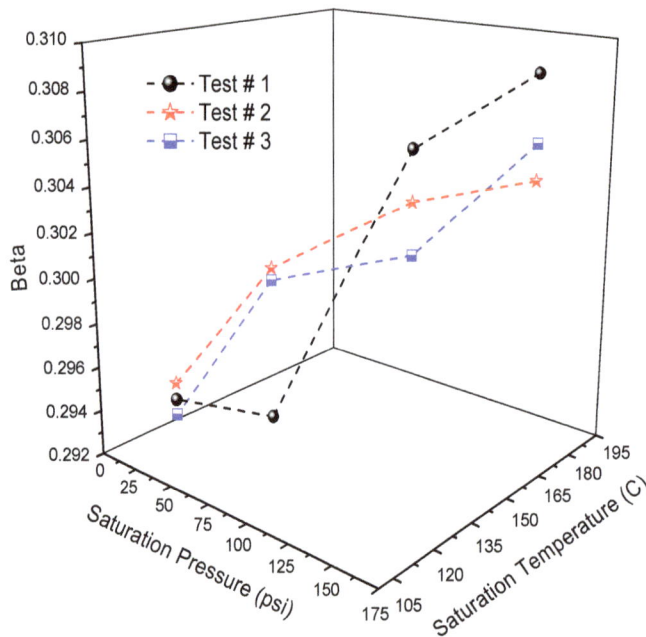

Figure 16. 3-D scatter plot of comparison of three tests of shape factor of heavy-oil/steam at different saturation temperature and pressure, (A): Test#1, (B): Test#2, (C): Test#3

6. Conclusions

In this study, we have experimentally investigated and visualized the effects of the saturated steam properties on the interfacial tension of heavy-oil/steam systems and its parameters using the pendant drop method. The experimental setup and procedures were specifically designed to simulate the initial reservoir and steam injection conditions in Duri Field (100 psia and 38 °C, and 0.75 psia/ft). It should be noted here that this experiment has not been completed yet to cover the entire saturation temperature and saturation pressure range. This is because of technical problems and the increase in heat loss at the elevated saturation temperatures. Nevertheless, the research findings shown in this paper can possibly be useful for references and for operating companies as an important source for understanding and visualizing the interfacial tension between steam and heavy oil under a specific reservoir condition. Based on the observations and analyses of the results, the main and important conclusions of this study can be briefly described as follows.

- Density calculation results show that the heavy-oil density is decreasing with temperature and pressure entirely because of the effect of temperature rather than pressure whereas the steam density is increasing with temperature and pressure mainly because of the effect of pressure instead of temperature.

- It also indicates that the saturated steam pressure, which corresponds to saturated steam temperature, is exactly the same to that found in the international steam tables.

- Exposing the heavy-oil pendant drop to steam during aging time period results in the maximum oil drop diameter (de) continues to decrease (the volume of the pendant drops tends to be smaller), while the diameter at the distance de from the drop apex (ds) remains relatively stable. For that reason, it is concluded that the shape factor (β) increases during aging time.

- The value of the radius of curvature at the oil drop apex (R_o) decreases during aging time largely because of a decrease in volume or in size of the oil pendant drop. As a result, the interfacial tension of the heavy-oil and steam increases.

- In order to capture and to visualize accurately the shape of the oil pendant drop used for interfacial tension measurement, the light intensity must be adjusted to the optimum condition before the measurement begins.

- Unstable digital images of the heavy-oil pendant drops which are associated with up-ward and down-ward movement during the aging time can also be used for indicating wettability alteration. This phenomenon maybe explained by the tension force between heavy-oil and needle and gravity force of the oil drops.

- Experimental observations show that, in general, steam injection affects the heavy-oil/steam system interfaces and the interfacial tension decreases when the aging time increases in all the systems.

- Combining all the test results of the equilibrium interfacial tension in 3D scatter plot, it can be concluded that the interfacial tension decreases when the saturation pressure and temperature increases in all the systems.

- The experimental observations show a clear trend of interfacial tension at different saturation pressures, different saturation temperatures, and different aging times. It may be useful as an important reference for understanding the heavy-oil/steam interactions.

7. Current and Future Work

In the present investigation, pendant drop experiments were conducted at elevated saturation pressure and temperature for steam injection conditions. The results showed a decrease in interfacial tension of a dead-heavy-oil/steam system. To gain a better understanding of the heavy-oil/steam interfacial tension in real reservoir conditions, this research should be continued on different live heavy-oil samples with the maximum saturation pressure of 500 psia and the maximum temperature of 242 °C. Manufacturing a good thermal insulator to reduce the heat loss during steam generation for evaluating higher saturation temperature and investigating the effect of different qualities of steam on the heavy-oil/steam interfacial tension is necessary.

Acknowledgement

The authors would like to thank to the Department of Energy and Mineral Resources Engineering of Sejong University, South Korea, the Department of Chemical Engineering of Hongik University, South Korea, and the Department of Petroleum Engineering of Bandung Institute of Technology, Indonesia, Chevron Pacific Indonesia.

Nomenclature

γ Surface or interfacial tension of liquid/liquid system (mN/m).

$\Delta\rho$ Mass density difference between oil drop and steam (g/cm^3).

g Gravity constant (cm/sec^2).

R_o Radius of curvature at the drop apex (mm).

β Shape factor.

S Ratio of ds and de.

ds Horizontal diameter of oil drop at a vertical distance of de (mm).

de Maximum diameter of oil drop (mm).

ρ_o Heavy-oil density (g/cm3).

ρ_v Steam density (g/cm3).

References

Ahmed, T. (1989). *Hydrocarbon Phase Behavior* (1st ed.). Gulf Publishing Company, Houston, Texas.

Donaldson, E. C., & Alam, W. (2008). *Wettability*. Gulf Publishing Company, Houston, Texas.

Drelich, J., Fang, Ch., & White, C. L. (2002). *Measurement of Interfacial Tension in Fluid-Fluid Systems*, Encyclopedia of Surface and Colloid Science, Marcel Dekker, Inc. 2002, 3152-3166.

Finn, K. H. (2012). *DROPimage Advanced Manual*, Advanced Edition, University of Oslo, Norway.

Guo, B., & Schechter, D. S. (1997). *A Simple and Accurate Method for Determining Low IFT from Pendant Drop Measurements*, International Symposium on Oilfield Chemistry, January 1997, Houston, Texas. http://dx.doi.org /10.2118/37216-MS.

Herd, M. D., Lassahn, G. D., Thomas, C. P., Bala, G. A., & Eastman, S. L. (1992). *Interfacial Tension of Microbial Surfactants Determined by Real-Time Video Imaging of Pendant Drops*, SPE/DOE Enhanced Oil Recovery Symposium, January, Tulsa, Oklahoma. http://dx.doi.org /10.2118/24206-MS.

Naser, M. A., Permadi, A. K., Bae, W., Ryoo, W. S., & Dang, S. T. (2015). *A Novel Experimental Method to Generate Steam within a Small Optical Cell for Measuring Interfacial Properties,* This paper has been accepted for publication on 13 April 2015 in the Arabian Journal of Science and Engineering, Dhahran, Saudi Arabia. http://dx.doi.org /10.1007/s13369-015-1659-0

Hjelmeland, O. S., & Larrondo, L. E. (1986). Experimental Investigation of the Effects of Temperature, Pressure, and Crude Oil Composition on Interfacial Properties, Society of Petroleum Engineers, July. http://dx.doi.org /10.2118/12124-PA.

Huygens, R. J. M., Boersma, D. M., Ronde, H., & Hagoort, J. (1995). Interfacial Tension Measurement of Oil/Water/Steam Systems Using Image Processing Techniques. *SPE Advanced Technology Series, 3*(March), 129-138. http://dx.doi.org /10.2118/24169-PA.

McCaffery, F. G. (1972). Measurement of Interfacial Tensions and Contact Angles at High Temperature and Pressure. *Journal of Canadian Petroleum Technology, 11*(July), 26-32. http://dx.doi.org /10.2118/72-03-03

Okasha. T. M., & Al-Shiwaish, A. J. A. (2010). *Effect of Temperature and Pressure on Interfacial Tension and Contact Angle of Khuff Gas Reservoir, Saudi Arabia.* SPE/DGS Saudi Arabia Section Technical Symposium and Exhibition, January, Al-Khobar, Saudi Arabia. http://dx.doi.org /10.2118/136934-MS.

PetroWiki (2013). Interfacial tension, Society of Petroleum Engineers Publishing PetroWikiWeb. Retrieved from http://petrowiki.org/Interfacial_tension

Rajayi, M., & Kantzas, A. (2011). Effect of Temperature and Pressure on Contact Angle and Interfacial Tension of Quartz/Water/Bitumen Systems. *Journal of Canadian Petroleum Technology, 50*(June), 61–67. http://dx.doi.org /10.2118/148631-PA

Rushing, J. A., Newsham, K. E., Van Fraassen, K. C., Mehta, S. A., & Moore, G. R. (2008). *Laboratory Measurements of Gas-Water Interfacial Tension at HP/HT Reservoir Conditions*, CIPC/SPE Gas Technology Symposium Joint Conference, January, Calgary, Alberta, Canada. http://dx.doi.org /10.2118/114516-MS.

Shariat, A., Moore, R. G., Mehta, S. A., Van Fraassen, K. C., & Rushing, J. A. (2012). *Gas/Water IFT Measurements Using the Pendant Drop Method at HP/HT Conditions: The Selected Plane vs. Computerized Image Processing Methods*, SPE Annual Technical Conference and Exhibition, January, San Antonio, Texas, USA. http://dx.doi.org /10.2118/159394-MS.

Vargaftik, N. B., Volkov B. N., & Voljak, L. D., (1983). International Tables of the Surface Tension of Water. *Journal of Phys and Chem, 12(October)*, 817-820.

Wang, W., & Gupta, A. (1995). *Investigation of the Effect of Temperature and Pressure on Wettability Using Modified Pendant Drop Method*, SPE Annual Technical Conference and Exhibition, Dallas, Texas, January. http://dx.doi.org /10.2118/30544-MS.

Wang. J., & Buckley. J. (2002). *Standard Procedure for Separating Asphaltenes from Crude Oils*, Petroleum Recovery Research Center, Socorro, New Mexico Tech. 2002. http://www.prrc.nmt.edu/groups/petrophysics/media/pdf/prrc_02-02.pdf

Wolfgang, W., & Hans-Joachim, K. (2008). *International Steam Tables - Properties of Water and Steam based on the Industrial Formulation IAPWS-IF97*, Springer.

Worthington, A. M. (1881). *On pendant drops*, Proceedings of the Royal Society of London 32, pp. 362-37.

Yang, D., & Gu, Y. (2004). *Visualization of Interfacial Interactions of Crude Oil-CO2 Systems under Reservoir Conditions*, SPE/DOE Symposium on Improved Oil Recovery, January, Tulsa, Oklahoma. http://dx.doi.org /10.2118/89366-MS.

Yaser, S., Mohammad, A., Hassan, K., & Ole, T., (2012). *Experimental Analyses of Athabasca Bitumen Properties and Field Scale Numerical Simulation Study of Effective Parameters on SAGD Performance*, Journal of Energy and Environment Research, 2(1), 140-154. http://dx.doi.org /10.5539/eer.v2n1p140.

Permissions

All chapters in this book were first published in EER, by Canadian Center of Science and Education; hereby published with permission under the Creative Commons Attribution License or equivalent. Every chapter published in this book has been scrutinized by our experts. Their significance has been extensively debated. The topics covered herein carry significant findings which will fuel the growth of the discipline. They may even be implemented as practical applications or may be referred to as a beginning point for another development.

The contributors of this book come from diverse backgrounds, making this book a truly international effort. This book will bring forth new frontiers with its revolutionizing research information and detailed analysis of the nascent developments around the world.

We would like to thank all the contributing authors for lending their expertise to make the book truly unique. They have played a crucial role in the development of this book. Without their invaluable contributions this book wouldn't have been possible. They have made vital efforts to compile up to date information on the varied aspects of this subject to make this book a valuable addition to the collection of many professionals and students.

This book was conceptualized with the vision of imparting up-to-date information and advanced data in this field. To ensure the same, a matchless editorial board was set up. Every individual on the board went through rigorous rounds of assessment to prove their worth. After which they invested a large part of their time researching and compiling the most relevant data for our readers.

The editorial board has been involved in producing this book since its inception. They have spent rigorous hours researching and exploring the diverse topics which have resulted in the successful publishing of this book. They have passed on their knowledge of decades through this book. To expedite this challenging task, the publisher supported the team at every step. A small team of assistant editors was also appointed to further simplify the editing procedure and attain best results for the readers.

Apart from the editorial board, the designing team has also invested a significant amount of their time in understanding the subject and creating the most relevant covers. They scrutinized every image to scout for the most suitable representation of the subject and create an appropriate cover for the book.

The publishing team has been an ardent support to the editorial, designing and production team. Their endless efforts to recruit the best for this project, has resulted in the accomplishment of this book. They are a veteran in the field of academics and their pool of knowledge is as vast as their experience in printing. Their expertise and guidance has proved useful at every step. Their uncompromising quality standards have made this book an exceptional effort. Their encouragement from time to time has been an inspiration for everyone.

The publisher and the editorial board hope that this book will prove to be a valuable piece of knowledge for researchers, students, practitioners and scholars across the globe.

List of Contributors

Yehuwdah E. Chad-Umoren
Department of Physics, University of Port Harcourt, Rivers State, Nigeria

Efe Ohwekevwo
Department of Physics, Rivers State University of Science and Technology, Port Harcourt, Rivers State, Nigeria

Vahid Alipour Tabrizy
Department of Reserve Replacement, ASG PTC, Statoil ASA, Norway

Aly A. Hamouda
Department of Petroleum Engineering, University of Stavanger, 4036 Stavanger, Norway

Mohammed N. Kajama
Centre for Process Integration and Membrane Technology (CPIMT), IDEAS Research Institute, School of Engineering, The Robert Gordon University, Aberdeen, AB10 7GJ, United Kingdom

Ngozi C. Nwogu
Centre for Process Integration and Membrane Technology (CPIMT), IDEAS Research Institute, School of Engineering, The Robert Gordon University, Aberdeen, AB10 7GJ, United Kingdom

Edward Gobina
Centre for Process Integration and Membrane Technology (CPIMT), IDEAS Research Institute, School of Engineering, The Robert Gordon University, Aberdeen, AB10 7GJ, United Kingdom

Clare Hall
Land Economy, Environment and Society Research Group, SRUC, United Kingdom

Fraser Allan
Land Economy, Environment and Society Research Group, SRUC, United Kingdom

Xiongwen Chen
Department of Biological & Environmental Sciences, Alabama A&M University, Normal, AL 35762, USA

Ayami Hayashi
Systems Analysis Group, Research Institute of Innovative Technology for the Earth, Kyoto, Japan

Keigo Akimoto
Systems Analysis Group, Research Institute of Innovative Technology for the Earth, Kyoto, Japan
Graduate School of Art and Science, The University of Tokyo, Tokyo, Japan

Takashi Homma
Systems Analysis Group, Research Institute of Innovative Technology for the Earth, Kyoto, Japan

Kenichi Wada
Systems Analysis Group, Research Institute of Innovative Technology for the Earth, Kyoto, Japan

Toshimasa Tomoda
Systems Analysis Group, Research Institute of Innovative Technology for the Earth, Kyoto, Japan

Rosy Lalnunsangi
Department of Zoology, Mizoram University, Aizawl, Mizoram, India

Dibyendu Paul
Department of Environmental Studies, North Eastern Hill University, Shillong, Meghalaya, India

Lalit Kumar Jha
Department of Environmental Studies, North Eastern Hill University, Shillong, Meghalaya, India

Ngozi C. Nwogu
Centre for Process Integration and Membrane Technology, IDEAS Research Institute, Robert Gordon University, United Kingdom

Mohammed N. Kajama
Centre for Process Integration and Membrane Technology, IDEAS Research Institute, Robert Gordon University, United Kingdom

Kennedy Dedekuma
Centre for Process Integration and Membrane Technology, IDEAS Research Institute, Robert Gordon University, United Kingdom

Edward Gobina
Centre for Process Integration and Membrane Technology, IDEAS Research Institute, Robert Gordon University, United Kingdom

Chin-Tsu Chen
Department of Commercial Design and Management, National Taipei University of Business, Taiwan

Jin-Li Hu
Institute of Business and Management, National Chiao Tung University, Taiwan

Shin-Lung Lin
Institute of Business and Management, National Chiao Tung University, Taiwan

Satoshi Kodama
Department of Chemical Engineering, Graduate School of Science and Engineering, Tokyo Institute of Technology, Tokyo, Japan

Kazuya Goto
Chemical Research Group, Research Institute of Innovative Technology for the Earth, Kyoto, Japan

Hidetoshi Sekiguchi
Department of Chemical Engineering, Graduate School of Science and Engineering, Tokyo Institute of Technology, Tokyo, Japan

George Yaw Obeng
Technology Consultancy Centre, College of Engineering, Kwame Nkrumah University of Science and Technology, Kumasi, Ghana

Ebenezer Nyarko Kumi
The Energy Center, College of Engineering, Kwame Nkrumah University of Science and Technology, Kumasi, Ghana

Boukary Ouédraogo
Université Ouaga2, Burkina Faso

Patrick Point
Université Montesquieu Bordeaux IV, France

Hu Wu
Department of Environmental Science and Technology, Tokyo Institute of Technology, Yokohama, Japan

Yafei Shen
Department of Environmental Science and Technology, Tokyo Institute of Technology, Yokohama, Japan

Noboru Harada
Department of Environmental Science and Technology, Tokyo Institute of Technology, Yokohama, Japan

Qi An
Department of Environmental Science and Technology, Tokyo Institute of Technology, Yokohama, Japan

Kunio Yoshikawa
Department of Environmental Science and Technology, Tokyo Institute of Technology, Yokohama, Japan

Ramkishore Singh
Department of Physics, Durban University of Technology, Durban 4000, South Africa

S P Singh
School of Energy & Environmental Studies, Takashashila Campus, Khandwa Road, Devi Ahilya University, Indore-452017, India

I J Lazarus
Department of Physics, Durban University of Technology, Durban 4000, South Africa

T. E. Oladimeji
Chemical Engineering Department, Covenant University, Ota, Ogun State Nigeria

J. A. Sonibare
Chemical Engineering Department, Obafemi Awolowo University, Ile-Ife, Nigeria

K. M. Odunfa
Mechanical Engineering Department, University of Ibadan, Ibadan, Nigeria

O. R. Oresegun
Chemical Engineering Department, Covenant University, Ota, Ogun State Nigeria

Perica Ilak
Faculty of Electrical Engineering and Computing, University of Zagreb, Croatia

Slavko Krajcar
Faculty of Electrical Engineering and Computing, University of Zagreb, Croatia

Madi Abdullah Naser
Department of Petroleum Engineering, Bandung Institute of Technology, Indonesia
Department of Energy and Mineral Resources Engineering, Sejong University, South Korea

Asep Kurnia Permadi
Department of Petroleum Engineering, Bandung Institute of Technology, Indonesia

Wisup Bae
Department of Energy and Mineral Resources Engineering, Sejong University, South Korea

Wonsun Ryoo
Department of Chemical Engineering, Hongik University, South Korea

Septoratno Siregar
Department of Petroleum Engineering, Bandung Institute of Technology, Indonesia